Sustainability in Plant and Crop Protection

Volume 12

Series Editor
Aurelio Ciancio, Sezione di Bari, Consiglio Nazionale delle Ricerche Istituto per la Protezione delle Piante, Bari, Italy

The series describes new perspectives recently developed in pest and disease management, based on innovative tools emerging from basic and applied research. The volumes will aim at interested readers involved in plant protection and crop management, for whom soil biodiversity, crop sustainability and, in general, organic approaches are fundamental. Different cropping systems will be treated by researchers involved in cutting edge studies worldwide. A number of basic issues including sustainability, life-cycle assessment, evolution, plant nutrition and organic cropping technologies will provide a common framework, within which different components of the crop production cycle will be focused on. These will range from roots and endophytes to pest and disease control, through the management of soil microbiome and fertility. These issues will be examined at the field and crop levels, including the effects of invasive species and climate changes on agroecosystems. Recent advancements in massive sequencing will represent the basis of dedicated volumes, dealing with transcriptomics and related approaches. They will illustrate the potentials and benefits of extensive DNA and RNA data analyses and studies, for practical purposes in crop protection, disease management and food production.

Contributions on any of the above cited topics are welcome. Potential Editors proposing a new volume are requested to contact the Series Responsible Editor and provide a short CV (400 words max) listing a selection of their most significant publications. In order to broaden the base of contributors and avoid redundancies, only one volume per Editor is allowed. Exceptionally, in case of many contributed chapters, a two-issues volume can eventually be considered.

More information about this series at http://www.springer.com/series/13031

Chrystel Y. Olivier • Tim J. Dumonceaux
Edel Pérez-López
Editors

Sustainable Management of Phytoplasma Diseases in Crops Grown in the Tropical Belt

Biology and Detection

Springer

Editors
Chrystel Y. Olivier
Agriculture and AgriFood Canada
Saskatoon, SK, Canada

Edel Pérez-López
Department of Biology
University of Saskatchewan
Saskatoon, SK, Canada

Tim J. Dumonceaux
Department of Veterinary Microbiology
University of Saskatchewan
Saskatoon, SK, Canada

Agriculture and Agri-Food Canada,
Saskatoon Research
and Development Centre
Saskatoon, SK, Canada

ISSN 2567-9805 ISSN 2567-9821 (electronic)
Sustainability in Plant and Crop Protection
ISBN 978-3-030-29652-0 ISBN 978-3-030-29650-6 (eBook)
https://doi.org/10.1007/978-3-030-29650-6

This Springer imprint is published by the registered company Springer Nature Switzerland AG.
The registered company address is: Gewerbestrasse 11, 6330 Cham, Switzerland

Preface

Phytoplasmas are devastating plant pathogens that are capable of infecting a wide array of plant species, including most crop plants. These are vectored by phloem-feeding insects, mostly leafhoppers, plant hoppers, and psyllids, which transmit the disease during feeding. Phytoplasma-infected plants exhibit symptoms of malformed plants and seeds, sometimes resulting in severe declines in the yield of field crops, fruits, and vegetables worldwide. Economic losses associated with phytoplasma infections are high every year due to their impacts on economically and ecologically important crops. In tropical countries, which commonly features agriculture among the top three income sources, the impact has been especially devastating. The lack of an effective cure means that appropriate management is required to reduce the impact of diseases caused by phytoplasmas. This includes rapid, rigorous, and sensitive methods to detect phytoplasma infections in plants and insects and the ability to accurately differentiate and classify phytoplasmas to facilitate epidemiological investigations of disease incidence and spread. Several books covering different aspects of this important plant pathogen have been published, but to date, this is the first book focusing on the management of phytoplasma diseases affecting plant hosts in selected tropical countries. These countries can be strongly impacted by phytoplasmas due to the importance of agriculture in their economies as well as the continuous growth cycles of insect vectors.

In 12 chapters contributed by experienced scientists worldwide, we have explored the management strategies employed by farmers in Mexico, Cuba, Argentina, Brazil, Colombia, China, and other tropical countries. We are sincerely grateful to all of the contributing authors and to Springer Nature, all of whom were key to the realization of this book. The editors also express special appreciation to Jannet Tam and Devvyn Murphy for their assistance in proofreading this book. We are confident that this book will be useful for plant pathologists, agronomists, entomologists, extension specialists, and farmers in general interested in the management of phytoplasma diseases.

Saskatoon, Saskatchewan, Canada

Chrystel Y. Olivier
Tim J. Dumonceaux
Edel Pérez-López

Contents

About the Editors

Chrystel Y. Olivier is a Research Scientist at Agriculture and Agri-Food Canada Saskatoon Research and Development Centre (AAFC) since 2001. Her research interests include phytoplasma disease epidemiology and management in field crops. Her lab worked extensively on aster yellow disease epidemiology and control in canola and cereal crops and on leafhopper populations associated with the disease.

Tim J. Dumonceaux has been a Research Scientist at the Agriculture and Agri-Food Canada Saskatoon Research and Development Centre (AAFC) since 2009. His research interests include the molecular diagnostics of plant and animal pathogens, the impacts of microbial communities on a variety of agricultural ecosystems, and the applications of white-rot fungi in the production of biofuels. His lab has developed a suite of tools for identifying and characterizing phytoplasmas based on the universal microbial barcode chaperonin-60 (cpn60), and he maintains an interest in applying these tools to the detection, characterization, and quantification of phytoplasma infections in plant and insect tissues.

Edel Pérez-López is a Postdoctoral Fellow at the University of Saskatchewan, Biology Department, working with clubroot and the soil-borne obligate parasite *Plasmodiophora brassicae*, although he keeps actively collaborating in research projects related to the identification and characterization of phytoplasmas in South America, Saudi Arabia, and Canada. He has been working with phytoplasmas affecting crops in Cuba, Mexico, Peru, Canada, and Saudi Arabia since 2012 and on the development of diagnostic methods to identify and characterize this group of plant pathogenic bacteria.

Chapter 1
The CpnClassiPhyR Facilitates Phytoplasma Classification and Taxonomy Using cpn60 Universal Target Sequences

Kevin Muirhead, Edel Pérez-López, Brian W. Bahder, Janet E. Hill, and Tim J. Dumonceaux

Abstract Phytoplasmas ('*Candidatus* Phytoplasma' spp.) are plant pathogenic bacteria that are transmitted by insects and cause developmental changes leading to altered floral morphologies and decreased seed set in a very wide variety of plant species, including most cultivated plants. Using rRNA and protein-encoding genes, 36 ribosomal groups and a similar number of species have been defined within the '*Candidatus* Phytoplasma' genus. The identification of phytoplasma strains infecting plants and insects has been facilitated by the availability of PCR primers that amplify a fragment of the 16S rRNA-encoding gene from any phytoplasma, the sequence of which is subjected to restriction fragment length polymorphism (RFLP) analysis and compared to RFLP patterns from reference sequences. An analogous classification scheme targeting a protein-encoding gene, chaperonin-60 (*cpn*60), has been described. In this work, we present software that automates the determination

K. Muirhead
Department of Microbiology, Immunology and Infectious Diseases, University of Calgary, Calgary, AB, Canada
e-mail: kevin.muirhead@ucalgary.ca

E. Pérez-López
Department of Biology, University of Saskatchewan, Saskatoon, SK, Canada
e-mail: epl733@mail.usask.ca

B. W. Bahder
University of Florida Fort Lauderdale Research and Education Center, Davie, FL, USA
e-mail: bbahder@ufl.edu

J. E. Hill
Department of Veterinary Microbiology, University of Saskatchewan, Saskatoon, SK, Canada
e-mail: janet.hill@usask.ca

T. J. Dumonceaux (✉)
Department of Veterinary Microbiology, University of Saskatchewan, Saskatoon, SK, Canada

Agriculture and Agri-Food Canada, Saskatoon Research and Development Centre, Saskatoon, SK, Canada
e-mail: tim.dumonceaux@canada.ca

C. Y. Olivier et al. (eds.), *Sustainable Management of Phytoplasma Diseases in Crops Grown in the Tropical Belt*, Sustainability in Plant and Crop Protection 12,
https://doi.org/10.1007/978-3-030-29650-6_1

of *cpn*60-based groups and subgroups. This software, the CpnClassiPhyR (http://cpnclassiphyr.ca), compares input *cpn*60 sequences to a reference database, performs RFLP-based similarity coefficient calculations, and facilitates phylogenetic analysis of the query sequence. We further describe the application of PCR primers that target phytoplasma *cpn*60 genes to generate new sequences from 16S groups II, IV, XI, XIV, and XV. We also used the CpnClassiPhyR to examine *cpn*60 sequences from reported genome sequences of various '*Ca.* Phytoplasma' species, providing results that are consistent with the 16S-based classification of these strains.

Keywords Classification · Phytoplasma · RFLP · cpn60 · Taxonomy

1.1 Introduction

Phytoplasmas ('*Candidatus* Phytoplasma' spp.), which were previously identified as mycoplasma-like organisms (Zreik, Carle, Bove, & Garnier, 1995), are plant pathogenic bacteria that infect a very wide array of plant species, affecting both cultivated and wild plants in all regions of the world (Bertaccini, Duduk, Paltrinieri, & Contaldo, 2014). Phytoplasmas are classified within the taxonomic class Mollicutes, and they are pleomorphic microorganisms that lack a typical bacterial cell wall (Namba, 2011). These phytopathogens propagate well within their plant and insect vector hosts, but they have not been reproducibly cultivated in axenic cultures. For this reason, traditional taxonomic classification criteria cannot be applied to phytoplasmas and the taxonomy of these microorganisms is under the criteria identified for uncultured microorganisms (Firrao et al., 2004). At least 25 species of '*Ca.* Phytoplasma' have been recognized based primarily on DNA sequence analysis of 16S rRNA-encoding genes and other housekeeping genes (Firrao et al., 2004). Phytoplasma taxonomy is further defined by the 16S rRNA gene through the use of restriction fragment length polymorphism (RFLP) analysis of the 16S rRNA F2nR2 fragment with a set of 17 endonucleases (Lee, Davis, & Gundersen-Rindal, 2000; Lee, Gundersen-Rindal, Davis, & Bartoszyk, 1998; Lee et al., 2004). This approach has identified more than 30 groups of phytoplasmas, designated 16SrI-16SrXXXVI, with each group containing subgroups designated by letters (Miyazaki et al., 2017; Naderali et al., 2017). The application of *in silico* RFLP analysis based on DNA sequence data as an alternative to the *in vitro* RFLP, along with the development of the online phytoplasma classification tool *i*PhyClassifier, has increased the accuracy and accessibility of phytoplasma classification based on 16S rRNA gene sequences (Pérez-López, Luna-Rodríguez, Olivier, & Dumonceaux, 2016; Wei, Lee, Davis, Suo, & Zhao, 2007; Zhao et al., 2009). The proliferation of new groups and subgroups of phytoplasma reported in the literature, however, has led to the recognition of a need for a system to formally register and recognize novel groups and subgroups based on specified criteria (Zhao & Davis, 2016).

While phytoplasma classification and taxonomy based on 16S rRNA-encoding gene sequences has proven to be very useful for studying the distribution and epidemiology of phytoplasma infections, there are limitations associated with their exclusive use. In many cases, rRNA-encoding gene sequences share high DNA sequence identity among closely related taxa, which limits the resolution of closely related but distinct strains. Furthermore, phytoplasma genomes contain two copies of the 16S rRNA-encoding gene (Wei, Davis, Lee, & Zhao, 2007; Zhao et al., 2009), and in some cases these two copies provide distinct RFLP typing results – a phenomenon referred to as 16S rRNA gene heterogeneity (Pérez-López & Dumonceaux, 2016; Zhao et al., 2009). These factors have led to the exploration of other gene sequences for the identification and taxonomic characterization of phytoplasmas. Single-copy, protein-coding genes including *rp* (Martini et al., 2007), *tuf* (Makarova et al., 2012), *secY* (Lee, Zhao, & Bottner, 2006), *rpoB* (Valiunas, Jomantiene, & Davis, 2013), and others have been explored for this use and these gene sequences typically provide a higher resolution of closely related taxa compared to 16S rRNA-encoding genes. However, accessing these genes from a sample infected with a completely unknown phytoplasma type can prove difficult, with complex PCR primer combinations or group-specific primers often required (Martini et al., 2007; Valiunas et al., 2013).

Another gene that has been exploited in phytoplasma detection, classification, and taxonomy is chaperonin-60 (*cpn*60, with synonyms *groEL* and *hsp60*), which encodes a 60 kDa protein with a canonical role in tertiary structure formation for proteins in nearly all bacteria (Hemmingsen et al., 1988), with the exception of certain Mollicutes (Clark & Tillier, 2010; Schwarz, Adato, Horovitz, & Unger, 2018). Phytoplasmas generally contain a single copy of *cpn*60 in their genomes, except for '*Ca.* Phytoplasma' pruni, group 16SrIII (Saccardo et al., 2012). For other groups of phytoplasmas, *cpn*60 has been shown to be a useful marker that provides improved resolution of closely related taxa compared to ribosomal RNA-encoding sequences (Contaldo, Mejia, Paltrinieri, Calari, & Bertaccini, 2012; Mitrović et al., 2011, 2015). A subsequence of *cpn*60 corresponding to nucleotides 274–828 of the *E. coli cpn*60 gene has been identified as a suitable barcode marker for the Domain Bacteria (Links, Dumonceaux, Hemmingsen, & Hill, 2012) and is accessible with a set of universal primers that amplify the *cpn*60 "universal target" (*cpn*60 UT) from bacteria and eukaryotes (Hill, Town, & Hemmingsen, 2006). These universal primers were modified based on available *cpn*60 sequences from phytoplasmas to develop a set of PCR primers and amplification conditions that facilitate the amplification of the *cpn*60 UT from a diverse range of phytoplasmas (Dumonceaux, Green, Hammond, Pérez-López, & Olivier, 2014).

Following the strategy previously used in the phytoplasma classification scheme based on the 16S rRNA gene, we developed a complementary, coherent system to classify phytoplasmas based on RFLP analysis of *cpn*60 UT sequences with seven endonucleases (Pérez-López, Olivier, Luna-Rodríguez, & Dumonceaux, 2016). This validated classification scheme generally allows a finer differentiation of phytoplasma strains within the same 16S rRNA RFLP subgroup, with the identification of complementary *cpn*60 UT groups and subgroups. Like 16S rRNA-based classification, the *cpn*60 UT-based RFLP analysis relies upon the calculation of a similarity coefficient (F) from the band sizes generated by *in silico* restriction digestion of DNA sequences. To facilitate the broader application of this complementary scheme for phytoplasma classification and taxonomy, we have developed the CpnClassiPhyR, an on-line tool that enables the identification and classification of phytoplasma strains based on the *cpn*60 UT sequence. In this work, we describe this publicly available resource and its application to the description of groups and subgroups of phytoplasmas for which *cpn*60 sequences have not previously been reported, corresponding to the 16S rRNA groups II, IV, XI, XIV, and XV. We also describe the use of the CpnClassiPhyR to identify novel groups and subgroups of phytoplasmas from reported genome sequences. Finally, we offer to the scientific community a collection of *cpn*60 UT plasmid clones corresponding to the broadest diversity of phytoplasmas that we have been able to obtain.

1.2 Results

1.2.1 Generation of Novel cpn60 *UT Sequences from Previously Unexplored Phytoplasmas*

The phytoplasma *cpn*60 UT primers successfully amplified the target gene from a very wide array of phytoplasma samples, including multiple subgroups within 16S rRNA groups I, II, IV, V, VII, and IX-XV (Fig. 1.1). Many of these groups were

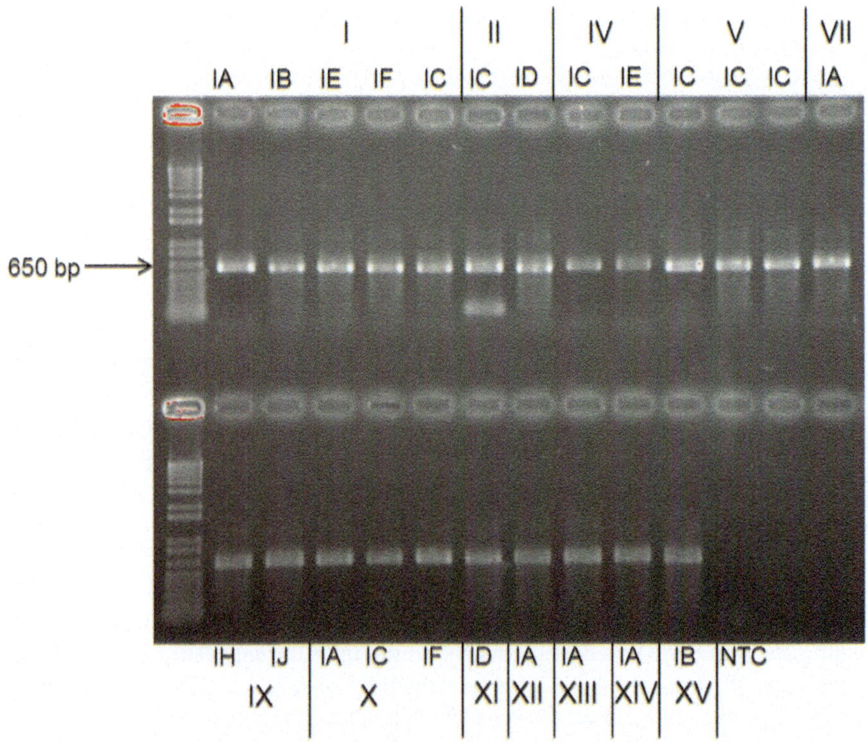

Fig. 1.1 Amplification of phytoplasmas using the *cpn*60 UT primers. Information on the phytoplasma strains used for amplification of each group/subgroup is provided in Table 1.1

previously reported (Dumonceaux et al., 2014), but amplicon from groups IV ('*Ca.* P.' palmae, '*Ca.* P.' cocostanzaniae), XI ('*Ca.* P.' oryzae) (Zhang et al., 2016), XIV ('*Ca.* P.' cynodontis), and XV ('*Ca.* P.' brasiliense) had not been previously generated with the phytoplasma *cpn*60 UT PCR primers. Overall, the *cpn*60 genes that were successfully amplified by these primers were highly diverse in sequence, with as low as 61% pairwise sequence identity between strains.

The application of the phytoplasma *cpn*60 UT primers to a wide variety of phytoplasmas collected from around the world has provided us with a collection of cloned *cpn*60 UT fragments from the phytoplasmas identified in Table 1.1 and Fig. 1.2. These are offered to the scientific community for use as plasmid standards, PCR controls, and a means to evaluate novel molecular diagnostics for analytical specificity. The plasmids have been deposited at addgene (www.addgene.org), with accession numbers provided at http://cpnclassiphyr.ca.

1.2.2 The CpnClassiPhyR: A Cpn(60-Based) Classi(Fier for) Phy(Toplasmas Using) R(FLP) Analysis

The CpnClassiPhyR is publicly available at http://cpnclassiphyr.ca. Users are directed to a welcome page containing general information regarding phytoplasmas and the purpose of the software. Along the left-hand side of the welcome page are several tabs, which are described in detail below.

CpnClassiPhyR RFLP analysis of input DNA sequences is performed here. The algorithm used by the software is described in Fig. 1.4. Input sequences are provided by the user and can be in either orientation (protein-coding or reverse-complemented), or can be full-length *cpn*60 genes, and can retain residual primer and cloning vector sequences. The CpnClassiPhyR first checks the length of the sequence and automatically discards those that are <450 bp. Next, the software determines if the sequence is likely derived from a phytoplasma by comparing its sequence using BLAST (Altschul, Gish, Miller, Myers, & Lipman, 1990) against a non-redundant reference database of *cpn*60 UT nucleotide sequences derived from '*Ca.* Phytoplasma' spp. obtained from cpnDB, the chaperonin database (Hill, Penny, Crowell, Goh, & Hemmingsen, 2004). A cutoff of 70% nucleotide sequence identity has been set for this determination, based on the sequence identity to any phytoplasma sequence returned by the *cpn*60 UT sequence of *A. laidlawii*. Sequences below 70% identity may include non-phytoplasma *cpn*60 sequences, or sequences that do not correspond to *cpn*60 at all. Users are recommended to check public databases to verify the identity of their sequence. BLAST results are returned for phytoplasma *cpn*60 sequences that are less than full-length (552 bp) but these sequences are not subjected to any further steps. Next, retained sequences are trimmed to the universal target and oriented in the protein-coding, reading frame 1 orientation. This trimmed, oriented sequence is made available to the user for download. The trimmed sequence is then subjected to *in silico* RFLP analysis using the previously described scheme (Pérez-López, Olivier, et al., 2016) and a report is provided to the user

Table 1.1 *cpn*60 clone panel. Plasmid DNA containing the cloned *cpn*60 UT sequence from each of the samples described below is available to the phytoplasma research community upon request

clone name	cpn60 UT classification	cpn60 UT GenBank accession	cpnDB ID[1]	16S classification	16S GenBank accession	Phytoplasma species	Host	country of origin	Literature references
O2L[2]	I-IA	KJ939998	b27096	16SrI-A	MH279536	'*Ca. P.* asteris'	Onion	Canada (Saskatchewan)	Dumonceaux et al. (2014)
SF1[3]	I-IB	KJ940013	b27105	16SrI-B	MH279534	'*Ca. P.* asteris'	Flax	Canada (Saskatchewan)	Dumonceaux et al. (2014)
BbSP	I-I(E/AI)AI	KU523402	b27454	16SrI-E/AI[4]	MH279523 (E); MH279522 (AI)	'*Ca. P.* asteris'	Blueberry	Canada (Nova Scotia)	Pérez-López, Olivier, et al. (2016)
CVB	I-IF	KJ939995	b27145	16SrI-F	MH279545	'*Ca.* P. asteris'	Periwinkle	Italy	Dumonceaux et al. (2014)
AY-col	I-IC	KJ939994	b27144	16SrI-R	MH279535	'*Ca.* P. asteris	Periwinkle	Italy	Dumonceaux et al. (2014)
R018	II-IC	MH279495	b31187	16SrII-C	MH266698	'*Ca.* P. aurantifolia'	Lime	Iran	This work
JbWB 190	II-IA	KY704478	b31189	16SrII-D	KY581664	'*Ca.* P. australasiae'	Jojoba	Saudi Arabia	Omar, Pérez-López, Al-Jamhan, and Dumonceaux (2017)
TagTall	IV-IC	MH922781	b32834	16SrIV-C	MH922778	'*Ca.* P. cocostanzaniae'	Coconut	Tanzania	This work
Sab4	IV-IE	MH922782	b32835	16SrIV-D	MH922779	'*Ca.* P. palmae'	Palm	USA (Florida)	This work
Palm A-DR	IV-IE	MH922783	b32836	16SrIV-E	MH922780	'*Ca.* P. palmae'	Coconut	Dominican Republic	This work

FD	V-IC	KJ939992	b27153	16SrV	NA[5]	‘*Ca.* P. ulmi’	Grape	France	Dumonceaux et al. (2014)
PGY	V-IC	KJ939991	b27154	16SrV-C	MH279539	‘*Ca.* P. ulmi’	Periwinkle	Germany	Dumonceaux et al. (2014)
RS	V-IC	KJ939990	b27155	16SrV-C	MH279544	‘*Ca.* P. rubi’	Periwinkle	Germany	Dumonceaux et al. (2014)
AshY	VII-IA	KJ939978	b27156	16SrVII-A	MH279540	‘*Ca.* P. fraxini’	Ash	France	Dumonceaux et al. (2014)
Cr	IX-IH	KJ939989	b27158	16SrIX-A	NA[6]	‘*Ca.* P.’ phoenicium’	Periwinkle	Cuba	Dumonceaux et al. (2014)
ChicBS	IX-IJ	KY986918	b31188	16SrIX-J	KY986922	‘*Ca.* P. phoenicium’	Chicory	Saudi Arabia	Pérez-López, Omar, Al-Jamhan, and Dumonceaux (2018)
AP	X-IA	KJ939977	b27159	16SrX-A	MH279542	‘*Ca.* P. mali’	Apple	Germany	Dumonceaux et al. (2014)
PYLR	X-IC	KJ940010	b27165	16SrX-C	MH279541	‘*Ca.* P. pyri’	Pear	USA	Dumonceaux et al. (2014)
ESFY	X-IF	KJ940007	b27166	16SrX-F	MH279543	‘*Ca.* P. prunorum’	Periwinkle	Italy	Dumonceaux et al. (2014)
SCWL	XI-ID	MH279493	b31184	16SrXI-D	KC295286	‘*Ca.* P. oryzae’	Sugarcane	China	This work
BN44948	XII-IA	KJ939979	b27168	16SrXII-A	MH279537	‘*Ca.* P. solani’	Periwinkle	Germany	Dumonceaux et al. (2014)

(continued)

Table 1.1 (continued)

clone name	cpn60 UT classification	cpn60 UT GenBank accession	cpnDB ID[1]	16S classification	16S GenBank accession	Phytoplasma species	Host	country of origin	Literature references
S31b-GP-MA	XIII-(A/I)I	KT444666	b27178	16SrXIII-A/I[7]	KT444665 (A); KT444664 (I)	'*Ca. P.* hispanicum'	Strawberry	Mexico (Michoacan)	Pérez-López and Dumonceaux (2016)
BGWL	XIV-IA	MH279492	b31185	16SrXIV-A	MH279538	'*Ca.* P. cynodontis'	Berumuda grass	Saudi Arabia	This work
PeruPBT	XV-IB	MH279494	b31186	16SrXV-B	KX810334	'*Ca.* P.' brasiliense	Papaya	Peru	This work and Wei et al. (2017)

In all cases except O2L and SF1, amplicon was generated using the phytoplasma *cpn*60 UT primers described on the "primer sets" tab of the CpnClassiPhyR

[1]chaperonin database ID (http://cpndb.ca)

[2]plasmid contains the 826 bp product of groELF/H280p primers (Dumonceaux et al., 2014). Contains *cpn*60 UT and flanking *cpn*60 sequences

[3]plasmid contains the 1.4 kb product of groEL F/R primers (Mitrović et al., 2011). Contains *cpn*60 UT and flanking *cpn*60 sequences

[4]Sample had F2nR2 sequence heterogeneity (Perez-Lopez et al., manuscript submitted). The sequence MH279522 was suggested as a new subgroup, 16SrI-AI

[5]insufficient template volume remains to generate F2nR2 sequence; 16Sr group information provided by S. Malembic-Maher (personal communication)

[6]NA, not available. The 16S sequence generated from this sample was too short for iPhyclassifier (714 bp), but showed 100% id to GenBank KF941131, which is 16SrIX-A

[7]sample had F2nR2 sequence heterogeneity (Pérez-López & Dumonceaux, 2016)

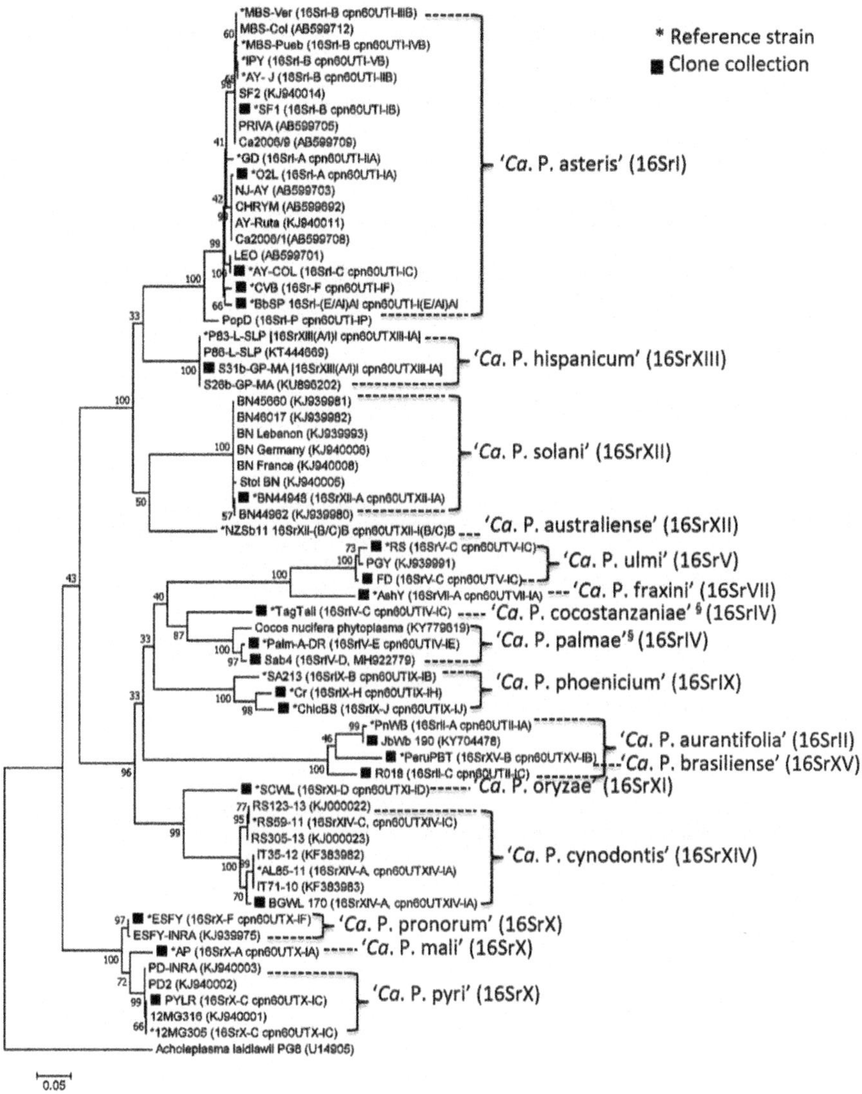

Fig. 1.2 The known diversity of phytoplasmas based on *cpn*60 UT sequences. This phylogenetic tree was constructed using the maximum likelihood method and incorporates *cpn*60 UT sequences generated by our group since 2014 and sequences retrieved from Genbank. Strains for which cloned *cpn*60 gene fragments are available to the scientific community (Table 1.1) are indicated (■). Reference strains for each group/subgroup (Table S1) are indicated (∗). The phylogenetic tree was bootstrapped 1000 times. Bar, 5 substitutions in 100 positions

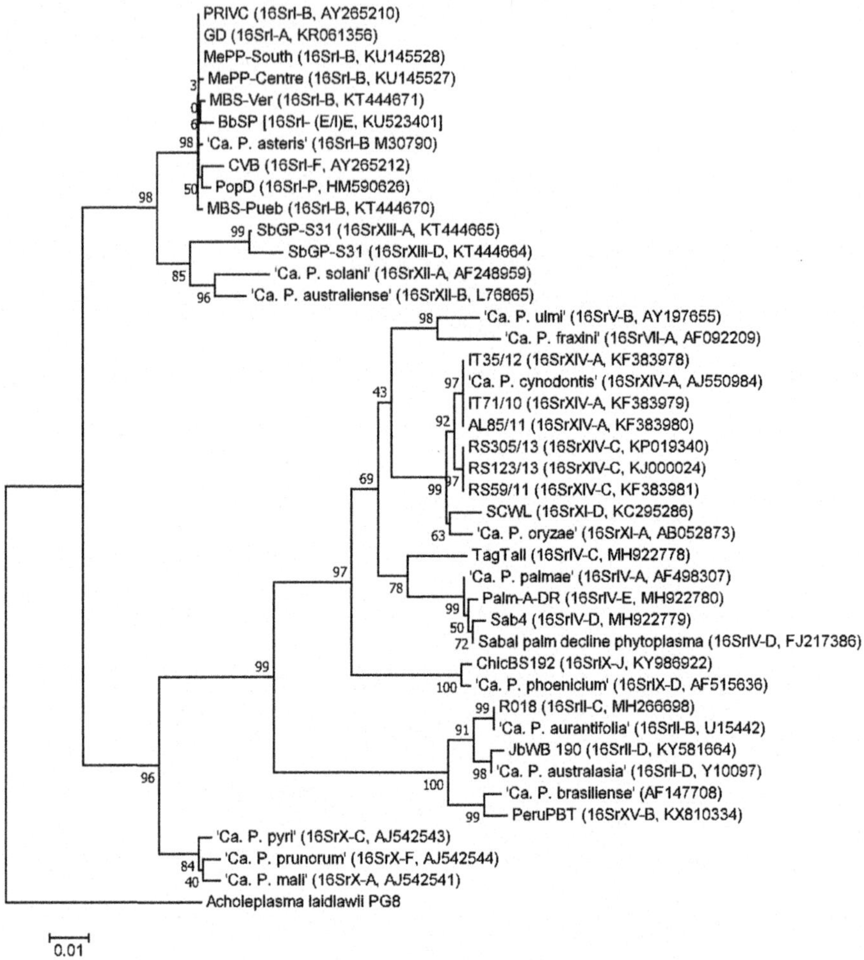

Fig. 1.3 Phylogenetic tree reconstructed through the maximum likelihood method of F2nR2 sequences from the phytoplasma strains examined in this study. The phylogenetic tree was bootstrapped 1000 times. Bar, 1 substitution in 100 positions

depending on the F-value determined for the nearest match. This is achieved by comparison to a reference database of *cpn*60 UT sequences from phytoplasmas. This table is dynamic and will be updated as new validated *cpn*60 UT sequences accrue. Samples that match a previously described group and subgroup ($F \geq 0.97$) are reported to the user. New subgroups are identified by samples with $F \geq 0.59$ and $F < 0.97$ compared to their closest match; similarly, new groups are identified by samples showing $F < 0.59$ to any previously reported group. When these criteria for group/subgroup inclusion are not met, users are provided the opportunity to register their sequence as a new group or subgroup (described below), and new, validated

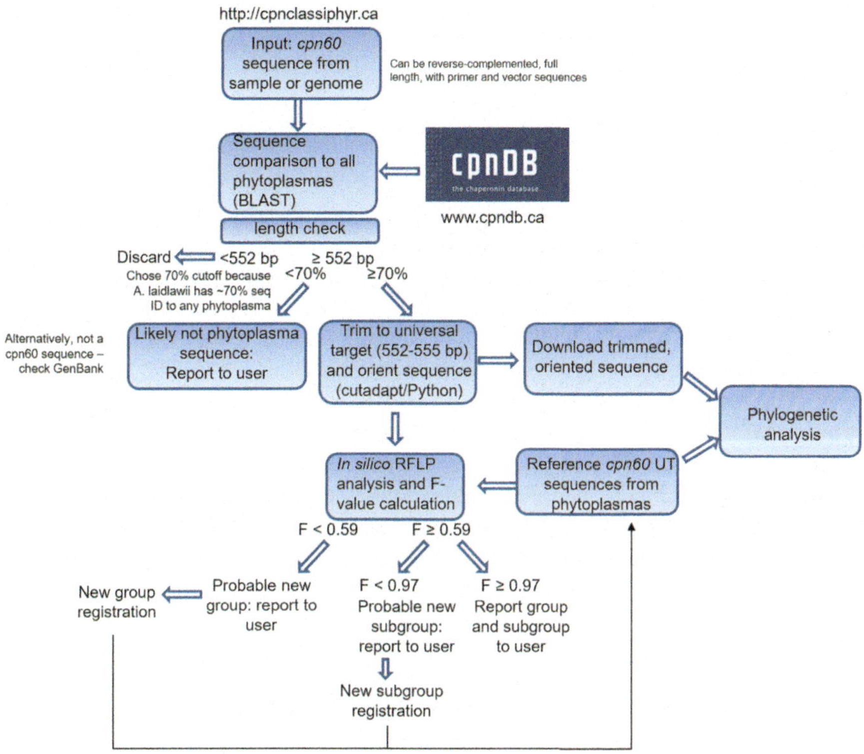

Fig. 1.4 Schematic diagram of the algorithm used by the CpnClassiPhyR for assigning group/subgroups to *cpn*60 UT sequences from phytoplasmas

sequences are added to the reference database to facilitate the identification of future phytoplasma query sequences.

Subgroups Detailed information is provided for each of the reference strains under the "subgroups" tab. The Subgroup Overview presents the known characteristics of all the reference strains in the database, including the *cpn*60 UT and 16S-based classifications, cpnDB identification number, the Genbank accession number, strain name and description including the '*Ca.* Phytoplasma' species, host information, sequence length and the virtual RFLP gel. When available, a link to the corresponding 16S rRNA-encoding sequence is also provided so that users can verify the 16S-based group/subgroup of all strains. The *cpn*60 UT sequences of all the reference strains can be downloaded from this tab. The second tab, "Number of Bands", shows the band sizes generated after the *in silico* digestion of the *cpn*60 UT sequences belonging to the reference strains with the seven endonucleases that form the *cpn*60 UT classification scheme. The third tab in this window is the Similarity Coefficient Matrix, which shows the F values determined by pairwise comparisons of all the *cpn*60 UT reference strains. The last tab is the Subgroups Tree, which allows the user to view the phylogenetic relationship of all the *cpn*60 UT reference sequences – this is described in more detail below.

Primer Sets The primer sets used to access the *cpn*60 UT from any phytoplasma have been previously described (Dumonceaux et al., 2014) and are also presented here along with detailed information on PCR setup and conditions. It is expected that these primers may need to be modified as new *cpn*60 sequences are described and this will be updated accordingly.

Publications A list of relevant publications that provide context for the CpnClassiPhyR in the scientific literature is provided.

Downloads The code that comprises the CpnClassiPhyR is provided as a Github link (https://github.com/kevmu/CpnClassiPhyR). Here the code that comprises the CpnClassiPhyR is made openly available to the scientific community.

cpnDB We provide users with a direct link to cpnDB, the publicly available chaperonin database (http://cpndb.ca) (Hill et al., 2004).

1.2.3 New Group/Subgroup Registration Service

The "dashboard" tab opens a login portal where registered users can manually enter information required to define a novel *cpn*60 UT-based group or subgroup of phytoplasma. Information is collected from users on the phytoplasma species, strain, and proposed *cpn*60 group/subgroup. In addition, users are required to include the 16S-based classification of the strain, along with the GenBank accession numbers of both the *cpn*60 UT sequence and the 16S F2nR2 sequence. Users are also asked to provide data on the host, the geographic origin of the sample, and if possible the literature reference for the sequence being submitted. Upon submission, the new sequence is automatically downloaded from NCBI (https://www.ncbi.nlm.nih.gov/) and subjected to analysis using the CpnClassiPhyR – a detailed report is sent by email to the user. If the CpnClassiPhyR results determine that the new sequence is a classified as a probable new group or subgroup, the new sequence and its associated information are added dynamically to the *cpn*60 reference dataset ("subgroups") used to calculate F values and phylogenetic trees to facilitate the identification of phytoplasma strains by all users.

1.2.4 CpnClassiPhyR Output

The analysis of a query sequence with CpnClassiPhyR results in an output window with five tabs. To illustrate the functionality of the software, we used the *cpn*60 sequence (full length) from the genome of the New Jersey Aster Yellows strain (NJ-AY; GenBank AB599703) (Sparks, Bottner-Parker, Gundersen-Rindal, & Lee, 2018). The first tab is the CpnClassiPhyR Overview, and it provides the major characteristics of the sequence such as its best match in cpnDB, the alignment length, the strandedness (+ or -) of the query sequence and the *E*-value with the best match. Users are also provided the option of downloading the trimmed, oriented *cpn*60 UT sequence in FASTA format (552 bp in this case). The second tab, Best Subgroup Matches, provides the key output of the software: the *cpn*60 UT-based RFLP classification of the best subgroup match of the query sequence, the similarity coefficient calculation, the F value, the *cpn*60 UT classification, and the RFLP patterns of the query sequence and the best match presented as virtual gels, which are downloadable. The results of the analysis can also be downloaded as a csv file. The third tab in this window is the "Number of Bands", where the software shows in a table the number of bands generated with each of the seven endonucleases (*Alu*I, *Bfa*I, *Hinf*I, *Hpa*I, *Mse*I, *Rsa*I, and *Taq*I) after analyzing the query sequence. The fourth tab in this window is the Similarity Coefficient Matrix, where the software shows the F values resulting from the comparison of the query sequence to all reference strains. In the present example, strain NJ-AY was identified as *cpn*60 UT I-IA, with an F-value of 1.00 compared to the reference strain for this subgroup, O2L. This is consistent with the identification of NJ-AY as a 16SrI-A strain (Sparks et al., 2018). The complete matrix of F-values generated in this tab can be downloaded by users as a csv file. The last tab, CpnClassiPhyR tree, shows the results of automated phylogenetic tree calculation of the input sequence in the context of the entire set of reference *cpn*60 UT sequences from phytoplasmas, with *Acholeplasma laidlawii* as an outgroup. Users can view an interactive tree (Fig. 1.5) with a link to CpnClassiPhyR_PhyD3_tree.php, a visualization tool powered by PhyD3 (https://phyd3.bits.vib.be/). Various options are provided for downloading phyloxml tree XML, SVG and PNG files associated with the phylogenetic tree. The input FASTA, MEGA options, summary, and CLUSTALW alignment files are also made available for download, allowing users to reproduce results.

Strain differentiation using the CpnClassiPhyR was also demonstrated using *cpn*60 sequences amplified from a symptomatic blueberry (*Vaccinium corymbosum*) plant in Nova Scotia, Canada, as well as from an insect collected in the same field. The insect was identified by CO1 barcoding analysis (Ratnasingham & Hebert, 2007) as *Penthimia americana* (CO1 sequence deposited to GenBank with accession number MF958490). This insect is not known to transmit phytoplasma infections in blueberry. Consistent with this, the *cpn*60 sequences amplified from the insect and the symptomatic plant in the same field were distinct, with discernable differences in digestion patterns produced by *Mse*I and *Rsa*I (Fig. 1.6). The

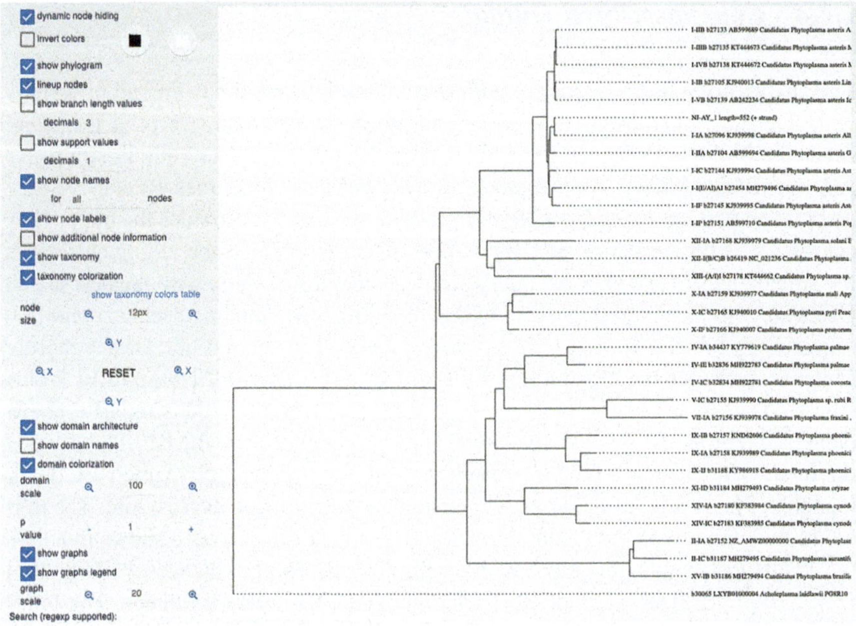

Fig. 1.5 Interactive phylogenetic analysis of the New Jersey Aster Yellows strain (NJ-AY; GenBank AB599703)

Fig. 1.6 Distinct restriction patterns with *Mse*I and *Rsa*I from *cpn*60 sequences amplified from infected plants (P) and a leafhopper (H) collected in the same field

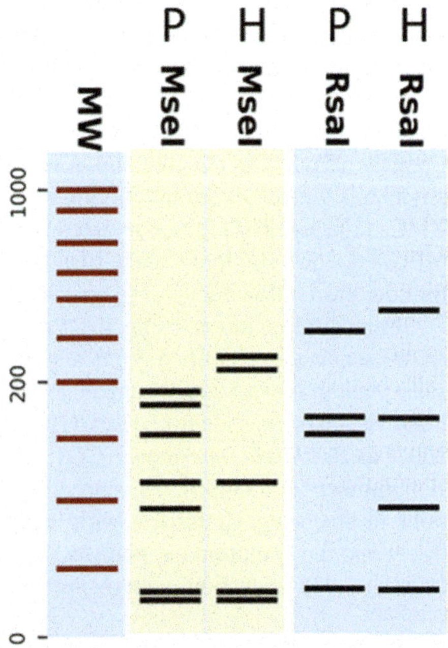

Table 1.2 CpnClassiPhyR output resulting from analysis of *cpn*60 sequences amplified from an infected plant and an insect identified as *Penthimia americana* collected in the same field in Nova Scotia, Canada, in 2016

Strain	Best match	Similarity coefficient calculation	F value	Classification
Blueberry-cpn60	I-I(E/AI)AI	F = (2 ∗ 25) / (25 + 25)	1.00	Exact match to Cpn60 UT I-I(E/AI)AI
Insect-cpn60	I-IC	F = (2 ∗ 23) / (23 + 23)	1.00	Exact match to Cpn60 UT I-IC

Table 1.3 Similarity coefficient matrix generated by the CpnClassiPhyR of the insect-derived *cpn*60 sequence compared to other group I phytoplasma sequences

cpn60 UT	I-I(E/AI)AI	I-IA	I-IB	I-IC	I-IF	I-IIA	I-IIB	I-IIIB	I-IP	I-IVB	I-VB
Blueberry-cpn60	1.00	0.79	0.76	0.79	0.82	0.61	0.61	0.64	0.85	0.76	0.82
Insect-cpn60	0.79	0.74	0.75	1.00	0.81	0.72	0.64	0.67	0.71	0.79	0.72

CpnClassiPhyR placed these sequences in distinct *cpn*60 subgroups, with the plant-derived sequence matching the known sequence of blueberry stunt phytoplasma (*cpn*60 I-I(E/AI)AI) and the insect sequence matching *cpn*60 I-IC (AY-col) (Table 1.2). Moreover, the similarity coefficient matrix generated by the CpnClassiPhyR showed that the insect-derived sequence had an F value that placed it in the same group as all other group I phytoplasma sequences while the sequence was clearly identified as *cpn*60 UT I-IC (Tables 1.2 and 1.3).

1.2.5 Application of the CpnClassiPhyR to the Identification of New Groups and Subgroups

Using the phytoplasma *cpn*60 universal primers, we generated and analyzed *cpn*60 UT sequences from five phytoplasma strains belonging to four 16S rRNA based groups. The strains we have added to the *cpn*60 reference database for this study are Peru-PBT (16SrXV-B) described by Wei et al. (2017); strain SCWL (16SrXI-D) from China (Zhang et al., 2016), strain BGWL (16SrXIV-A) from Saudi Arabia (Omar, 2016), and strain R018 (16SrII-C) from Iran. In addition, the primers were applied to the phytoplasma strains TagTall (16SrIV-C), obtained from a Tagnanan Tall coconut variety in Tanzania, Sab4 (16SrIV-D), and Palm-A-DR (16SrIV-E), which was derived from a coconut palm in the Dominican Republic. No *cpn*60 sequences have previously been reported for 16S groups XI, XIV, and XV, and only other subgroups within groups 16SrIV (GenBank KY779619), 16SrXIV (Mitrović et al., 2015) and 16SrII (Chung, Chen, Lo, Lin, & Kuo, 2013) have been identified. All available information regarding the strains used in this study is provided in Table 1.1. For each of the strains included in this work, the 16S rRNA-encoding gene and *cpn*60 UT were amplified, cloned, and sequenced.

```
   TGRTTITCICCAAAACCAGGIGCATT    - H280p
   TGRTKITCICCAAAACYWGGIGCWTC    - D0318
845TGATTATCTCCGAATCCTGGAGCTTC820 - 16SrIV
              *  *
```

Fig. 1.7 Alignment of the downstream primer landing site of the 16SrIV 'Cocos nucifera' phytoplasma *cpn*60 sequence (GenBank KY779619) with the phytoplasma *cpn*60 UT primers. The sites of mismatch are indicated (*)

The results of phylogenetic analysis of phytoplasma strains based on *cpn*60 UT sequences was consistent with the 16S rRNA gene-based classification of the strains (Figs. 1.2 and 1.3). The BGWL strain belonging to the 16SrXIV group branched with the other strains within this group, and the new sequence from R018 clustered with Peanut Witches' Broom (Group 16SrII). The *cpn*60 UT sequence amplified from the strain PeruPBT, a member of the 16SrXV group ('*Ca.* P.' brasiliense), branched with the strains within the 16SrII group, as previously reported (Harrison et al., 2014). Similarly, the new *cpn*60 UT sequences from the strains in group 16SrIV reported in this study were closely related to strains belonging to the group 16SrV and 16SrVI, as previously reported (Harrison et al., 2014). While amplicon was successfully generated with the phytoplasma *cpn*60 primers using genomic DNA from 16SrIV-infected samples, examination of the amplified sequences and of a full-length *cpn*60 gene from 'Cocos nucifera' phytoplasma (GenBank KY779619) revealed two mismatches in the downstream primer landing site for 16SrIV phytoplasmas (Fig. 1.7). This resulted in the CpnClassiPhyR initially failing to trim correctly the full-length *cpn*60 gene to the UT; however the script has been adjusted to accommodate this observation.

In this study we have expanded the number of described *cpn*60 UT groups from nine to twelve (Fig. 1.2), which represents a third of the phytoplasma 16Sr groups that have been reported to date (Naderali et al., 2017). The known *cpn*60-based diversity of phytoplasmas was similarly expanded to include sixteen '*Ca.* Phytoplasma' species (Fig. 1.2). The phylogenetic tree obtained using the full diversity of known phytoplasma *cpn*60 sequences (Fig. 1.2) showed a clear differentiation of the three major phytoplasma subclades previously described (Chung et al., 2013; Hogenhout et al., 2008; Zhao, Davis, Wei, Shao, & Jomantiene, 2014; Zhao, Wei, Davis, & Lee, 2010), and corresponded well with the 16S rRNA-based phylogenetic tree (Fig. 1.2, S4).

Including these new sequences, the reference *cpn*60 UT database for the CpnClassiPhyR now includes 30 reference strains, with 11 subgroups within the *cpn*60 UT-I group, 2 subgroups within the *cpn*60 UT-II group, 2 subgroups within the *cpn*60 UT-IV, 1 subgroup within group *cpn*60 UT-V, 1 subgroup within group *cpn*60 UT-VII, 3 subgroups within *cpn*60 UT-IX, 3 subgroups within *cpn*60 UT-X, 1 subgroup within *cpn*60 UT-XI, 2 subgroups within *cpn*60 UT-XII, 1 subgroup within *cpn*60 UT-XIII, 2 subgroups within *cpn*60 UT-XIV, and 1 subgroup within group *cpn*60 UT-XV. This reference dataset is readily expandable as new *cpn*60 UT sequences accumulate, either from genomic sequencing efforts or through the application of existing or adapted *cpn*60 phytoplasma-specific primers.

1.2.6 Discrepancies

The *cpn*60 UT RFLP patterns generated by strains 16SrIV-D (Sab4) and 16SrIV-E (Palm A-DR) were identical, so the RFLP scheme as currently described cannot discern these subgroups. Despite this, the *cpn*60 UT sequences from these strains were slightly more divergent than the corresponding 16S sequences – the strains shared 99.3% sequence identity at F2nR2 compared to 99.0% sequence identity at *cpn*60. Conversely, RFLP using F2nR2 sequences was unable to discern 16SrXIV-A (AL85/11) from 16SrXIV-C (RS59/11), while these strains were readily differentiated into distinct subgroups using *cpn*60 UT RFLP.

1.2.7 Application of the CpnClassiPhyR to cpn*60 Sequences Extracted from Genomes*

To determine the *cpn*60 UT group and subgroup classification for publicly available phytoplasma genomes, *cpn*60 and, where available, 16S rRNA-encoding gene sequences were extracted from 17 phytoplasma genomes, including a variety of 16S-based groups and subgroups (Table 1.4). Wherever possible we analyzed one or both 16S rRNA-encoding genes (depending on their availability within the genome sequence) using iPhyClassifier along with the full-length *cpn*60 gene using CpnClassiPhyR. In several cases, published genomes did not contain 16S rRNA-encoding genes, or the genes were of insufficient length for classification using the *i*PhyClassifier (e.g. '*Ca.* P.' solani strains 284/09 and 231/09; '*Ca.* P.' phoenicium strains SA213 and CHiP; '*Ca.* P.' oryzae strain Mbita1; Table 1.4). However, *cpn*60 genes were found in all analyzed genomes (Table 1.4). In nearly all cases in which 16S rRNA-encoding genes were available, the classification provided by the CpnClassiPhyR was consistent with the 16S rRNA group/subgroup classification. In the case of '*Ca.* P.' phoenicium strain CHiP, which lacks 16S rRNA-encoding gene sequence data, the *cpn*60 UT sequence extracted from the genome suggested that the strain may belong to a new *cpn*60 subgroup, with F = 0.93 to group IX-IJ (Table 1.4). In the absence of any known 16S rRNA-encoding gene sequences from this strain, we have not added it to the phytoplasma *cpn*60 UT RFLP reference database.

Several of the genome sequences also displayed clear evidence of 16S rRNA-encoding gene heterogeneity, including two strains of '*Ca.* P.' australiense (NZSb11 and NC_010544) with identical *cpn*60 sequences that placed them in a clade with '*Ca.* P.' solani', also a group 16SrXII phytoplasma (Fig. 1.2). Due to this clear evi-

Table 1.4 CpnClassiPhyR results using *cpn*60 sequences from published phytoplasma genomes

Species	strain	16S group-subgroup (F)	*cpn*60 UT group-subgroup (F)	accession no.
'*Ca.* P.' solani	284/09	ND[a]	XII-IA (1.00)	GCA_000970375.1
'*Ca.* P.' solani	231/09	ND[a]	XII-IA (1.00)	GCA_000970395.1
'*Ca.* P.' phoenicium	SA213	16SrIX[b]	IX-IB (1.00)	GCA_001189415.1
'*Ca.* P.' phoenicium	CHiP	ND[a]	IX-IJ (0.93)	GCA_002968365.1
'*Ca.* P.' australiense	NZSb11	16SrXII-B (0.98); 16SrXII-C (1.00)	XII-I(B/C)B (1.00)	CP002548.1
'*Ca.* P.' australiense	NC_010544	16SrXII-B (1.00); 16SrXII-C (1.00)	XII-I(B/C)B (1.00)	AM422018.1
Echinacea purpurea witches'-broom phytoplasma	NCHU2014	II-A (1.00)	II-IA (1.00)	NZ_LKAC00000000
'*Ca.* P.' oryzae	Mbita1	16SrXI[c]	X-IF (0.5)	NZ_LTBM00000000.1
'*Ca.* P.' asteris	New Jersey Aster yellows	16SrI-A	I-IA (1.00)	MAPF00000000.1
'*Ca.* P.' asteris	Onion yellows OY-M	16SrI-B (1.00)	I-IB (1.00)	AP006628.2
'*Ca.* P.' asteris	Maize bushy stunt phytoplasma	16SrI-B (1.00)	I-IIIB (1.00)	NZ_CP015149.1
'*Ca.* P.' asteris	TW1	16SrI-A (0.97); 16SrI-B (1.00)	I-IB (1.00)	QGKT00000000
'*Ca.* P.' mali		16SrX-A (1.00)	X-IA (1.00)	NC_011047.1
'*Ca.* P.' asteris	Wheat blue dwarf phytoplasma	16SrI-R (1.00); 16SrI-S (1.00)	I-IC (1.00)	NZ_AVAO00000000.1
'*Ca.* P.' asteris	Rice orange leaf phytoplasma	16SrI-B (1.00)	I-IIIB (1.00)	NZ_MWKN00000000.1
'*Ca.* P.' aurantifolia		ND[a]	II-IC (1.00)	NZ_MWKN00000000.1
Peanut witches' broom phytoplasma		16SrII-A (1.00)	II-IA (1.00)	NZ_AMWZ00000000

[a]ND, not determined. No 16S rRNA-encoding genes were found in the genomic sequence

[b]Subgroup undetermined. A single 16S rRNA-encoding gene was found in the genomic sequence, which was of insufficient length for classification using the iPhyClassifier. An accession number is provided for a 16S rRNA-encoding gene sequence for SA213 (KM275491), but this sequence types as 16SrIX-D while the strain is identified as 16SrIX-B (Quaglino et al., 2015)

[c]Subgroup undetermined. No 16S rRNA-encoding genes were found in the genomic sequence and the strain was identified only to the group level using 16S amplicon analysis (Fischer et al., 2016)

dence of 16S rRNA-encoding gene sequence heterogeneity in multiple strains, the *cpn*60 sequence extracted from these genomes was used as a reference to define the novel *cpn*60 subgroup XII-I(B/C)B, according to the suggested nomenclature for heterogeneous subgroups (Wei, Lee, et al., 2007). In other cases, the single *cpn*60 sequence from the genome provided clarity to the heterogeneous results provided by the two copies of the 16S rRNA-encoding gene (e.g. '*Ca.* P.' asteris strain TW1 – Table 1.4). The wheat blue dwarf phytoplasma, which contains two annotated 16S rRNA-encoding gene sequences identified as 16SrI-R and 16SrI-S using iPhyClassifier (Table 1.4), was identified as *cpn*60-I-IC using CpnClassiPhyR (Table 1.4).

The *cpn*60 UT sequence from '*Ca.* P.' oryzae strain Mbita1 (group 16SrXI) (Fischer et al., 2016) was different from all previously reported *cpn*60 UT sequences, with a maximum F value of 0.5 with a reference strain from *cpn*60 group X, which suggests the possibility that this strain belongs to a new group. This result was observed despite the fact that the CpnClassiPhyR reference database contains a sequence from group *cpn*60-XI (SCWL). Since the 16S rRNA-encoding gene sequences for this strain are not publicly available, it is difficult to determine if this strain should be designated as a new *cpn*60 group so we have not included the *cpn*60 UT sequence from this strain as a reference in the *cpn*60 UT classification scheme used by the CpnClassiPhyR.

1.3 Discussion

The taxonomic classification of phytoplasmas using RFLP analysis of target gene sequences is an established practice that has provided useful information for the detection, classification, and surveillance of phytoplasmas worldwide. While RFLP-based classification schemes have been described for several loci including *rp* (Martini et al., 2007), *cpn*60 (Pérez-López, Olivier, et al., 2016) and others, to date only RFLP analysis of 16S rRNA-encoding gene sequences has been provided to users through an interactive online calculation tool, the *i*PhyClassifier (Zhao et al., 2009).

We provide a phytoplasma classification portal, the CpnClassiPhyR, which is an open access, user-friendly platform for the identification of phytoplasma strains based on *cpn*60 UT sequences. The CpnClassiPhyR performs similarity coefficient calculations, provides RFLP patterns as virtual gel images, and facilitates phylogenetic analysis of an input phytoplasma *cpn*60 UT sequence in the context of the known diversity of *cpn*60 sequences derived from phytoplasmas. The *cpn*60-based classification scheme we have provided is firmly rooted in the existing 16S RFLP-based classification scheme, and is meant as a complementary tool providing robust taxonomic information for phytoplasmas that can help to resolve closely related strains and provide clarity in the context of strains featuring heterogeneous 16S rRNA-encoding loci. The registration of new *cpn*60 reference sequences as we have implemented it here includes mandatory information on 16S group and subgroup, and ideally (acknowledging that this is not always possible) links to the 16S rRNA-encoding gene sequences themselves so that users can verify the taxonomic affiliation of the strain under analysis.

Universal PCR primers that amplify either the 16S rRNA (Gundersen-Rindal & Lee, 1996) or *cpn*60 (Dumonceaux et al., 2014) taxonomic marker genes from phytoplasmas have been described, which facilitates access to these gene sequences from a sample containing a phytoplasma of unknown taxonomic affiliation. However, while the primers targeting the F2nR2 subsequence of 16S rRNA are known to be effective on all described phytoplasma taxa, this is not the case for the phytoplasma *cpn*60 primers. We have demonstrated the efficacy of these primers on approximately 33% of the known diversity of phytoplasmas, corresponding to a very broad range of phytoplasma taxa, but have not had access to samples infected with several other phytoplasma strains. The ability of the primers to amplify the *cpn*60 UT from phytoplasmas containing mismatches has been demonstrated here using group 16SrIV. Nevertheless, it is reasonable to assume that *cpn*60 sequences from phytoplasmas other than those described here will become available as a result of ongoing and planned genome sequencing efforts. The CpnClassiPhyR has been designed to handle this easily, providing the ability to enter full-length *cpn*60 gene sequences that are automatically trimmed to the universal target sequence prior to analysis. However it must be noted that one very important group of phytoplasmas, the 16SrIII group ('*Ca.* P.' pruni) appears not to contain a *cpn*60 gene sequence in its genome (Saccardo et al., 2012), so a *cpn*60-based classification scheme will not apply to this group. At present this is the only known exception; however it remains to be seen if all other taxonomic groups of phytoplasmas contain *cpn*60 genes.

The reference database used by the CpnClassiPhyR (described in the "subgroups" tab) can easily be expanded to include reference strains corresponding to new *cpn*60 UT groups and subgroups as they are described. The site has been designed to accommodate the accumulation of novel *cpn*60 reference sequences, and the routine determination of paired 16S rRNA and *cpn*60 sequences from the same sample will improve the quality of the reference dataset and thereby the taxonomic classification of unknown strains. We also recommend that users investigate the possibility of 16S rRNA-encoding gene heterogeneity, which is a reasonably common phenomenon (Table 1.4). This could be determined by routinely cloning amplicons generated using the phytoplasma 16S-targeted primers (never sequencing amplicons directly) and examining the sequence of many clones in order to evaluate this possibility. Phytoplasma taxonomic classifications will be improved by using both 16S rRNA-encoding gene sequences (iPhyClassifier) and *cpn*60 gene sequences through the CpnClassiPhyR. In turn, improved taxonomic classification of phytoplasmas facilitates strain tracking and phytoplasma surveillance, which will provide benefits to agriculture worldwide.

1.4 Materials and Methods

1.4.1 Generation of cpn60 *and 16S rRNA Amplicon from Phytoplasmas*

Due to the lack of a central repository for samples corresponding to identified phytoplasma strains, samples containing phytoplasma DNA were obtained from scientists worldwide through e-mail requests. Samples from outside of Canada were received as DNA extracts from identified sources, and the method of DNA extraction was typically unknown. Samples from within Canada were received as DNA or as infected plant material, and in the latter case DNA was extracted using a commercial kit (Plant DNA mini kit, Qiagen, Toronto ON). Amplicon corresponding to the *cpn*60 UT was generated from DNA extracts as described in (Dumonceaux et al., 2014) and on the "primers" tab of the CpnClassiPhyR web page. A dilution series was performed on each sample in order to overcome the effects of PCR inhibition. Generation of amplicon corresponding to the F2nR2 subsequence of the 16S rRNA-encoding gene was achieved using nested PCR with primers P1/P7 (Deng & Hiruki, 1991; Schneider, Seemüller, Smart, & Kirkpatrick, 1995) in the first round and R16F2n/R16R2 in the second round (with a 1:30 dilution of the first round PCR products) as described (Gundersen-Rindal & Lee, 1996). Positive reactions were determined through the visualization in 1% agarose gel of a band of ~650 bp (actual size is 604–607 bp) corresponding to the *cpn*60 UT (Dumonceaux et al., 2014) or a band of ~1.25 kb corresponding to the F2nR2 fragment (the sequence located between the annealing sites of primers R16F2n/R16R2) (Wei, Davis, et al., 2007). In all cases, *cpn*60 or F2nR2 amplicon was ligated into the vector pGEM-T Easy (Promega, Madison WI) and used to transform competent *E. coli* TOP10 cells (ThermoFisher Scientific, Waltham MA). The sequence of one or more selected clones was determined. Stocks of *E. coli* containing plasmids containing phytoplasma *cpn*60 gene fragments were prepared in 15% glycerol and stored at −80 °C. From these stocks, plasmid DNA was prepared using a commercial kit (QiaPrep Spin miniprep kit, Qiagen, Toronto, ON). Dilutions of the miniprep DNA samples from each phytoplasma strain described in Table 1.1 and Fig. 1.2 are available to the scientific community from the corresponding author by request.

1.4.2 CpnClassiPhyR Code

The software is a command-line based utility written in the Python programming language using Biopython (Cock et al., 2009). The CpnClassiPhyR workflow is composed of two main programs: generate_subgroup_metadata.py and phyto_RFLP_digest.py which share a common python class CpnClassiPhyR.py to perform the *in silico* RFLP digests, similarity coefficient calculations and generate virtual gel images. *In silico* RFLP digests are performed using the restriction enzymes AluI

(AG/CT), BfaI (C/TAG), HinfI (G/ANTC), HpaI (GTT/AAC), MseI (T/TAA), RsaI (GT/AC) and TaqI (T/CGA), according to the described scheme (Pérez-López, Olivier, et al., 2016). Regular expressions are used to search for the recognition sequence and to cut at the position of the restriction site. The number of fragments (bands) and the length of each sequence fragment (band sizes) is calculated for each enzyme. The total number of bands are calculated by adding the number the fragments of each enzyme. The similarity coefficient or F value is calculated using the formula, $F = 2Nxy/Nx + Ny$, where Nx and Ny are the total number of bands for strain x and strain y, respectively. Nxy is calculated by counting the number of bands common between strain x and strain y. Each list of band sizes is sorted in ascending order and the unique bands are retained. Fragments of the same size are considered as the same band. A similarity coefficient matrix is generated using the F value calculated for each strain x and strain y pair. Virtual agarose gel images are generated using BandWagon-0.1.3 (https://edinburgh-genome-foundry.github.io/BandWagon/ref.html).

The generate_subgroup_metadata.py program generates the subgroup metadata and JSON (Javascript Object Notation) data files. JSON and CSV (comma-separated value) files document restriction fragment sequences, the number of bands and band sizes for each restriction enzyme, the similarity coefficient matrix, and virtual gel images for each subgroup classification. JSON data files are generated for use as a database within the phyto_RFLP_digest.py program. Output files are organized in the subgroups metadata folder which contains the data, json and gels subfolders. The data folder contains the num_bands_subgroups.csv and similarity_coefficient_matrix_subgroups.csv. The num_bands_subgroups.csv file lists the number of bands and band sizes generated by each enzyme for each subgroup reference sequence. The similarity_coefficient_matrix_subgroups.csv file displays the F value for each strain x and strain y pair enzyme for each subgroup reference sequence. The json directory contains the subgroup_RFLP_digests.json and subgroup_sc_matrix.json files. The subgroup_ subgroup_RFLP_digests.json file lists the fragment nucleotide sequences, fragment sizes, number of bands for each restriction enzyme and the full nucleotide and peptide sequences and the length of the sequences. The subgroup_sc_matrix.json file lists the similarity coefficient of each strain x and strain y pair. The gels folder contains the *in silico* RFLP virtual agarose gels for each group/subgroup sequence. The program generates the CpnClassiPhyR metadata and JSON data files. JSON and CSV files document restriction fragment sequences, the number of bands and band sizes for each restriction enzyme, the similarity coefficient matrix, and virtual gel images for each input sequence.

The phyto_RFLP_digest.py program requires a fasta file as input and an absolute directory path for the output directory. The input fasta formatted sequence(s) can consist of full length and/or trimmed *cpn60* universal target nucleotide sequences in the anti-sense and sense strand orientation. If the input sequence is less than 450 bps the sequence is documented as being too short for further analysis are not considered for further analysis. The CpnClassiPhyR class uses a BLAST nucleotide database consisting of non-redundant *cpn*60 UT sequences from phytoplasma strains obtained from cpnDB (cpndb.ca). A blastn alignment is performed against this custom database using the BLAST+ software package (Camacho et al., 2009). The CpnClassiPhyR class module searches the results of the first 50 hits for matches to phytoplasma. Sequences with ≥70% nucleotide sequence identity to any of the phytoplasma reference sequences are retained for further assessment. The best blast hit of each alignment is used for determining the closest reference sequence, strandedness, and start and end position of the *cpn*60 UT sequence. If the sequence is in the antisense orientation the reverse complement of the sequence is derived prior to *in silico* RFLP analysis. If the input sequence is greater than or equal to 552 bps plus the length of the forward and reverse *cpn*60 UT primers (Dumonceaux et al., 2014), cutadapt (Martin, 2011) (version 1.16 + 16.g6819c78) is used to remove the primer sequences to obtain the trimmed and oriented *cpn*60 UT sequence. The sequences are then subjected to an *in silico* RFLP digest and F values are calculated and compared to each *cpn*60 UTgroup/subgroup reference sequence (Fig. 1.4). If $F < 0.59$ the input sequence is flagged as being a probable new c*pn*60 group. If $F > = 0.59$ and $F < 0.97$, the input sequence is suggested as a probable new subgroup of the closest matching group. Sequences with $F \geq 0.97$ are identified and the group and subgroup reported. F values are also reported as a similarity coefficient matrix for input sequences in comparison to all the *cpn*60 reference sequences.

1.4.3 Phylogenetic Analysis

Phylogenetic trees are automatically calculated for input sequences in the context of phytoplasma *cpn*60 reference sequences using the generate_subgroup_trees.py and generate_cpnclassiphyr_trees.py programs. Phylogenetic analysis is performed using the MEGA-CC 10 software package (Kumar, Stecher, Li, Knyaz, & Tamura, 2018; Kumar, Stecher, Peterson, & Tamura, 2012) and automated in Python using the Bio.Phylo module of BioPython. Phylogenetic trees are rooted using the outgroup sequence *Acholeplasma laidlawii* PG8R10 (cpnDB b30065, GenBank LXYB01000004) and converted to newick and phyloxml tree file formats. Trimmed and oriented *cpn60* UT nucleotide sequences are aligned to the *cpn60* UT group and subgroup reference database using the CLUSTALW algorithm and phylogenetic trees are generated using the Neighbor-Joining method and Tamura-Nei model with bootstrapping 1000 times.

For phylogenetic analysis of the sequences determined in this study, 62 *cpn*60 UT sequences representing samples amplified using the phytoplasma *cpn*60 universal primers as well as sequences retrieved from Genbank were aligned using the CLUSTALW option of the MEGALIGN program. Phylogenetic analysis was conducted for *cpn*60 UT sequences using the Maximum-likelihood method with MEGA v6.0 (Tamura, Stecher, Peterson, Filipski, & Kumar, 2013), and bootstrapping 1000 times. *Acholeplasma ladlawii* strain PG-8A (U14905) was used as outgroup to root the tree. The same analysis was performed with the F2nR2 fragment of the 16S rRNA-encoding gene.

1.4.4 Examination of Genome Sequences for 16S rRNA and cpn60 Gene Sequences

Publicly available genome sequences from phytoplasmas were downloaded from GenBank (accession numbers provided in Table 1.4). Genomes were queried for the presence of 16S rRNA-encoding genes using command-line BLASTn (Altschul et al., 1990) with 16S rRNA gene sequences from related phytoplasma strains as query. Similarly, *cpn*60 genes were found in the genome sequences using BLASTn with *cpn*60 sequences from related phytoplasma strains. In addition, the PGAP annotation of the genome sequence was searched for the presence of genes annotated as 16S rRNA or *cpn60*/*hsp*60/*groEL*. Wherever possible, one or both 16S rRNA gene sequences retrieved from genomes was typed using the iPhyclassifier (Zhao et al., 2009) to determine the group/subgroup of the strain.

1.5 Concluding Remarks

In addition to '*Candidatus* Phytoplasma' species, phytoplasma strains have been classified into ribosomal groups and subgroups using RFLP analysis of a fragment of 16S ribosomal RNA-encoding genes. Complementary classifications using other genes can provide finer resolution of phytoplasmas. We have developed a web-based interface, the CpnClassiPhyR, for classifying phytoplasma strains using the sequence of a single-copy, protein-coding gene, chaperonin-60 (*cpn*60). Classification of phytoplasmas using both ribosomal RNA-encoding sequences and *cpn*60 will improve phytoplasma strain tracking, benefiting agricultural production systems worldwide.

Acknowledgements This work was supported by Agriculture and Agri-Food Canada A-base Project 1668 – Sustainable Small Fruit Production. Establishment of this tool was enabled in part by support provided by the University of Saskatchewan (https://www.usask.ca/), Westgrid (https://www.westgrid.ca/), and Compute Canada (www.computecanada.ca). The funders had no role in study design or execution, or in the preparation and editing of the manuscript or decision to publish.

References

Altschul, S. F., Gish, W., Miller, W., Myers, E. W., & Lipman, D. J. (1990). Basic local alignment search tool. *Journal of Molecular Biology, 215*(3), 403–410.

Bertaccini, A., Duduk, B., Paltrinieri, S., & Contaldo, N. (2014). Phytoplasmas and phytoplasma diseases: A severe threat to agriculture. *American Journal of Plant Sciences, 5*(12), 1763.

Camacho, C., Coulouris, G., Avagyan, V., Ma, N., Papadopoulos, J., Bealer, K., … Madden, T. L. (2009). BLAST+: Architecture and applications. *BMC Bioinformatics, 10*(1), 421.

Chung, W.-C., Chen, L.-L., Lo, W.-S., Lin, C.-P., & Kuo, C.-H. (2013). Comparative analysis of the peanut witches'-broom phytoplasma genome reveals horizontal transfer of potential mobile units and effectors. *PLoS One, 8*(4), e62770.

Clark, G. W., & Tillier, E. R. (2010). Loss and gain of GroEL in the Mollicutes. *Biochemistry and Cell Biology, 88*(2), 185–194.

Cock, P. J., Antao, T., Chang, J. T., Chapman, B. A., Cox, C. J., Dalke, A., ... Wilczynski, B. (2009). Biopython: Freely available Python tools for computational molecular biology and bioinformatics. *Bioinformatics, 25*(11), 1422–1423.

Contaldo, N., Mejia, J. F., Paltrinieri, S., Calari, A., & Bertaccini, A. (2012). Identification and GroEL gene characterization of green petal phytoplasma infecting strawberry in Italy. *Phytopathogenic Mollicutes, 2*(2), 59–62.

Deng, S., & Hiruki, C. (1991). Amplification of 16S rRNA genes from culturable and nonculturable mollicutes. *Journal of Microbiological Methods, 14*(1), 53–61.

Dumonceaux, T. J., Green, M., Hammond, C., Pérez-López, E., & Olivier, C. Y. (2014). Molecular diagnostic tools for detection and differentiation of phytoplasmas based on chaperonin-60 reveal differences in host plant infection patterns. *PLoS One, 9*(12), e116039.

Firrao, G., Andersen, M., Bertaccini, A., Boudon, E., Bove, J., Daire, X., ... Gibb, K. S. (2004). '*Candidatus* Phytoplasma', a taxon for the wall-less, non-helical prokaryotes that colonize plant phloem and insects. *International Journal of Systematic and Evolutionary Microbiology, 54*(4), 1243–1255.

Fischer, A., Santana-Cruz, I., Wambua, L., Olds, C., Midega, C., Dickinson, M., ... Jores, J. (2016). Draft genome sequence of "Candidatus Phytoplasma oryzae" strain Mbita1, the causative agent of Napier grass stunt disease in Kenya. *Genome Announcements, 4*(2), e00297–e00216.

Gundersen-Rindal, D., & Lee, I.-M. (1996). Ultrasensitive detection of phytoplasmas by nested-PCR assays using two universal primer pairs. *Phytopathologia Mediterranea, 34*, 144–151.

Harrison, N., Davis, R. E., Oropeza, C., Helmick, E. E., Narvaez, M., Eden-Green, S., ... Konan Konan, J. L. (2014). '*Candidatus* phytoplasma palmicola': A novel taxon associated with a lethal yellowing-type disease (LYD) of coconut (*Cocos nucifera* L.) in Mozambique. *International Journal of Systematic and Evolutionary Microbiology, 64*(6), 1890–1899.

Hemmingsen, S. M., Woolford, C., van der Vies, S. M., Tilly, K., Dennis, D. T., Georgopoulos, C. P., ... Ellis, R. J. (1988). Homologous plant and bacterial proteins chaperone oligomeric protein assembly. *Nature, 333*(6171), 330.

Hill, J. E., Penny, S. L., Crowell, K. G., Goh, S. H., & Hemmingsen, S. M. (2004). cpnDB: A chaperonin sequence database. *Genome Research, 14*(8), 1669–1675.

Hill, J. E., Town, J. R., & Hemmingsen, S. M. (2006). Improved template representation in cpn60 polymerase chain reaction (PCR) product libraries generated from complex templates by application of a specific mixture of PCR primers. *Environmental Microbiology, 8*(4), 741–746.

Hogenhout, S. A., Oshima, K., Ammar, E. D., Kakizawa, S., Kingdom, H. N., & Namba, S. (2008). Phytoplasmas: Bacteria that manipulate plants and insects. *Molecular Plant Pathology, 9*(4), 403–423.

Kumar, S., Stecher, G., Li, M., Knyaz, C., & Tamura, K. (2018). MEGA X: Molecular evolutionary genetics analysis across computing platforms. *Molecular Biology and Evolution, 35*(6), 1547–1549.

Kumar, S., Stecher, G., Peterson, D., & Tamura, K. (2012). MEGA-CC: Computing core of molecular evolutionary genetics analysis program for automated and iterative data analysis. *Bioinformatics, 28*(20), 2685–2686.

Lee, I.-M., Davis, R. E., & Gundersen-Rindal, D. E. (2000). Phytoplasma: Phytopathogenic mollicutes. *Annual Reviews in Microbiology, 54*(1), 221–255.

Lee, I.-M., Gundersen-Rindal, D. E., Davis, R. E., & Bartoszyk, I. M. (1998). Revised classification scheme of phytoplasmas based on RFLP analyses of 16S rRNA and ribosomal protein gene sequences. *International Journal of Systematic and Evolutionary Microbiology, 48*(4), 1153–1169.

Lee, I.-M., Gundersen-Rindal, D. E., Davis, R. E., Bottner, K., Marcone, C., & Seemüller, E. (2004). '*Candidatus* Phytoplasma asteris', a novel phytoplasma taxon associated with aster yellows and related diseases. *International Journal of Systematic and Evolutionary Microbiology, 54*(4), 1037–1048.

Lee, I.-M., Zhao, Y., & Bottner, K. (2006). SecY gene sequence analysis for finer differentiation of diversestrains in the aster yellows phytoplasma group. *Molecular and Cellular Probes, 20*(2), 87–91.

Links, M. G., Dumonceaux, T. J., Hemmingsen, S. M., & Hill, J. E. (2012). The chaperonin-60 universal target is a barcode for bacteria that enables de novo assembly of metagenomic sequence data. *PLoS One, 7*(11), e49755.

Makarova, O., Contaldo, N., Paltrinieri, S., Kawube, G., Bertaccini, A., & Nicolaisen, M. (2012). DNA barcoding for identification of 'Candidatus Phytoplasmas' using a fragment of the elongation factor Tu gene. *PLoS One, 7*(12), e52092.

Martin, M. (2011). Cutadapt removes adapter sequences from high-throughput sequencing reads. *EMBnet Journal, 17*(1), 10–12.

Martini, M., Lee, I.-M., Bottner-Parker, K. D., Zhao, Y., Botti, S., Bertaccini, A., … Khan, A. (2007). Ribosomal protein gene-based phylogeny for finer differentiation and classification of phytoplasmas. *International Journal of Systematic and Evolutionary Microbiology, 57*(9), 2037–2051.

Mitrović, J., Kakizawa, S., Duduk, B., Oshima, K., Namba, S., & Bertaccini, A. (2011). The groEL gene as an additional marker for finer differentiation of '*Candidatus* Phytoplasma asteris'-related strains. *Annals of Applied Biology, 159*(1), 41–48.

Mitrović, J., Smiljković, M., Seemüller, E., Reinhardt, R., Hüttel, B., Büttner, C., … Duduk, B. (2015). Differentiation of '*Candidatus* Phytoplasma cynodontis' based on 16S rRNA and groEL genes and identification of a new subgroup, 16SrXIV-C. *Plant Disease, 99*(11), 1578–1583.

Miyazaki, A., Shigaki, T., Koinuma, H., Iwabuchi, N., Rauka, G. B., Kembu, A., … Maejima, K. (2017). '*Candidatus* Phytoplasma noviguineense', a novel taxon associated with Bogia coconut syndrome and banana wilt disease on the island of New Guinea. *International Journal of Systematic and Evolutionary Microbiology, 68*(1), 170–175.

Naderali, N., Nejat, N., Vadamalai, G., Davis, R. E., Wei, W., Harrison, N. A., … Zhao, Y. (2017). '*Candidatus* Phytoplasma wodyetiae', a new taxon associated with yellow decline disease of foxtail palm (*Wodyetia bifurcata*) in Malaysia. *International Journal of Systematic and Evolutionary Microbiology, 67*(10), 3765–3772.

Namba, S. (2011). Phytoplasmas: A century of pioneering research. *Journal of General Plant Pathology, 77*(6), 345–349.

Omar, A. F. (2016). Association of 'Candidatus Phytoplasma cynodontis' with Bermuda grass white leaf disease and its new hosts in Qassim province, Saudi Arabia. *Journal of Plant Interactions, 11*(1), 101–107.

Omar, A. F., Pérez-López, E., Al-Jamhan, K. M., & Dumonceaux, T. J. (2017). First report of a new jojoba (Simmondsia chinensis) witches'-broom disease in Saudi Arabia and its association with infection by a 'Candidatus Phytoplasma australasiae'-related phytoplasma strain. *Plant Disease, 101*(8), 1540.

Pérez-López, E., & Dumonceaux, T. J. (2016). Detection and identification of the heterogeneous novel subgroup 16SrXIII-(A/I) I phytoplasma associated with strawberry green petal disease and Mexican periwinkle virescence. *International Journal of Systematic and Evolutionary Microbiology, 66*(11), 4406–4415.

Pérez-López, E., Luna-Rodríguez, M., Olivier, C. Y., & Dumonceaux, T. J. (2016). The underestimated diversity of phytoplasmas in Latin America. *International Journal of Systematic and Evolutionary Microbiology, 66*(1), 492–513.

Pérez-López, E., Olivier, C. Y., Luna-Rodríguez, M., & Dumonceaux, T. J. (2016). Phytoplasma classification and phylogeny based on in silico and in vitro RFLP analysis of cpn60 universal target sequences. *International Journal of Systematic and Evolutionary Microbiology, 66*(12), 5600–5613.

Pérez-López, E., Omar, A. F., Al-Jamhan, K. M., & Dumonceaux, T. J. (2018). Molecular identification and characterization of the new 16SrIX-J and cpn60 UT IX-J phytoplasma subgroup associated with chicory bushy stunt disease in Saudi Arabia. *International Journal of Systematic and Evolutionary Microbiology, 68*(2), 518–522.

Quaglino, F., Kube, M., Jawhari, M., Abou-Jawdah, Y., Siewert, C., Choueiri, E., … Lova, M. M. (2015). 'Candidatus Phytoplasma phoenicium'associated with almond witches'-broom disease: From draft genome to genetic diversity among strain populations. *BMC Microbiology, 15*(1), 148.

Ratnasingham, S., & Hebert, P. D. (2007). BOLD: The barcode of life data system (http://www.barcodinglife.org). *Molecular Ecology Notes, 7*(3), 355–364.

Saccardo, F., Martini, M., Palmano, S., Ermacora, P., Scortichini, M., Loi, N., & Firrao, G. (2012). Genome drafts of four phytoplasma strains of the ribosomal group 16SrIII. *Microbiology, 158*(11), 2805–2814.

Schneider, B., Seemüller, E., Smart, C., & Kirkpatrick, B. (1995). Phylogenetic classification of plant pathogenic mycoplasma-like organisms or phytoplasmas. In S. Razin & J. Tully (Eds.), *Molecular and diagnostic procedures in mycoplasmology* (Vol. 1, pp. 369–380). San Diego, CA: Academic Press.

Schwarz, D., Adato, O., Horovitz, A., & Unger, R. (2018). Comparative genomic analysis of mollicutes with and without a chaperonin system. *PLoS One, 13*(2), e0192619.

Sparks, M. E., Bottner-Parker, K. D., Gundersen-Rindal, D. E., & Lee, I.-M. (2018). Draft genome sequence of the New Jersey aster yellows strain of 'Candidatus Phytoplasma asteris'. *PLoS One, 13*(2), e0192379.

Tamura, K., Stecher, G., Peterson, D., Filipski, A., & Kumar, S. (2013). MEGA6: Molecular evolutionary genetics analysis version 6.0. *Molecular Biology and Evolution, 30*(12), 2725–2729.

Valiunas, D., Jomantiene, R., & Davis, R. E. (2013). Evaluation of the DNA-dependent RNA polymerase β-subunit gene (rpoB) for phytoplasma classification and phylogeny. *International Journal of Systematic and Evolutionary Microbiology, 63*(10), 3904–3914.

Wei, W., Davis, R. E., Lee, M., & Zhao, Y. (2007). Computer-simulated RFLP analysis of 16S rRNA genes: Identification of ten new phytoplasma groups. *International Journal of Systematic and Evolutionary Microbiology, 57*(8), 1855–1867.

Wei, W., Lee, I., Davis, R. E., Suo, X., & Zhao, Y. (2007). Virtual RFLP analysis of 16S rDNA sequences identifies new subgroups in the clover proliferation phytoplasma group. *Bulletin of Insectology, 60*(2), 349.

Wei, W., Pérez-López, E., Davis, R. E., Bermúdez-Díaz, L., Granda-Wong, C., Wang, J., & Zhao, Y. (2017). 'Candidatus Phytoplasma brasiliense'-related strains associated with papaya bunchy top disease in northern Peru represent a distinct geographic lineage. *Crop Protection, 92*, 99–106.

Zhang, R.-Y., Li, W.-F., Huang, Y.-K., Wang, X.-Y., Shan, H.-L., Luo, Z.-M., … Yin, J. (2016). Group 16SrXI phytoplasma strains, including subgroup 16SrXI-B and a new subgroup, 16SrXI-D, are associated with sugar cane white leaf. *International Journal of Systematic and Evolutionary Microbiology, 66*(1), 487–491.

Zhao, Y., & Davis, R. E. (2016). Criteria for phytoplasma 16Sr group/subgroup delineation and the need of a platform for proper registration of new groups and subgroups. *International Journal of Systematic and Evolutionary Microbiology, 66*(5), 2121–2123.

Zhao, Y., Davis, R. E., Wei, W., Shao, J., & Jomantiene, R. (2014). Phytoplasma genomes: Evolution through mutually complementary mechanisms, gene loss and horizontal acquisition. In *Genomics of plant-associated bacteria* (pp. 235–271). Berlin, Germany: Springer.

Zhao, Y., Wei, W., Davis, R. E., & Lee, I. M. (2010). Recent advances in 16S rRNA gene-based phytoplasma differentiation, classification and taxonomy. In P. G. Weintraub & P. Jones (Eds.), *Phytoplasmas: Genomes, plant hosts and vectors* (pp. 64–92). Oxfordshire, UK: CABI.

Zhao, Y., Wei, W., Lee, M., Shao, J., Suo, X., & Davis, R. E. (2009). Construction of an interactive online phytoplasma classification tool, iPhyClassifier, and its application in analysis of the peach X-disease phytoplasma group (16SrIII). *International Journal of Systematic and Evolutionary Microbiology, 59*, 2582–2593.

Zreik, L., Carle, P., Bove, J., & Garnier, M. (1995). Characterization of the mycoplasmalike organism associated with witches'-broom disease of lime and proposition of a *Candidatus* taxon for the organism,"*Candidatus* phytoplasma aurantifolia". *International Journal of Systematic and Evolutionary Microbiology, 45*(3), 449–453.

Chapter 2
Epidemiology of Non-culturable Phloem-Limited Pathogens of Citrus; Case Study Phytoplasma

Mohammad Djavaheri, Maryam Ansari, and M. Hossein Borhan

Abstract Iran is one of the countries with an important citrus industry. Iranian citrus production suffers from various diseases such as non-culturable phloem-limited diseases (NCPLDs) that caused historical losses on citrus production. Citrus Tristeza Virus (CTV), Witches' broom disease of lime (WBDL) and Citrus greening (HLB) are the most important NCPLDs in Iran. These three NCPLDs have been observed one after another on citrus orchards across the country. All three diseases are present on the far south Persian Gulf citrus belt with CTV being limited to the Caspian Sea citrus belt. '*Candidatus* Phytoplasma aurantifolia' is associated with WBDL in citrus cultivation areas located in the Persian Gulf and neighboring countries. WBDL is limited to a few Middle Eastern countries including Iran and has therefore been largely studied in Iran. This chapter reviews the research that has been conducted on this disease in addition to the current challenges that WBDL poses on citriculture. The similarities between WBDL and two other NCPLDs in Iran are discussed, with focus on the epidemiology of NCPLDs that could help coordinate different practices to effectively manage these diseases.

Keywords Non-culturable phloem-limited pathogens · Citrus · Phytoplasma · Iran

Abbreviations

*C*Pa	*Candidatus* Phytoplasma aurantifolia
*C*Las	*Candidatus* Liberibacter asiaticus
*C*Laf	*Candidatus* Liberibacter africanus
*C*Lam	*Candidatus* Liberibacter americanus
CTV	Citrus Tristeza Virus

M. Djavaheri (✉) · M. H. Borhan
Agriculture-Agrifood Canada, Saskatoon, SK, Canada
e-mail: mohammad.djavaheri@canada.ca; hossein.borhan@canada.ca

M. Ansari
College of Agriculture, Shiraz University, Shiraz, Iran

C. Y. Olivier et al. (eds.), *Sustainable Management of Phytoplasma Diseases in Crops Grown in the Tropical Belt*, Sustainability in Plant and Crop Protection 12,
https://doi.org/10.1007/978-3-030-29650-6_2

HLB	Citrus greening/Huanglongbing
NCPLDs	non-culturable phloem-limited diseases
WBDL	Witches' broom disease of lime

2.1 Introduction: Citrus and Citriculture in Iran

Citrus industry and cultivation has a history of over 300 years in Iran (Ebrahimi, 2002). Citrus varieties with improved traits have been imported for years from other countries such as Spain, Italy, Japan etc.. Encyclopaedia Iranica (2011) describes local and imported citrus varieties that are now present in Iran. To support the citrus producers, the government of Iran established the Citrus & Subtropical Fruits Research Institute (CSFRC) in 1930. CSFRC has expanded over the last 80 years and has now research stations throughout the country.

The imported varieties have been tested for their adaptation to various climates in Iran and citrus varieties that were selected for commercial use were grown in three citriculture belts i.e., Caspian Sea belt (Mediterranean climate), central and Persian Gulf belts (diverse climates from hot and dry to tropical) (Fig. 2.1). Citrus rootstocks and scions in different belts are summarized in Ebrahimi (2002). Citrus production includes fruits from orange to mandarin, clementine, lime, lemon and some unique varieties like Persian sweet lemon (*Citrus limetta* Risso, likely a mutant of acid lemon) and Bakraee (a local *Citrus reticulata* Blanco hybrid). The Mazandaran province, located on the Caspian Sea belt, produces near 47% of citrus (Alipour, HoseinBeyki, Jahed, Rahnama, & Sharifnia, 2013). The significant turnover for citrus industry encouraged the producers to increase citrus production and in 2017, the aim of five million tons was reached (Young Journalist Club, 2017). At present, Iran is ranked as one of the major citrus-producing countries in the world (Barzegar, Sohi, & Rahimian, 2006).

2.2 Challenging Diseases of Citrus in Iran

Citrus pathogens cause economic damages ranging from yield loss to decline and death of infected fruiting trees. The pathogens threatening citrus cultivation in Iran were either imported with new commercial varieties (Ebrahim-Nesbat & Nienhaus, 1978; Ebrahimi, 2002) or found their way into citrus trees by adaptation of their insect vector to feed on citrus (Barzegar et al., 2006; Bové et al., 2000). Iranian citrus trees have been shown to be naturally infected by fungal, bacterial, nematode, viral and viral-like, as well as mollicute pathogens (Akbarpour & Banihashemi, 1998; Barzegar et al., 2006; Ebrahim-Nesbat & Nienhaus, 1978; Ebrahimi, 2002; Faghihi, Salehi, Bagheri, & Izadpanah, 2009; Serra et al., 2014; Tanha Maafi & Damadzadeh, 2008). Three puzzling phloem-limiting pathogens received most attention among all other viral and viral-like diseases of citrus plantation in Iran: the Citrus tristeza virus, the Witches' broom disease of lime and the huanglongbing/citrus greening disease. Those three diseases are heavily affecting the Iranian citrus

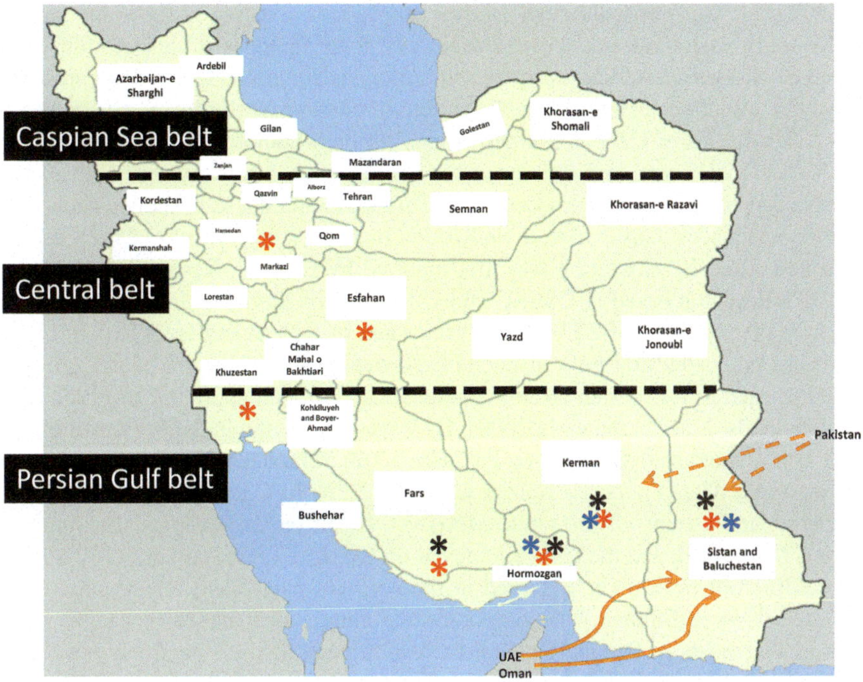

Fig. 2.1 Distribution of WBDL (Red asterisks), and HLB (Blue asterisks for infected plants and Black asterisks for infected psyllid vectors) on citrus trees in southern Iran. Iran has three major citriculture areas designated Caspian Sea belt, Central belt and Persian Gulf belt. WBDL (solid brown lines) is suggested to have entered to the country from southwest neighboring countries i.e., Oman, and/or the United Arab Emirates. However, HLB (Dashed brown line) is thought to be originated from Pakistan. For WBDL and HLB, the port of entry to Iran is suggested to be the far south east province of Sistan-and-Baluchestan. (Map source: https://lizardpoint.com/geography/iran-quiz.php)

industry, especially the production of Mexican lime, presenting new challenges to research and citriculture.

Plant pathology labs at the University of Mazandaran (in the Northern Province of Mazandaran) and Shiraz University (located in the south west part of the country-Fars province) have long been addressing different challenges posed by citrus pathogens. Government-funded research institutes under the Agricultural Research, Education and Extension Organization (AREEO) including Iranian Research Institute of Plant Protection (IRIPP), and Agricultural biotechnology research institute of Iran (ABRII) have had collaboration with these two universities. Besides these research institutes/labs, the certification of the germplasm and quarantine is implemented by the Plant Protection Organization (PPO) of Iran's Ministry of Agriculture.

Citrus tristeza virus (CTV) is the most prevalent virus and viral-like disease in citrus groves of Iran. CTV spreads through the infected budwood and by aphids

(Barzegar et al., 2006); although for over 30 years the existing aphid with potential CTV transmission did not show such activity. CTV includes strains with different degrees of severity. Additionally, the scion/rootstock combination, and the presence of active insect vectors determine the extent of damage by this virus. The presence of severe strains of CTV in northern Iran with high homology to strains found in Japan and California, and activity of the aphid vector, *Aphis gossypii* Glover, has been a cause for concern for a long time. This strain could lead to sudden decline of the infected trees. Citrus tristeza was found repeatedly in orchards of the Caspian Sea belt since 1978, when it was first reported by Ebrahim-Nesbat and Nienhaus (1978). Research based in CSFRC and University of Mazandaran was initiated to address the problems of CTV and its vector and to suggest strategies to control the diseases and slow down its dispersion. However, with time, the citrus growers became used to CTV as an endemic pathogen and did not apply the proposed control methods. Later in the early 2000s, CTV was also found in the Persian Gulf citrus belt (Izadpanah, Shafiee, & Pakniat, 2002). Three aphid species including *A. gossypii*, *Myzus persicae* Sulzer, and *A. craccivora* C.L. Koch were shown to experimentally transmit the virus. Moreover, based on the serological activity of two CTV antisera and the CTV symptoms on mexican lime, southern CTV strains were different from those observed in the northern part. More importantly, CTV plant hosts were different with sweet and sour orange, local mandarin, and Mexican lime being hosts in Persian Gulf citrus belt as compared to mandarin and sweet orange in the North.

Witches' broom disease of lime (WBDL) was first found on a few Mexican lime trees in rural communities located in south east Iran (Bové et al., 2000), then quickly spread to other areas that were devoid of the disease. WBDL gradually reduces the tree productivity over the years and eventually kills the trees roughly 5 years after infection.

Almost a decade later, symptoms of sudden decline occurred in southern Iran on sweet orange and Bakraee production (local citrus rootstock) (Faghihi et al., 2009; Golmohammadi et al., 2016; Salehi, Faghihi, Khanchezar, Bagheree, & Izadpanah, 2012). PCR-based methods revealed that a new emerging disease caused by '*Candidatus* Liberibactor asiaticus', was associated with the decline of the infected trees. Resistance for the citrus greening disease has yet to be found. Therefore, Citrus greening posed a new threat to citrus cultivation especially in the commercial citriculture in southern Iran.

Lack of implementation of adequate control practices has resulted in further spread of CTV, WBDL and HLB in Iran. Only recently, IPPO has been able to stop the dispersal of WBDL and HLB to new areas by implementing strict quarantine measures. The presence of CTV, WBDL, HLB, and viroids on the citrus trees in Iran is a unique situation. For example, there are examples of co-occurrence of those pathogens on the same infected trees, which provide great opportunities to study co-infection of plants with two or more pathogens belonging to different groups (Saberi, Alavi, Safaie, Moslemkhany, & Azadvar, 2017; Serra et al., 2014). Also, WBDL being present only in few Middle Eastern countries including Iran, most of the research on WBDL has been developed and conducted in Iran. Therefore, this review discusses substantial challenges caused by three groups of non-culturable phloem-limited pathogens on the Iranian citrus industry, with a greater focus on the two most challenging pathogens for the citriculture i.e., WBDL and citrus greening. The diseases are discussed in the order of their occurrence and first report in various citrus growing orchards and the introduction and geographical spread of the diseases are discussed from an epidemiology perspective. Concluding remarks summarize research-based strategies to combat the diseases.

2.2.1 Witches' Broom Disease of Lime (WBDL), Host Range, and Plant Response to the Pathogen

Mexican acid lime (*Citrus aurantifolia* (Christm.) Swingle) is an economically important crop for growers in the southern provinces of Iran (Sistan-Balouchestan, Hormozgan, Kerman, and Fars). Iran cultivate over 40,000 hectares of acid lime and produces nearly 40% of the world production. However, its cultivation is threatened by several prokaryotic diseases such as witches' broom disease of lime (WBDL), citrus bacterial canker, citrus stubborn, and citrus greening.

WBDL-infected citrus express symptoms of shortened internodes and proliferation of small pale green leaves, known as witches' brooms (Bové et al., 2000). In the advanced stages, the infected tree becomes almost non-productive and dies quickly (in five to six years). The damage of WBDL to Iranian citriculture is estimated to be over $ US 200,000,000 (Mardi et al., 2011). WBDL strongly impacted Iran's citrus industry due to restrictions on export and transport of seedlings and rootstocks from infested areas. The Iranian Witches' Broom Disease of Lime Network (IWBDLN) was established in 2007, in order to attract attention of all national and international researchers to control the disease. Because, research on NCPLDs in citrus orchards requires long time, immediate focus of IWBDLN funding was set on prevention i.e., quarantine measures, chemical control of insect vector(s) and eradication programs (Mardi et al., 2011).

'*Candidatus* Phytoplasma aurantifolia' (*C*Pa), a cell-wall-less, non-cultivable prokaryote, is the causal agent of WBDL. *C*Pa like many other devastating pathogens of citrus is strictly phloem-restricted, and biotrophic. In Iran, WBDL naturally occurs on sour lime (*C. aurantifolia*), Bakraee (*C. reticulata* hybrid), citron (*C. medica*) and grapefruit (*C. paradise*) and limequat (*C. aurantifolia* hybrid) (Bagheri, Faghihi, Salehi, & Khanchezar, 2010; Bové et al., 2000; Djavaheri & Rahimian, 2004; Faghihi, Bagheri, Askari Seyahooei, Pezhman, & Faraji, 2017). Surprisingly, Ghayeb Zamharir, Paltrinieri, Hajivand, Taheri, and Bertaccini (2017) traced back *C*Pa (16SrII-B) from grapevines in Markazi, a central province of Iran with almost no citrus cultivation. Hashemi-Tameh, Bahar, and Zirak (2014) also reported *C*pa from a few scattered commercial apple orchards in the Isfahan province with no commercial citrus cultivation. Additionally, *C*Pa infection of a dozen plants including other citrus species as well as herbaceous plants to WBDL has been reported in greenhouse experiments (Faghihi et al., 2017). Disease symptoms on herbaceous plants consist of typical phytoplasma disease symptoms i.e., yellowing, internode shortening, witches' broom, virescence, phyllody, and plant death. Based on the 16S rRNA sequences, *C*Pa and the phytoplasma strains detected in carrot, alfalfa, sesame, egg-plant and tomato big buds are related. Studies on *C*Pa in Iran described the pathogen as a homogenous microorganism (Faghihi et al., 2017; Salehi, Izadpanah, Siampour, Bagheri, & Faghihi, 2007), as they all cluster within the 16SrII group. The genome of *C*Pa comprises over 440 genes and is publicly available on NCBI website (Foissac & Carle, 2017).

The insect vector of *C*Pa in Iran and other countries around Persian Gulf is *Hishimonus phycitis* Distant (Bové et al., 2000). PCR amplification of a conserved region of rDNA belonging to *C*Pa and identified in infected *H. phycitis* provided the first indication that this leafhopper was the natural vector for WBDL. The leafhopper *H. phycitis*, has a very limited ability for disease transmission. In greenhouse experiments, *H. phycitis* was able to transmit the disease to Bakraee (*C. reticulata* hybrid), but failed to transmit to sour lime and other susceptible plants. Recent detection of the disease in grapevine and apple, in two central regions of Iran that are far from each other and from WBDL-infected citrus regions, raises the question of whether other vectors can actively transmit *C*Pa.

Studies on IWBDL have mainly been focused on the pathogen host range, the vector identity and its transmission abilities. Recently, the IWBDLN incorporated over 10 programs carried out at different institutes, to look for strategies to control the disease (Mardi et al., 2011). Next to conventional practices (like nutrition, agronomy, breeding), -Omics-based technologies including transcriptomics and proteomics of lime infected by witches' broom were implemented to answer some key questions about IWBDL: (I) What is the best method for early and accurate detection of the pathogen? (II) What are the resistance mechanisms used by plants to fight *C*Pa? What are the disease resistance signaling pathways employed by the plant to restrict *C*Pa establishment (early response), growth (midterm response) and propagation (late response)?

A proteomic approach identified 55 proteins in the leaves of infected Mexican lime trees with a significant response to *C*Pa (Taheri et al., 2011). Out of the 55 proteins, 39 were involved in plant oxidative stress, photosynthesis, metabolism, and stress responses. Monavarfeshani et al. (2013) reported nearly 200 proteins that were more abundant in infected lime trees and a similar number of proteins to be less abundant in *C*Pa-infected plants as compared to healthy plants. In addition to stress and defense-related responses, proteins involved in cellular processes like sugar metabolism, growth and development, signal transduction, photosynthesis, cell cycle, and cell wall organization showed alteration in infected plants. The authors detected similar defense-related proteins in both susceptible and resistant hosts of WBDL. Ehya et al. (2013) suggested the involvement of miRNAs in modulating the response of Mexican lime to infection by CPa.

Transcriptome analysis on Mexican lime trees (Mardi, Farsad, Gharechahi, & Salekdeh, 2015) revealed over 2800 transcripts that were significantly altered as compared to healthy plants. Genes with plausible role in plant-pathogen interaction, including cell wall biogenesis and degradation, sucrose metabolism, secondary metabolism, hormone (JA, auxin, ethylene, and BRs) biosynthesis and signalling, amino acid and lipid metabolism were mainly up-regulated in infected plants, however some cellular processes like ubiquitin proteolysis and oxidative phosphorylation pathways were down-regulated.

These results were in agreement with Zafari, Niknam, Musetti, and Noorbakhsh (2012) who showed that some of the enzymes with roles in plant defense including polyphenol oxidase (PPO), peroxidase (POX) and superoxide dismutase (SOD), have higher activity in *C*Pa-infected leaves of Mexican lime, suggesting that antioxidant enzymes are activated upon infection by phytoplasma. Furthermore, the content of total proteins and sugars as well as chlorophyll and plant growth hormone –auxin- was shown to be significantly reduced in the *C*Pa-infected leaves.

A graduate student project, conducted by Hosseinzadeh Ghasemabad (unpublished data, 2012) studied the possibility of transmission of *C*Pa to Arabidopsis. *Arabidopsis thaliana* (L.) Heynh. is extensively used in various plant-parasite interactions. Availability of the whole genome sequences and genetic resources in the form of lines/mutants with over or under-expressed genes, make *A. thaliana* an ideal candidate for the plant-pathogen interaction research. Since Arabidopsis is susceptible to various species of Dodder (Birschwilks, Sauer, Scheel, & Neumann, 2007), this parasitic vine was used as a transmission tool. *C*Pa was first transmitted to periwinkle by grafting from naturally infected trees, then transmitted to Arabidopsis via dodders (Fig. 2.2). In as early as 5 days, transmission of *C*Pa to Arabidopsis was confirmed using appropriate primers and PCR tests (Ansari, 2016). Transcriptional

Fig. 2.2 Attachment of dodder to the plants and transmission of WBDL to Arabidopsis from various infected hosts. Transmission from infected periwinkle (**a** and **b**), and symptomatic seedlings of key lime (**c**). Arrows show the dodder stems around Arabidopsis stem

changes of a few disease-related genes in dodder, as well as in *C*Pa-infected Arabidopsis were compared to that of healthy control plants. The differential expression of a few marker genes for SA- and JA-dependent defenses revealed that both pathways are involved in early response of Arabidopsis to WBDL. The dodder was able to transmit *C*Pa directly from infected Mexican lime to Arabidopsis within the first 7 days of being infested by *C*Pa-infected dodder (Ansari, 2016).

2.2.2 Greening or Huanglongbing, Host Range, and Plant Response to the Pathogen

Huanglongbing (HLB) or citrus greening was recently reported from most provinces in southern citrus belt of Iran (Faghihi et al., 2009; Salehi et al., 2012). HLB infection has been accounted for the rapid decline of sweet orange (*Citrus sinensis* cv. Valencia) in the southern part of the country. The pathogen infection causes mottling of the leaves resembling Zn deficiency, malformation of the fruits and eventually kills the infected tree. Symptoms of greening disease have previously been observed in Asia, Oceania, Africa and South and North Americas (Bové, 2006). Worldwide dispersion of the pathogen and its economic impact on citrus industry has led to significantly more research as compared to other pathogens of citrus.

Three non-culturable phloem-limited gram-negative bacterium of the genus *Candidatus* Liberibacter (*C*L) are associated with HLB symptoms. HLB-associated bacteria have been classified into three species *C*L asiaticus (*C*Las), *C*L africanus (*C*Laf) and *C*L americanus (*C*Lam) (Bové, 2006). All citrus cultivars are susceptible to HLB and insect vectors have been reported on various citrus species/cultivars. Sechler et al. (2009) reported an artificial medium for the cultivation of *C*Las, *C*Laf, and *C*lam, but attempts to replicate this protocol failed. Sap-sucking psyllids have been reported to act as natural insect vector for HLB disease. Similar to WBDL, grafting and dodder have been used to transmit *C*L bacterium (Bové et al., 2000; Salehi et al., 2012).

Interestingly, both HLB and its vectors are temperature-sensitive (Bové, 2006). *C*las and its vector *Diaphorina citri* Kuwayama are both heat tolerant with an optimum growth of over 30 °C while *C*laf and its vectore *Trioza erytreae* Del Guercio show an optimum growth at 22–25 °C and both vectors can transmit *C*Las and *C*Laf (Bové, 2006). *D. citri* and *C*Las have long been present in the Indian subcontinent located on the east side of Iran. *D. citri*, was first observed in Iran in 1997 in areas close to eastern neighbor country, Pakistan (Bové et al., 2000). Since then, high populations of *C*Las -infected *D. citri* have been found in citrus plantations of Hormozgan, Fars, Sistan-Baluchestan and Kerman provinces in southern Iran (Salehi et al., 2012).

Wang et al. (2016) profiled transcriptome of two naturally infected grapefruit varieties that were HLB- tolerant or susceptible. Their results indicated a clear role for basal resistance mechanisms being involved in the tolerance of these trees after being exposed to the pathogen for over 10 years. They found genes, functionally associated with transcription factors, abiotic stress, hormone signaling, cell wall, beta- glucanase, proteolysis, glutathione S-transferase, lipids, Pathogenesis Related (PR)-genes (DMR6-like and NPR1-like genes), and secondary metabolism being differentially expressed between the HLB tolerant and the HLB susceptible citrus trees.

2.3 Practical Measures to Control Phloem-Limiting Pathogens of Citrus; Case Study Phytoplasma

Citrus trees are usually grafted on sour orange (*Citrus aurantium* L.) rootstock and most commercial citrus species including orange, lime, and mandarin, as well as the sour orange rootstock are sensitive to CTV, WBDL and HBL. Control of tristeza disease could be achieved by replacing sour orange by rootstocks that are tolerant to the disease (Barzegar et al., 2006; Bové & Ayres, 2007). Most growers from the major citrus-growing regions, where CTV is endemic, have adapted to the presence of this disease. For HLB and WBDL however, the only known control method is to prevent healthy trees from becoming infected (Najafinia & Azadvar, 2016). Therefore, HLB and WBDL are a more serious problem than tristeza disease.

Due to the difficulty of eliminating directly the pathogens, control of HBL and WBDL is mostly preventive and start with strict quarantine measures to impair further national and international spread of *C*Pa and its vector. Other control methods rely mainly on lowering inoculum pressure in infected areas (Bové et al., 2000). Pathogen elimination could be achieved through early detection of the pathogen associated with removal of infected trees and chemical treatments against insect vectors. Breeding for resistence/tolerance, implementing cross-protection and the use of antibiotics and genetic engineering are part of the tool box used to control the diseases.

2.3.1 WBDL and HLB Origins and Foreign and Internal Quarantine for Effective Sanitation Practices in Citrus Orchards

WBDL has long been advocated to be imported from Oman and/or the UAE. The geographical distance and natural barriers between Oman/UAE and the locations where WBDL was observed in Iran support this hypothesis (Fig. 2.1). Mexican (key) limes (*Citrus aurantifulia*) and bitter orange (*Citrus aurantium*) have been imported to Iran from the Arab peninsula about 1200 years ago (Alipour et al., 2013). However, the disease has been recorded on the trees in Oman in the 1980s and in Iran 10 years later. Additionally, there is high similarity between phytoplasma strains associated with some of the herbaceous plants and *C*Pa, raising a plausible hypothesis that *C*Pa is likely transmitted from diseased weedy plants to citrus trees rather than from infected trees. The possibility of infection of lime trees from herbaceous hosts may practically results in a high risk of re-infection in the areas where eradication of symptomatic trees and quarantine measures have taken place. *C*Pa has recently been found on *Amaranthus* in Mexico (Ochoa-Sánchez, Parra-Cota, Aviña-Padilla, Délano-Frier, & Martínez-Soriano, 2009), although it has not yet been recorded on citrus in the Americas.

There are puzzling scenarios for the introduction of CTV and HLB to Iran as well. Undoubtedly, CTV has been introduced to the Caspian Sea citrus belt with imported infected germplasm from other countries (Barzegar et al., 2006; Ebrahim-Nesbat & Nienhaus, 1978; Ebrahimi, 2002). However, the origin of CTV disease in southern citrus growing regions where CTV has been around much longer than in the northern part of Iran is unknown. Interestingly, the vector of citrus HLB, *D. citri*, has been found on lime trees and annual crops and weeds, raising the question about the herbaceous origin of HLB. The natural spread of both HLB and WBDL has mainly been thought to be vector-dependent. It has been suggested that the low-paced spread of WBDL was due to the low efficiency of its vector, *H. physitis* and that the fast spread of HLB could indicate the efficacy of its psyllid vector, *D. citri*.

Successful quarantine actions implemented by IPPO for WBDL have stopped its further spread to northern parts of the Fars province and other citrus-growing provinces in Iran. Internal quarantine for WBDL has been a success story to slow down the spread of the disease; however its effectiveness could be questioned.

2.3.2 Accurate and Early Detection of the Pathogen

Detection of the pathogen is the key step in efficient management of plant pathogens especially for non-culturable pathogens. WBDL have a long incubation period, multiplying in their host plant without causing any visible symptoms while remaining transferrable for over 5 years in naturally-occurring disease orchards. Additionally, symptoms of some citrus diseases like HLB are influenced by temperature (Bové, 2006). Therefore, in addition to the latent period of the disease, environmental cues add to the complexity of diagnosing the disease.

Amiri, Izadpanah, Hamzeh Zarghani, and Salehi (2016) compared field as well as artificially graft-infected lime trees in the greenhouse and showed that a high population of the *C*Pa was present in the leaves, stems, crown and roots of the plants that were already showing symptoms of witches' brooms. Interestingly, the authors showed that the concentration of phytoplasma was the highest in August as compared to that on March and December, suggesting that the *C*Pa is well-adapted to high temperatures. Authors recommend that surveys are best to be conducted in summer from the crown bark in symptomless tree or from the green bark of stems and veins of terminal leaves in young seedlings and grafted plants.

Rad et al. (2012) developed a quantum dot (QD)-based nano-biosensor for the detection of phytoplasma by targeting immunodominant membrane protein (IMP) of *C*Pa. The sensor showed high specificity and detected *C*Pa propagules found in concentrations as low as 5000 *C*Pa per mL. Despite recent advances to accurately detect *C*Pa in the earliest time post infection, PCR-based techniques used to amplify phytoplasma rDNA remain the easiest and most reliable tool. The method is cost-effective and accurate as evident by its wide use in recent publications (Faghihi et al., 2017; Ghayeb Zamharir et al., 2017).

2.3.3 Vector Control by Chemical Spray

A great achievement for all three NCPLDs of citrus is the discovery of their natural insect vector. Control of the insect vector has been shown to be the most practical measure to control the pathogens that rely on their specialized vector for transmission and spread. The presence of the insect vector before observing the pathogen is alarming and regular surveys are needed to make sure that the disease does not spread to non-infected regions. Occurrence and epidemics of both WBDL and HLB on citrus have been reported a few years after their insect vectors were detected in non-infected citrus orchards (Bové, 2006; Bové et al., 2000; Salehi et al., 2012). CTV infection was limited only to mandarin trees in Caspian Sea citrus belt before its aphid vector became capable of transmission to other citrus cultivars. This is an example that underlines the importance of controlling the insect vectors as a crucial step to slow down the spread of the pathogen.

2.3.4 Eradication of Trees with Symptoms

Removal of plants with typical symptoms is a known method used to lower the disease pressure. This practice could stop further spread of the pathogen by lowering the inoculum pressure in infected areas in addition to keep active sap-sucking vectors away from the source of the pathogen (Lee, Davis, & Gundersen-Rindal, 2000).

Eradication has been a major challenge for growers and the government in Iran. The citriculture where the WBDL has first been observed in the far south east of the country is in extremely poor and remote areas and is very limited. In those places, lime trees are the only source of living for the growers. These issues posed a great challenge for the government to convince growers of the potential harm of the infected individual trees for their uninfected trees and other farmers in the region. It took years for the IWBDLN to allocate funds and convince local producers to adopt the eradication practices for trees infected with WBDL. Unfortunately, the delay in eradication likely contributed to the spread of the disease to other citrus growing areas and to the new cultivars of citrus.

2.3.5 Control of *C*Pa by Injecting Surfactant or Antibiotic Tetracycline

Antibiotics have been previously used for the treatment of plants infected with pathogens including bacteria (Hu, Jiang, & Wang, 2017; Sundin & Wang, 2018). For pathogens that are living inside the plants, accessibility of the antibiotics is important and antimicrobial injection in the tree is an efficient method. Using trunk injection, Askari et al. (2011) tested the effectiveness of a bacterially-produced surfactin as well as tetracycline in reducing the population of *C*Pa in artificially infected seedlings of key lime. A quantitative real-time PCR assay revealed that each of the two antimicrobial compounds reduced the number of *C*Pa over time. Furthermore, they showed that surfactin and tetracycline had an additive effect in the reduction of *C*Pa quantities in infected tissues. Although these results are positive, it is not clear how long the inhibitory effect would last in the antibiotic-treated trees. Moreover, this method is expensive and not practical in areas with large number of diseased plants.

2.3.6 Cross-Protection

There are areas in the Persian Gulf citrus belt where all three NCPLDs are present individually or mixed, in citrus trees. Saberi et al. (2017) showed a mixed infection of sweet lemon with a phytoplasma (belonged to peanut witches'broom 16SrII group) and HLB. Serra et al. (2014) showed CTV and citrus viroids both being present on Mexican lime in the same tree. Cross-protection has previously been used to reduce the damage caused by severe strains of CTV by manually infecting the trees with mild strains of the same virus (R. F. Lee & Keremane, 2013). Foroutan, Rahimian, Kazemitabar, Mardi, and Alavi (2009) reported a reduction in disease symptoms as well as reduced density of *C*Pa where a CTV strain was experimentally inoculated on the trees prior to *C*Pa infection on the same trees.

2.3.7 Breeding Programs and Horticultural Best Practices

Citrus trees in infected areas are constantly under environmental stresses. In the Persian Gulf citrus belt, very high temperatures as well as drought and lack of nutrients predispose citrus trees to diseases like WBDL and its vector. Using proper combination of rootstock and grafts has been successfully applied to control tristeza infection (Barzegar et al., 2006; Bové & Ayres, 2007). This method however has not yet been shown to be successful in controlling the WBDL as most commercial cultivars including local rootstocks (like Bakraee) are not resistance to *C*Pa. Another major barrier for this method in general is the compatibility of the rootstock and scion (Ebrahimi, 2002).

Breeding of citrus cultivars toward production of varieties with higher resistance to external stresses have been suggested (Omura & Shimada, 2016). Breeding programs are labor-, cost, and time- intensive; however they seem to be one of the most efficient ways to maintain favourable traits of commercial citrus varieties with improved plant tolerance to pathogens and abiotic stress. Asadi and Golmohammadi (2016) generated over 100 hybrid citrus plants by crossing five male parents and 4 females. Over 40% of the progenies were found to be free of *C*Pa and they did not develop any symptoms after being artificially infected by the pathogen. The hybrids are being evaluated for agronomic traits and promising candidates might be soon available to growers.

2.3.8 New Tools Through Implementation of Genetic Engineering

Shahryari, Safarnejad, Shams-Bakhsh, Schillberg, and Nölke (2013) generated a recombinant antibody to be used for the suppression of WBDL. A single-chain variable fragment antibody (scFvIMP6) was produced against the immunodominant membrane protein (IMP) of the witches' broom phytoplasma. Their results indicated a high activity for scFvIMP6 in planta thus being a good candidate for the detection of *C*Pa. Authors suggested that this antibody could likely be used to suppress witches' broom in lime trees. Dalio et al. (2017) have extensively reviewed citrus plant immunity mechanisms on most of the citrus diseases. Authors suggest application of molecular-based approaches as part of an integrated strategy, to prevent these diseases. Using model plants could also help understanding the mechanisms underlying disease resistance against citrus pathogens (Ansari, 2016; Hosseinzadeh Ghasemabad, 2012).

2.4 Concluding Remarks

Non-culturable phloem-limited diseases (NCPLDs) are threatening citrus industry in Iran and around the world. This review summarized the impact of three NCPLDs on citrus cultivation in Iran and the socioeconomic challenges that they pose. Citrus industry in Iran suffers from other **phloem-restricted** pathogens such as *Spiroplasma citri*, the causal agent of stubborn disease, as well as some **xylem-restricted** bacteria including *Xylella fastidiosa*, the causal agent of citrus variegated chlorosis. Banihashemian, Taheri, Alian, Bové, and Duran-Vila (2013) revealed that around 50% of the samples from two major citrus producing regions were infected with Citrus bark cracking viroid (CBCVd), Citrus bent leaf viroid (CBLVd), Citrus exocortis viroid (CEVd), Citrus dwarfing viroid (CDVd), Hop stunt viroid (HSVd), and Citrus viroid V (CVd-V) often in mixed infections. CTV, HLB and WBDL have had a major impact on the production of citrus in Iran; however other pathogens of citrus also contribute to the yield loss and decline of the trees.

So far, HLB and WBDL have been limited to the Persian Gulf citrus belt. Zagros and Alborz mountains are a physical barrier for natural migration of the insect vectors to northern citrus belt. However, extra caution is needed to avoid introducing these diseases to disease-free areas. Saberi et al. (2017) showed for the first time that sour orange (*Citrus aurantium* L.) which is mainly grown in northern citrus belt of Iran has been naturally infected (in Kerman province, central south) with HLB. Some psyllid vectors and strains of HLB are well-suited to grow in high altitudes, and this will require extra precautions from citrus growers to control the disease. Additionally, occurrence of *C*Pa on grapevine and apple in central Iran and previously observed in a few grapevine samples from South Africa (Botti & Bertaccini, 2006; Ghayeb Zamharir et al., 2017; Hashemi-Tameh et al., 2014) may indicate that the pathogen is likely to be more dispersed than currently known. Sanitation, fast and accurate detection, and vector control are still the best and most-effective practices to control NCPLDs.

References

Akbarpour, K., & Banihashemi, Z. (1998). Distribution and severity of citrus gummosis in Fars province, Iran. *Iranian Journal of Plant Pathology, 34*, 18–31.

Alipour, H., HoseinBeyki, A., Jahed, M., Rahnama, H., & Sharifnia, M. (2013). A review on citrus production and export marketing strategies in mazandaran province, Iran. *Middle-East Journal of Scientific Research, 14*(10), 1375–1380.

Amiri, M., Izadpanah, K., Hamzeh Zarghani, H., & Salehi, M. (2016). *Detection of 'Candidatus Phytoplasma aurantifolia' in lime trees during the latent period of the disease*. Paper presented at the 22th Iranian plant protection congress (August 26–30, 2016), College of Agriculture and Natural Resources, University of Tehran, Karaj, Iran. p. 87.

Ansari, M. (2016). *Transmission of WBDL to Arabidopsis from infected Mexican limes using dodder. Special problem course*. Shiraz, Iran: University of Shiraz.

Asadi, A. A., & Golmohammadi, M. I., (2016, Brazil). Cross breeding to produce new acid accessions tolerant to Witches' broom disease of lime (WBDL) in Iran. Citrus Research & Technology, p. 37.

Askari, N., Jouzani, G., Mousivand, M., Foroutan, A., Nazari, A. H., Abbasalizadeh, S., … Mardi, M. (2011). Evaluation of antiphytoplasma properties of surfactin and tetracycline towards lime witches broom disease using real-time PCR. *Journal of Microbiology and Biotechnology, 21*(1), 81–88.

Bagheri, A., Faghihi, M., Salehi, M., & Khanchezar, A. (2010). *First report of natural infection of grapefruit trees to lime witches broom phytoplasma*. Paper presented at the 19th Iranian plant protection congress (July 31–August 3, 2010), Iranian Research Institute of Plant Protection, Teheran, Iran, Vol. 1, p. 409.

Banihashemian, S., Taheri, H., Alian, Y. M., Bové, J., & Duran-Vila, N. (2013). Complex mixtures of viroids identified in the two main citrus growing areas of Iran. *Journal of Plant Pathology, 95*(3), 647–654.

Barzegar, A., Sohi, H. H., & Rahimian, H. (2006). Characterization of *Citrus* tristeza virus isolates in northern Iran. *Journal of General Plant Pathology, 72*(1), 46–51.

Birschwilks, M., Sauer, N., Scheel, D., & Neumann, S. (2007). *Arabidopsis thaliana* is a susceptible host plant for the holoparasite *Cuscuta* spec. *Planta, 226*(5), 1231–1241.

Botti, S., & Bertaccini, A. (2006). First report of phytoplasmas in grapevine in South Africa. *Plant Disease, 90*(10), 1360–1360.

Bové, J. (2006). Huanglongbing: A destructive, newly-emerging, century-old disease of citrus. *Journal of Plant Pathology, 88*(1), 7–37.

Bové, J., & Ayres, A. J. (2007). Etiology of three recent diseases of citrus in Sao Paulo state: Sudden death, variegated chlorosis and huanglongbing. *IUBMB Life, 59*(4–5), 346–354.

Bové, J., Danet, J., Bananej, K., Hassanzedeh, N., Taghizadeh, M., Salehi, M., ... Garnier, M. (2000). Witches' broom disease of lime (WBDL) in Iran. In J. Da Graça, R. Lee, & R. Yokomi (Eds.), *Proceedings of the fourteenth conference of the International Organization of Citrus Virologists* (pp. 207–212). Riverside, CA: International Organization of Citrus Virologists.

Dalio, R. J., Magalhaes, D. M., Rodrigues, C. M., Arena, G. D., Oliveira, T. S., Souza-Neto, R. R., ... Maximo, H. J. (2017). PAMPs, PRRs, effectors and R-genes associated with citrus–pathogen interactions. *Annals of Botany, 119*(5), 749–774.

Djavaheri, M., & Rahimian, H. (2004). Witches'-broom of bakraee (*Citrus reticulata* hybrid) in Iran. *Plant Disease, 88*(6), 683–683.

Ebrahimi, Y. (2002). Citrus industry in Iran. In K. Djelouah, A. D'Onghia, & C. Roistacher (Eds.), *Proceedings of the Mediterranean research network on certification of citrus (MNCC): 1998–2001* (Vol. 43, pp. 31–35): Bari, Italy: CIHEAM.

Ebrahim-Nesbat, F., & Nienhaus, F. (1978). Occurrence of citrus tristeza virus in Iran. *Journal of Plant Diseases and Protection, 85*(5), 308–312.

Ehya, F., Monavarfeshani, A., Fard, E. M., Farsad, L. K., Nekouei, M. K., Mardi, M., ... Salekdeh, G. H. (2013). Phytoplasma-responsive microRNAs modulate hormonal, nutritional, and stress signalling pathways in Mexican lime trees. *PLoS One, 8*(6), e66372.

Encyclopaedia Iranica. (2011). Citrus fruits. Retrieved March 24, 2019, from Ehsan Yarshater Center for Iranian Studies, Columbia University. http://www.iranicaonline.org/articles/citrus-fruits

Faghihi, M., Bagheri, A., Askari Seyahooei, M., Pezhman, A., & Faraji, G. (2017). First report of a '*Candidatus* Phytoplasma aurantifolia'-related strain associated with witches'-broom disease of limequat in Iran. *New Disease Reports, 35*, 24–24.

Faghihi, M., Salehi, M., Bagheri, A., & Izadpanah, K. (2009). First report of *citrus* huanglongbing disease on orange in Iran. *Plant Pathology, 58*(4), 793–793.

Foissac, X., & Carle, P. (2017). 'Candidatus Phytoplasma aurantifolia' strain WBDL, whole genome shotgun sequencing project. Retrieved April 24, 2019, from NCBI https://www.ncbi.nlm.nih.gov/nuccore/NZ_MWKN00000000.1

Foroutan, A., Rahimian, H., Kazemitabar, S. K., Mardi, M., & Alavi, S. M. (2009). *Interaction between citrus tristeza virus and phytoplasma associated with witches'broom disease of lime.* Paper presented at the 6th national biotechnology congress of Iran (August 13–15, 2009), Tehran, Iran.

Ghayeb Zamharir, M., Paltrinieri, S., Hajivand, S., Taheri, M., & Bertaccini, A. (2017). Molecular identification of diverse '*Candidatus* Phytoplasma' species associated with grapevine decline in Iran. *Journal of Phytopathology, 165*(7–8), 407–413.

Golmohammadi, M., Bani Hashemian, S., Mattos, D., Jr., Carlos, E., Novelli, V., Alves de Azevedo, F., ... Zaccheo, P. (2016). Decline of citrus trees in southern Iran and its relation to Huanglongbing and bacterial blight. *Citrus Research & Technology, 37*, 114.

Hashemi-Tameh, M., Bahar, M., & Zirak, L. (2014). '*Candidatus* Phytoplasma asteris' and '*Candidatus* Phytoplasma aurantifolia', new phytoplasma species infecting apple trees in Iran. *Journal of Phytopathology, 162*(7–8), 472–480.

Hosseinzadeh Ghasemabad, M. (2012). *Transmission of WBDL to Arabidopsis usind Dodder.* MSc, University of Shiraz, Shiraz, Iran.

Hu, J., Jiang, J., & Wang, N. (2017). Control of citrus huanglongbing via trunk injection of plant defense activators and antibiotics. *Phytopathology, 108*(2), 186–195.

Izadpanah, K., Shafiee, V., & Pakniat, A. (2002). The status of Citrus tristeza virus in the Fars and Bushehr provinces of Iran – Short Communications, Fifteenth IOCV Conference. *International Organization of Citrus Virologists Conference (1957–2010), 15*(15), 332–334.

Lee, I.-M., Davis, R. E., & Gundersen-Rindal, D. E. (2000). Phytoplasma: Phytopathogenic mollicutes. *Annual Reviews in Microbiology, 54*(1), 221–255.

Lee, R. F., & Keremane, M. (2013). Mild strain cross protection of tristeza: A review of research to protect against decline on sour orange in Florida. *Frontiers in Microbiology, 4*(259), 1–11.

Mardi, M., Farsad, L. K., Gharechahi, J., & Salekdeh, G. H. (2015). In-depth transcriptome sequencing of Mexican lime trees infected with '*Candidatus* Phytoplasma aurantifolia'. *PLoS One, 10*(7), 1–23.

Mardi, M., Khayam Nekouei, S., Farsad, L. K., Ehya, F., Shabani, M., Shafiee, M., … Hosseini Salekdeh, G. (2011). Witches' broom disease of Mexican lime trees: Disaster to be addressed before it will be too late. *Bulletin of Insectology, 64*(Suppl), S205–S206.

Monavarfeshani, A., Mirzaei, M., Sarhadi, E., Amirkhani, A., Khayam Nekouei, M., Haynes, P. A., … Salekdeh, G. H. (2013). Shotgun proteomic analysis of the Mexican lime tree infected with "Candidatus Phytoplasma aurantifolia". *Journal of Proteome Research, 12*(2), 785–795.

Najafinia, M., & Azadvar, M. (2016). Witches broom disease of lime and its management. *Indian Phytopathology, 69*, 330–332.

Ochoa-Sánchez, J. C., Parra-Cota, F. I., Aviña-Padilla, K., Délano-Frier, J., & Martínez-Soriano, J. P. (2009). *Amaranthus* spp.: A new host of '*Candidatus* Phytoplasma aurantifolia'. *Phytoparasitica, 37*(4), 381–384.

Omura, M., & Shimada, T. (2016). Citrus breeding, genetics and genomics in Japan. *Breeding Science, 66*(1), 3–17.

Rad, F., Mohsenifar, A., Tabatabaei, M., Safarnejad, M., Shahryari, F., Safarpour, H., … Fotokian, M. (2012). Detection of '*Candidatus* phytoplasma aurantifolia' with a quantum dots FRET-based biosensor. *Journal of Plant Pathology, 94*(3), 525–534.

Saberi, E., Alavi, S. M., Safaie, N., Moslemkhany, C., & Azadvar, M. (2017). Bacterial pathogens associated with citrus huanglongbing-like symptoms in southern Iran. *Journal of Crop Protection, 6*(1), 99–113.

Salehi, M., Faghihi, M., Khanchezar, A., Bagheree, A., & Izadpanah, K. (2012). Distribution of citrus huanglongbing disease and its vector in southern Iran. *Iranian Journal of Plant Pathology, 48*(2), 195–208.

Salehi, M., Izadpanah, K., Siampour, M., Bagheri, A., & Faghihi, S. (2007). Transmission of '*Candidatus* Phytoplasma aurantifolia'to Bakraee (*Citrus reticulata* hybrid) by feral *Hishimonus phycitis* leafhoppers in Iran. *Plant Disease, 91*(4), 466–466.

Sechler, A., Schuenzel, E., Cooke, P., Donnua, S., Thaveechai, N., Postnikova, E., … Schaad, N. (2009). Cultivation of '*Candidatus* Liberibacter asiaticus','*Ca.* L. africanus', and '*Ca.* L. americanus' associated with huanglongbing. *Phytopathology, 99*(5), 480–486.

Serra, P., Hashemian, S. M. B., Fagoaga, C., Romero, J., Ruiz-Ruiz, S., Gorris, M. T., … Duran-Vila, N. (2014). Virus-viroid interactions: Citrus Tristeza virus enhances the accumulation of Citrus dwarfing viroid in Mexican lime via virus-encoded silencing suppressors. *Journal of Virology, 88*(2), 1394–1397.

Shahryari, F., Safarnejad, M., Shams-Bakhsh, M., Schillberg, S., & Nölke, G. (2013). Generation and expression in plants of a single-chain variable fragment antibody against the immunodominant membrane protein of '*Candidatus* phytoplasma aurantifolia'. *Journal of Microbiology and Biotechnology, 23*(8), 1047–1054.

Sundin, G. W., & Wang, N. (2018). Antibiotic resistance in plant-pathogenic bacteria. *Annual Review of Phytopathology, 56*, 161–180.

Taheri, F., Nematzadeh, G., Zamharir, M. G., Nekouei, M. K., Naghavi, M., Mardi, M., … Salekdeh, G. H. (2011). Proteomic analysis of the Mexican lime tree response to '*Candidatus* Phytoplasma aurantifolia' infection. *Molecular BioSystems, 7*(11), 3028–3035.

Tanha Maafi, Z., & Damadzadeh, M. (2008). Incidence and control of the citrus nematode, *Tylenchulus semipenetrans* Cobb, in the north of Iran. *Nematology, 10*(1), 113–122.

Wang, Y., Zhou, L., Yu, X., Stover, E., Luo, F., & Duan, Y. (2016). Transcriptome profiling of Huanglongbing (HLB) tolerant and susceptible citrus plants reveals the role of basal resistance in HLB tolerance. *Frontiers in Plant Science, 7*, 933.

Young Journalist Club. (2017). Citrus fruit output to rise by 7%. *Financial Tribune*.

Zafari, S., Niknam, V., Musetti, R., & Noorbakhsh, S. N. (2012). Effect of phytoplasma infection on metabolite content and antioxidant enzyme activity in lime (*Citrus aurantifolia*). *Acta Physiologiae Plantarum, 34*(2), 561–568.

Chapter 3
Occurrence and Distribution of Phytoplasma Diseases in Iran

Majid Siampour, Keramatollah Izadpanah, Mohammad Salehi, and Alireza Afsharifar

Abstract The study of phytoplasmas in Iran dates back to three decades ago when the phytoplasmal nature of sesame phyllody was established. In this period, diverse phytoplasmas have been identified in association with diseases in a broad range of host plant species across at least 26 families. The body of reports on the incidence of these diseases is growing in this country both for the number of phytoplasma strains identified and the number of plant species they affect. Although Iran is a country with significant climate variation, phytoplasm diseases have been reported throughout the country nearly in all provinces. To date, phytoplasmas identified in Iran belong to nine 16Sr groups including 16SrI, 16SrII, 16SrVI, 16SrVII, 16SrIX, 16SrX, 16SrXII, 16SrXIV and 16SrXXX. Phytolasmas of 16SrII, 16SrVI, 16SrIX and 16SrXII groups are the most reported phytoplasmas from a broad range of plants. Phytoplasmas of 16SrII group are especially most prevalent in southern half of the country, while those of the other three groups have wider geographical distribution. On the other hand, phytoplasmas of 16SrXIV and 16SrXXX groups are somewhat restricted both in geographical distribution and host range. Some of the phytoplasma diseases in Iran — such as almond witches' broom and witches' broom disease of lime are highly destructive, with socio-economic impact.

Keywords Phytoplasma · Iran · Almond witches' broom · Witches' broom disease of lime

---

M. Siampour (✉)
Department of Plant Protection, College of Agriculture, Shahrekord University, Shahrekord, Iran
e-mail: Siampour@sku.ac.ir

K. Izadpanah · A. Afsharifar
Plant Virology Research Center, College of Agriculture, Shiraz University, Shiraz, Iran
e-mail: Izadpana@shirazu.ac.ir; afshari@shirazu.ac.ir

M. Salehi
Plant Protection Research Department, Fars Agricultural and Natural Resources Research and Education Center, Shiraz, Iran
e-mail: salehi_abarkoohi@yahoo.com

C. Y. Olivier et al. (eds.), *Sustainable Management of Phytoplasma Diseases in Crops Grown in the Tropical Belt*, Sustainability in Plant and Crop Protection 12,
https://doi.org/10.1007/978-3-030-29650-6_3

3.1 Introduction

Phytoplasmas are pathogens associated with diseases in several hundred plant species causing devastating plant diseases worldwide. For example, in Italy, apple proliferation disease caused by '*Candidatus* Phytoplasma mali'-related strains resulted in an estimated loss of €100 million in apple crop in 2001 (Strauss, 2009). Phytoplasma classification system has been mainly based on the sequence diversity of the 16S rRNA gene by which several phytoplasma strains have been classified within the genus '*Candidatus* Phytoplasma'. With the current IRPCM rules, over 40 '*Candidatus* Phytoplasma' species have been so far described (Miyazaki et al., 2017). RFLP analyses of the 16S rRNA gene have classified phytoplasma strains in different RFLP groups and subgroups (Zhao & Davis, 2016). To date, phytoplasmas have been classified into 33 16Sr groups, with each group being classified in several 16S rRNA subgroups (Zhao & Davis, 2016). The use of less conserved genes including *sec*Y and ribosomal protein genes (*rp*) has revolutionized the ability to discriminate between closely related starins and resulted in a finer phytoplasma strain differentiation (Lee, Bottner-Parker, Zhao, Davis, & Harrison, 2010; Martini et al., 2007).

The geographical distribution of phytoplasma strains depends on many factors which include the phytoplasma host range, the feeding behavior of insect vectors (polyphagous, oligophagous or monophagous), and the human activity such as transferring and using potentially infected plant material for planting or grafting (EPPO, 2017; Hogenhout et al., 2008). For instance, phytoplasmas within the aster yellows group (16SrI, '*Ca.* P. asteris') have vast geographical distribution. They are well known by having the largest host range which comprises over 30 Hemiptera species identified as vectors—mostly polyphagous—and more than 80 plant species (Hogenhout et al., 2008). On the other hand, phytoplasmas of 16Sr IV, 16SrVII or 16SrX groups have a more restricted geographical distribution since they affect far fewer plant species and are vectored by fewer species within the Cicadellidae and Psyllidae families (Foissac & Wilson, 2010).

Iran has multiple climate zones ranging from arid and semi-arid to subtropical along the Caspian Sea and forests in the northern parts. The mean temperature values range from 5 °C to 10 °C in January, the coldest month, and above 30 °C in July, the hottest month (Soltani et al., 2016). The very variable climate conditions are suitable for the growth of diverse plant species and crops, as well as the establishment of several pests and diseases. More than 8000 plant species from 167 botanical families and 1200 genera have been recorded from Iran (Heshmati, 2013). The most economically important crops cultivated in Iran are cereals (wheat, rice, barley, maize, etc.), followed by sugar beet, sugarcane, rapeseed, alfalfa, cotton, fruits and nut crops, and vegetables. To date, a number of phytoplasma diseases have been reported, many of which present a major risk of serious epidemics that may have significant socio-economic impacts. Vector-host plant interactions play an important role in limiting or expanding phytoplasma dispersal (Weintraub, 2007). Insect vectors play the most important role in the emergence of new phytoplasma diseases and their distribution is highly influenced by climate. It has been documented that milder winters will lead to an increase of phytoplasma survival not only from migration of their leafhopper vectors but also from their weedy hosts, and consequently, phytoplasma epidemics will become more important in higher latitudes (R. Jones & Barbetti, 2012). Hence, Iran's diverse of climate conditions can be seen to influence the movement of insect vectors between warmer and colder areas where more suitable environments favour phytoplasmas to thrive and spread into new geographical regions.

In this report, we have reviewed and summarized the characteristics and distribution of phytoplasma diseases reported from Iran in the past 30 years of the phytoplasma research in the country. Sesame phyllody was the first phytoplasma disease identified in Iran by electron microscopy, transmission assays, host range and symptomatology studies (M. Salehi & Izadpanah, 1992), followed by alfalfa witches' broom, almond and peach witches' broom, witches' broom disease of lime and tomato big bud (Salehi & Izadpanah, 1993, 1995a, 1995b, 2001; Salehi, Izadpanah, & Ebrahim-Nesbat, 1995; M. Salehi, Izadpanah, & Rahimian, 1997). The advance of polymerase chain reaction (PCR) and sequencing technologies applied to phytoplasma research in Iran (M. Salehi, Izadpanah, & Behjatnia, 2000) have unveiled the association of phytoplasmas belonging to nine 16Sr groups with diseases in plants from more than 25 botanical families (Table 3.1; Fig. 3.1). Some of these phytoplasma diseases were economically destructive for Iran. For example, almond witches' broom and witches' broom disease of lime have caused significant economic losses with socio-economic impact. Some others, like alfalfa witches' broom, had wide geographical distribution and were found in almost all major alfalfa growing areas of Iran.

3.2 Phytoplasma Diseases in Poaceae

Poaceae is a large plant family that includes about 600 genera of more than 9000 species. Bermuda grass (*Cynodon dactylon* L. Pers, Poaceae) is a cosmopolitan Poaceae weed widespread in tropical regions of the world (Arocha & Jones, 2010), originated from countries in the Middle East including Turkey, Afghanistan, Pakistan and Iran (Mohammadi Farsani, Etemadi, Sayed-Tabatabaei, & Talebi, 2011). A disease known as Bermuda grass white leaf (BGWL), associated with phytoplasmas of 16SrXIV group '*Ca*. P. cynodontis', was reported from Australia, Sudan, Italy and Cuba (Arocha et al., 2005; Marcone, Ragozzino, & Seemüller, 1997; Rao et al., 2008; Snehi et al., 2008). BGWL is present in several provinces of southern Iran (Table 3.1) (Marcone, Schneider, & Seemüller, 2004; M. Salehi, Heydarnejad, & Izadpanah, 2005; M. Salehi, Izadpanah, Siampour, & Taghizadeh, 2009). The leafhopper *Exitianus capicola* Still (Hemiptera: Cicadellidae) was confirmed as the vector of BGWL phytoplasma in Iran (M. Salehi, Izadpanah, Siampour, & Taghizadeh, 2009). BGWL symptoms included chlorosis and whitening of leaves, bushy growth, proliferation of axillary buds, small leaves and plant

Table 3.1 List of plant species reported to be infected by phytoplasmas in Iran. N/A refers to 'not available'

Plant Family	Common name	Scientific name	Main symptoms	16Sr group or subgroup	Distribution(Provinces)	GenBank Acc. No	References
Fabaceae							
	Sophora	*Sophora alopecuroides*	Yellowing, little leaf	XII-A	Tehran	KF263685	(Allahverdi, Rahimian, & Babaeizad, 2014b)
	Alfalfa	*Medicago sativa*	Witches' broom	II (II-C and II-D)	Fars, Yazd, Kerman,Esfahan, Khorasan, Khuzistan, Hamedan,Bushehr	DQ233655	Esmailzadeh-Hosseini, Khodakaramian, et al., 2016a)
	Alfalfa	*Medicago sativa*	Yellowing	XII-A	Esfahan, yazd, Kohguiluieh-Boyerahmad, East Azarbayijan, Qom, Markazi, Kordestan, Fars, Hamedan, Kermanshah, Chahramahal- Bakhtiari	KT763372	(Esmailzadeh-Hosseini, Khodakaramian, et al., 2016a; Esmailzadeh-Hosseini, Khodakaramian, Salehi, & Bertaccini, 2016b; S. Hosseini, Bahar, & Zirak, 2013a)
	Alfalfa	*Medicago sativa*	Witches' broom	I	West Azarbayjan	FJ788514	(Rashidi et al., 2010)
	Alfalfa	*Medicago sativa*	Witches' broom	VI-A	East Azerbaijan, Zajan	KU240021	(Esmailzadeh-Hosseini, Khodakaramian, et al., 2016b)
	Clover	*Trifolium sp.*	Little leaf	II	Esfahan	JQ710445	(S. Hosseini, Bahar, & Zirak, 2013b)
	Cowpea	*Vigna unguiculata*	Phyllody	VI-A	East Azerbaijan	KC633092	N/A
	Cooba	*Acacia salicina*	Little leaf	II	Khuzistan	KX088460	N/A
	Black locust	*Robinia pseudoacacia*	Witches' broom	IX	Khorasan razavi, Fars	KX553989	(Karimzade, Siampour, Zakiaghl, & Mehrvar, 2018)

	Faba bean	*Vicia faba*	Phyllody	II-D, II-C	Fars, Bushehr	KP869129, KP869128	(M. Salehi, R. Rasoulpour, et al., 2016)
Poaceae							
	Maize	*Zea mays*	Yellowing and dwarfing	VI-A	West Azerbaijan	KT807468	(Zibadoost, Rastgou, & Tazehkand, 2016)
	Johnsongrass	*Sorghum halepense*		VI-A*	West Azerbaijan	KT807469	(Zibadoost & Rastgou, 2016)
	Bermuda grass	*Cynodon dactylon*	White leaf	XIV-A	Fars, Boushehr, Chaharmahal-Bakhtiari, Hormozgan, Kerman, Khuzistan, Sistan- Baluchistan	EF444485	(M. Salehi, Izadpanah, Siampour, & Taghizadeh, 2009)
	Sugarcane	*Saccharum officinarum*	White leaf	II	Khuzistan	DQ471316	(M. Salehi, Ghasemi, Taker-Khani, & Izadpanah, 2001)
Asteraceae							
	Horseweed	*Conyza canadensis*		VI-A	West Azerbaijan	KT807471	(Zibadoost & Rastgou, 2016)
	Lettuce	*Lactuca sativa*	Phyllody	IX-D*	Fars	DQ889748	(M. Salehi, Izadpanah, Nejat, et al., 2007)
	Lettuce	*Lactuca sativa*	Leaf deformation	I	Esfahan	JQ015290	(Sichani et al., 2014)
	Wild lettuce	*Lactuca serriola*	Phyllody	IX-D*	Fars	DQ889749	(M. Salehi, Izadpanah, Nejat, et al., 2007)
	Prickly lettuce	*Lactuca serriola*	Phyllody	I	Esfahan	HM626108	(Sichani et al., 2014)
	Prickly lettuce	*Lactuca serriola*	Witches' broom	I	Markazi		(Babaie et al., 2007)

(continued)

Table 3.1 (continued)

Plant Family	Common name	Scientific name	Main symptoms	16Sr group or subgroup	Distribution(Provinces)	GenBank Acc. No	References
	Blanket flower	*gaillardia sp*	Phyllody	I	Markazi	HQ286479	(Sichani et al., 2014)
	Callistephus chinensis	*China aster*	Yellowing	I	Markazi	HQ2864	(Sichani et al., 2014)
	Niger seed	*Guizotia abyssinica*	Phyllody	I	Esfahan/Yazd	JQ015289	(Sichani et al., 2014)
	Sunflower	*Helianthus annuus*	Witches' broom	II (mix of VI and II)	Fars,, Esfahan, Kerman	JQ015289, GQ861502	(M. Salehi, Esmailzadeh-Hosseini, et al., 2015; Tazehkand et al., 2010)
	Pot marigold	*Calendula officinalis*	Phyllody	II-D	Yazd, Fars	KU297202	(Esmailzadeh-Hosseini, Salehi, Mirchenari, & Bertaccini, 2016)
	Black-eyed susan	*Rudbeckia hirta*	Phyllody	I	Markazi	DQ855286	(Babaie et al., 2007)
	Safflower	*Carthamus tinctorius*	Phyllody	VI-C	Fars, Yazd	DQ88948	(M. Salehi, Izadpanah, & Siampour, 2008)
Rubiaceae							
	Madder	*Rubia tinctorum*		VI-D	West Azerbaijan	KT807472	(Zibadoost & Rastgou, 2016)
Brasicaseae							
	Cabbage	*Brassica oleracea*	Yellows	VI-A	Fars	EF592606	(M. Salehi, Izadpanah, & Siampour, 2007)
	Rapeseed	*Brassica napus*	Phyllody	I	Fars, Ghazvin, Esfahan, Kerman, Yazd	FJ594960	(M. Salehi, Izadpanah, et al., 2011)
	Wallflower	*Erysiumu cheiri*	Phyllody	II	Kerman	GQ861503	(Tazehkand et al., 2010)

Amaranthaceae							
	Garden beet	*Beta vulgaris*	Witches' broom	II	Yazd	DQ302722	(Mirzaie, Esmailzadeh-Hosseini, Jafari-Nodoshan, & Rahimian, 2007)
	Sugar beet	*Beta vulgaris*	Witches' broom	II	Yazd	DQ195211	(M. Salehi & Izadpanah, 2005)
	Spinach	*Spinacia oleracea*	Yellowing	I	Kerman	GQ861505	(Tazehkand et al., 2010)
	Cockscomb	*Celosia cristata*	Flat stem, yellowing	VI	Khorasan Razavi, Markazi	JF508519	(Babaie et al., 2007)
	Cockscomb	*Celosia cristata*	Fasciation	II	Khuzistan	KX088458	(Babaie et al., 2007)
	Egyptean sea blite	*Suaeda aegyptiaca*	Witches' broom	VI	Hormozgan	KY411138	(Seyahooei, Hemmati, Faghihi, & Bagheri, 2017)
Apocynaceae							
	Periwinkle	*Catharanthus roseus*	Little leaf	I	Fars	DQ266089	NA
	Periwinkle	*Catharanthus roseus*	Phyllody	IX	Fars	KC332292	NA
	Periwinkle	*Catharanthus roseus*	Phyllody	VI	Behshahr	KC661072	(Fattahi, Salehi, Sharzehi, & Esmailzadeh-Hosseini, 2016)
	Periploca	*Periploca aphylla*	Witches' broom	XXX	Hormozgan	GQ249159	(Faghihi et al., 2010)
Amarylidaceae							
	Onion	*Allium cepa*	Flower malformation	I	Esfahan	HM626107	(Sichani et al., 2014)
	Scallion	*Allium cepa*	NA	I	Esfahan, Yazd	NA	(Sichani et al., 2014)

(continued)

Table 3.1 (continued)

Plant Family	Common name	Scientific name	Main symptoms	16Sr group or subgroup	Distribution(Provinces)	GenBank Acc. No	References
Rosaceae							
	Peach	*Prunus persica*	Yellows	IX-C	Mazandaran, West Azerbaijan,	KF932287	NA
	Peach	*Prunus persica*	Yellows	IX-B	Kurdistan	GU338000	(Ghayeb Zamharir, 2014)
	Peach	*Prunus persica*	Witches broom	VI	Esfahan	FJ204395	(Zirak et al., 2010)
	Peach	*Prunus persica*	Yellows	XII	East Azerbaijan,	FJ204394	(Zirak et al., 2010)
	Peach	*Prunus persica*	Yellows	II	Esfahan, Chaharmahl-Bakhtiari	FJ204393	(Zirak et al., 2010)
	Plum	*Prunus domestica*	Yellow leaf stunt	X-F	Mazandaran	KF739403	(Allahverdi et al., 2014a)
	Plum	*Prunus domestica*	Little leaf, leaf roll	II	Esfahan	FJ204400 (IR)	(Zirak et al., 2009b)
	Plum	*Prunus domestica*	Little leaf, yellowing	XII	Chaharmahal-bakhtiari	FJ409624	(Zirak et al., 2009b)
	Almond	*Prunus amygdalus*	Witches' broom	IX-C	East Azerbaijan, Fars, Chaharmahal-Bakhtiari,	DQ195209	(M. Salehi et al., 2006)
	Almond	*Prunus amygdalus*		IX-B	Fars, Kerman, Kohgyluieh-Boyerahmad, Kurdistan,	AF515637	(Verdin et al., 2003)
	Almond	*Prunus amygdalus*	Proliferation	II-C	Esfahan	FJ184381	(Zirak et al., 2009a)
	Almond	*Prunus amygdalus*	Proliferation	VI-D	Esfahan	FJ184380	(Zirak et al., 2009a)
	Almond	*Prunus amygdalus*	Yellowing, little leaf	XII	Esfahan	FJ184384	(Zirak et al., 2009b)

	Pear	*Pyrus sp.*	Decline	X-C '*Ca*.P.pyri'	Fars, Khorasan, Esfahan, Tehran, Esfahan	DQ471321	(M. Salehi, Izadpanah, Taghavi, et al., 2008)
	Pear	*Pyrus sp.*		X-B, '*Ca*. P. prunoroum	Gilan	KC902808	(Hashemi-Tameh et al., 2014b)
	Pear	*Pyrus sp.*		I	Esfahan	KC902811	(Hashemi-Tameh et al., 2014b)
	Apricot	*Prunus armeniaca*	Chlorotic leaf roll	IX-B	Fars	KY014991	(EPPO, 2017)
	Apricot	*Prunus armeniaca*	Leaf roll	II	Fars	KU962122	NA
	Gf677	*Prunus amygdalus x Prunus persica*	Witches' broom	IX-B	Fars	JX445141	(M. Salehi, Haghshenas, et al., 2011)
	Apple	*Malus pumila*	Yellowing, leaf roll, little leaf	I	Esfahan	KC902800	(Hashemi-Tameh et al., 2014a)
	Apple	*Malus pumila*	Resetting, little leaf	II	Esfahan	KC902794	(Hashemi-Tameh et al., 2014a)
	Wild almond	*Prunus scoparia*	Witches' broom	IX-B	Fars, Kohguiluyeh-Boyer Ahmad, Esfahan, Kerman	KM235725	(M. Salehi, Salehi, et al., 2015)
	Sweet cherry	*Prunus avium*	Leaf roll, witches broom	I	Esfahan	FJ204397	(Zirak et al., 2010)
	Sweet cherry	*Prunus avium*	Rosetting	II	Esfahan	FJ204398	(Zirak et al., 2010)

(continued)

Table 3.1 (continued)

Plant Family	Common name	Scientific name	Main symptoms	16Sr group or subgroup	Distribution(Provinces)	GenBank Acc. No	References
Cucurbitaceae							
	Cucumber	*Cucumis sativus*	Phyllody	II-D	Yazd, Fars, Kerman	KR822804	(M. Salehi, Siampour, et al., 2015)
	Cucumber	*Cucumis sativus*	Phyllody	VI-A	West Azerbaijan, East Azerbaijan,Khorasan Razavi, Yazd	JF508516, KR633068	(Esmailzadeh-Hosseini, Salehi, Mirchenari, Tarizeh, & Gholampoor, 2015)
	Squash	*Cucumis pepo*	Phyllody	II	Yazd	KR822805	(M. Salehi, Siampour, et al., 2015)
Solanaceae							
	Potato	*Solanum tuberosum*	Potato purple top	I, VI and XII	Esfahan, Chaharmahal-Bakhtiari, Tehran, Kermanshah, Hamedan, East Azarbayjan, Zanjan	FJ427296(VI) EU661607 (XII) FJ427297 (I) KX844627(XII, zanjan)	(P. Hosseini, Bahar, Madani, & Zirak, 2011)
	Tomato	*Solanum lycopersicum*	Big bud	VI-A	Kermanshah, Fars, Khorasan Razavi, East Azerbaijan, West Azerbaijan, Zanjan	JF508507	(Jamshidi et al., 2014)
	Tomato	*Solanum lycopersicum*	Big bud	IX-C	North Khorasan	JF508510	(Jamshidi et al., 2014)
	Tomato	*Solanum lycopersicum*	Witches' broom, Big bud,	II-D	Bushehr	KC182528	(E. Salehi, Salehi, et al., 2014)
	Tomato	*Solanum lycopersicum*	Little leaf	I	Ardebil, Markazi	GU797557	(Sichani et al., 2014)

	Tomato	*Solanum lycopersicum*	Big bud	XII-A	Fars	KX098490	(M. Salehi & Esmailzadeh-Hosseini, 2016)
	Eggplant	*Solanum melongena*	Big bud	IX-C, IX-D*	Hormozgan, Fars, Bushehr	JX464669, JX483702	(Tohidi et al., 2015)
	Eggplant	*Solanum melongena*	Big bud	II-D	Fars	JX083373	(Siampour et al., 2013)
	Eggplant	*Solanum melongena*	Big bud	XII-A	Zanjan	KX844625	NA
	Black nightshade	*Solanum nigrum*	Witches' broom	II	Hormozgan	GQ866886	(Samavi et al., 2012)
	Indian night shade	*solanum surattense*	Witches' broom	II	Hormozgan	GU550504	(Samavi et al., 2012)
	Petunia	*Petuni hybrida*	Witches' broom	II-D	Sistan-Baluchestan	JX570935	(Faghihi et al., 2014)
	Bell pepper	*Capsicum annuum*	Big bud	II-D	Hormozgan	KR706443	(Faghihi, Taghavi, Safaei, Siampour, & Najafabadi, 2016)
Apiaceae							
	Carrot	*Daucus carota*	Witches' broom	II-C	Yazd	KP840501	(M. Salehi, S. Esmailzadeh-Hosseini, E. Salehi, et al., 2016)
	Parsley	*Petroselinum crispum*	Witches' broom	II-D	Yazd	JX083374	(M. Salehi, Hosseini, Salehi, & Bertaccini, 2016)
Pedaliaceae							
	Sesame	*Sesamum indicum*	Phyllody	II-D	Khuzistan, Fars, Bushehr, Esfahan, Yazd	KF607108	Salehi et al. (2017)

(continued)

Table 3.1 (continued)

Plant Family	Common name	Scientific name	Main symptoms	16Sr group or subgroup	Distribution(Provinces)	GenBank Acc. No	References
	Sesame	*Sesamum indicum*		VI-A	Bushehr, Fars, KhorasanRazavi	KT265697	Salehi et al. (2017)
	Sesame	*Sesamum indicum*	Phyllody	IX-C	Fars, Khorasan Razavi	KF774193	Salehi et al. (2017)
Vitaceae							
	Grapevine	*Vitis vinifera*	Yellows	XII-A	Hamedan, Esfahan, Chaharmaha-Bakhtiari, Fars, Lorestan, Khorasan Razavi, Markazi, Yazd, Qazvine	KF130967	(E. Salehi, Taghavi, et al., 2014)
	Grapevine	*Vitis vinifera*	Yellows	I	Chaharmaha-Bakhtiari	KJ637210	(Mirchenari et al., 2015)
	Grapevine	*Vitis vinifera*	Yellows	II	Markazi,Qazvine		(Ghayeb Zamharir et al., 2017)
	Grapevine	*Vitis vinifera*	Scorch/ yellows	VII	Qazvine	KX267564	(Ghayeb Zamharir et al., 2017)
	Grapevine	*Vitis vinifera*	Yellows	IX-C	Fars, Qazvine	KX011516	(Ghayeb Zamharir et al., 2017; E. Salehi, Salehi, Taghavi, & Izadpanah, 2016)
Elaeagnaceae							
	Russian olive	*Elaeagnus angustifolia*	Witches' broom	I	East Azerbaijan, Tehran	FJ788515	(Rashidi et al., 2010)
Celastraceae							
	Japanese spindle	*euonymus japonicas*	Witches' broom	XII	Tehran	GQ273961	(Rashidi et al., 2010)
Rutaceae							
	Citron	*Citrus medica*	Witches' broom	II	Kerman	JN967029	(Faghihi, Bagheri, Askari Seyahooei, Pezhman, & Faraji, 2017)

	Grapefruit	*Citrus paradisi*	Witches' broom	II-B	Hormozgan, kerman	JF518980	(Faghihi et al., 2017)
	Limequat	*Citrofortunella floridana*	Witches' broom	II-B	Hormozgan	KY171947	(Faghihi et al., 2017)
	Lime	*Citrus aurantiifolia*	Witches' broom	II-B	Sistan and Bluchestan, Hormozgan, Kerman	KY412987	(Siampour et al., 2015)
	Bakraee	*Citrus reticulata hybrid*	Witches' broom	II-B	Hormozgan, Kermaan	–	(Djavaheri & Rahimian, 2004)
	Bitter orange	*Citrus aurantium*	Little leaf	II	Khuzistan	KX354934	NA
Cannabaceae							
	Hemp	*Cannabis sativa*	Witches' broom	XII, IX-C	Yazd, Fars, Fars	JF441274	(Sichani, Bahar, & Zirak, 2011)
Myrtaceae							
	Lemon bottlebrush	*callistemon citrinus*	Little leaf	II	Khuzistan	KX354936	NA
	Eucalyptus	*Eucalyptus globulus*	Leaf roll	II-D	Khuzistan	KX088462	(Azimi, Farokhi-Nejad, & Mehrabi-Koushki, 2017a)
	Eucalyptus	*Eucalyptus globulus*	Little leaf	XII-A	Fars	KX685876	NA
	Eucalyptus	*Eucalyptus globulus*	Little leaf	I	Fars	KT626569	(M. Salehi, Esmailzadeh-Hosseini, & Salehi, 2016)

(continued)

Table 3.1 (continued)

Plant Family	Common name	Scientific name	Main symptoms	16Sr group or subgroup	Distribution(Provinces)	GenBank Acc. No	References
Lythraceae							
	Pomegranate	*Punica granatum*	Little leaf	II-D	Fars	KT265695	(M. Salehi, Esmailzadeh-Hosseini, Rasoulpour, Salehi, & Bertaccini, 2016)
Combretaceae							
	Buttonwood	*Conocarpus erectus*	Little leaf	II	Khuzistan	KX088465	(Azimi, Farokhi-Nejad, & Mehrabi-Koushki, 2017b)
Oleaceae							
	Olive	*Olea europaea*	Little leaf	VII	Gilan	KX298840	(Ghayeb Zamharir & Razavi, 2016)
Sapotaceae							
	Sapodilla	*Manilkara zapota*	Flattened stem	II*	Hormozgan	KY171945	(Bagheri et al., 2017
Arecaceae							
	Date palm	*Phoenix dactylifera*	Yellowing	VII	Khuzistan	KX347967	(Ghayeb Zamharir et al., 2016)
Cupressaceae							
	Mediterranean cypress	*Cupressus sempervirens*	Witches' broom	II	Tehran	KU647632	(Bahari, Alavi, Shams-Bakhsh, & Saberi, 2016)
Berberidaceae							
	Barberry	*Berberis vulgaris*	Flatten stem	II	South Khorasan	KT879852	NA

*Shows the variants of the related subgroup

Fig. 3.1 Symptoms associated with phytoplasma diseases reported from Iran. (**a**) Symptoms of TBB on a tomato plant in the Zanjan province infected with a 16SrVI phytoplasma. (**b**) Witches' broom symptoms on a tomato plant in Bushehr province infected with a 16SrII phytoplasma. (**c**) Wild lettuce plant infected with a 16SrIX pytoplasma. D. Alfalfa witches' broom in the Yazd province infected with a 16SrII phytoplasma. (**e** and **f**) Symptoms of witches' broom, respectively, on almond and wild almond infected with '*Ca.* P. phoenicium' (16SrIX-B) in Fars province. (**g** and **h**). Witches' broom symptoms caused by '*Ca.* P. aurantifolia, on lime and limequat, respectively. (**i**). Petunia infected with a 16SrII phytoplasma reported from Sistan-Baluchistan

death (Marcone et al., 2004; M. Salehi, Izadpanah, Siampour, & Taghizadeh, 2009). Other phytoplasma strains causing white leaf in gramineous species such as brachiara grass (*Brachiaria distachya* L. Stapf, Poaceae), annual blue grass (*Poa annua* L., Poaceae) and growfoot grass (*Dactyloctenium aegyptum* L. Willd, Poaceae) were reported as closely related to the BGWL phytoplasma (16SrXIV-A subgroup)

(Marcone et al., 2004). Two different 16SrXIV phytoplasmas, subgroup 16SrXIV-A and 16SrXIV-B, have been identified in association with BGWL disease in Fars province (strains Firouzabad and Jouyom) distinguished by a HinfI restriction site on the 16S rRNA gene sequence. BGWL phytoplasmas have been also detected in two dicotyledonous plant species in Germany and Australia: *Cirsium arvense* L. Scop (Asteraceae) and *Galactia tenuifolia* Kallar (Fabaceae), and in the leafhopper *Psammotettix cephalothes* Herrich-Schäffer (Hemiptera: Cicadellidae) (Marcone et al., 2004).

Sugarcane white leaf (SCWL) is another threatening phytoplasma disease reported from south of Iran (Biabani, Ghasemi, Salehi, & Rahimian, 2009). Investigations from other countries have shown that the SCWL phytoplasma belongs to the 16SrXI-B subgroup ('*Ca.* P. oryzae'). The symptoms of the disease in Iran were similar to those described elsewhere and included white or striped leaf blades and stunting of the infected plants, brittle leaves and death of plants (Arocha & Jones, 2010; M. Salehi et al., 2000). In Iran, SCWL phytoplasma was detected by serological and PCR analyses from several commercial sugarcane cultivars such as CP-48, CP-69 and CP-71 (Biabani et al., 2009; M. Salehi et al., 2000). ELISA, dot immunobinding assay (DIBA) and tissue print immunobinding assay (TPIA) confirmed the association of a phytoplasma with SCWL disease in Khuzistan, the most important province for sugarcane cultivation in Iran (Biabani et al., 2009). There is no record on the 16S rRNA gene sequence of the SCWL phytoplasma in Iran nor information about its phylogenetic relationship to other SCWL strains in the world. However, the sequence of the spacer region of a Khuzistan isolate (Acc. no. DQ471316) was most similar (96.1% identity) to that of phytoplasmas within the 16SrII taxonomic group ('*Ca.* P. aurantifolia') (Biabani et al., 2009). *Matsumuratettix hiroglyphicus* Matsumura (Hemiptera: Cicadellidae) and *Yamatotettix flavovittatus* Matsumura (Hemiptera: Cicadellidae) were reported as the vectors of the SCWL phytoplasma in Thailand and Taiwan (Hanboonsong, Ritthison, Choosai, & Sirithorn, 2006). The presence and the role of these leafhopper species in SCWL development is yet not known in Iran.

3.3 Phytoplasma Diseases in Vegetable Crops

Phytoplasmas are known to cause devastating diseases in a wide number of vegetable species around the world. In Iran, phytoplasma diseases have been reported from tomato, potato, cucumber, squash, carrot, faba bean, parsley, lettuce and cabbage, where symptoms of phyllody, big bud, and witches' broom are among the most commonly described .

Among vegetable crops in Iran, the highest number of phytoplasma diseases has been reported on tomato. This country ranked the seventh major tomato producer in the world. Tomato big bud (TBB) is the most important tomato disease in several provinces including Khorasan Razavi, North Khorasan, East Azerbaijan, Kermanshah, Kurdistan, Fars, Esfahan, Ardabil, West Azarbaijan, Lorestan and

Zanjan (Jamshidi, Jafarpour, Rouhani, & Salehi, 2014; M. Salehi & Esmailzadeh-Hosseini, 2016; M. Salehi & Izadpanah, 1995b; Sichani, Bahar, & Zirak, 2014). Phytoplasmas of 16SrI, 16SrII, 16SrVI ('*Ca.* P. trifolii'), 16SrIX ('*Ca.* P. phoenicium') and 16SrXII (Stolbur) (Table 3.1) have been associated with TBB in Iran. TBB symptoms include unusual flowers with enlarged calyxes, virescence, proliferation and witches' broom (Fig. 3.1, Table 3.1). Phytoplasmas of 16SrVI (subgroup -A) were found in tomato fields of six different provinces of Iran (West and East Azerbaijan, Kermanshah, Zanjan, Fars and Khorasan Razavi). A 16SrVI-A strain was transmitted from infected tomato to tomato, eggplant and periwinkle using grafting and dodder (*Cuscuta* sp. L., Convolvulaceae) (E. Salehi, Salehi, & Masoumi, 2016).

Tomato witches' broom (ToWB) has been reported from Bushehr province in southern Iran (E. Salehi, Salehi, Taghavi, & Izadpanah, 2014). ToWB was associated with a phytoplasma of 16SrII-D subgroup. Symptoms of ToWB included small deformed leaves, stunting, witches' broom, proliferation of the axillary buds (Fig. 3.1), and not flowering. The ToWB phytoplasma was transmitted to eggplant by graft inoculation where characteristic symptoms of little leaf, shortening of internodes, yellowing, floral virescence, phyllody and witches' broom developed (E. Salehi, Salehi, et al., 2014). The association of phytoplasmas of 16SrII-D subgroup was also reported from tomato in Australia (Pearce, Scott, & Pethybridge, 2011), Egypt (Omar & Foissac, 2012) and India (Singh et al., 2012). Phytoplasmas of 16SrII were also associated with diseases of several other vegetable crops including carrot, faba bean, parsley, cucumber, squash and bell pepper in Iran (Table 3.1). In this country, cucumber and squash phyllodies, associated with a 16SrII-D phytoplasma, were two important phytoplasma diseases in Yazd province and transmitted by *Orosius albicinctus* Distant (Hemiptera: Cicadellidae). Both phytoplasma diseases were also transmitted to periwinkle, alfalfa, cucumber, carrot, sesame, sunflower, pot marigold, eggplant, squash, tomato and parsley by *O.albicinctus* (M. Salehi, Siampour, Esmailzadeh-Hosseini, & Bertaccini, 2015). Moreover, phytoplasmas of 16SrII-D caused cucumber phyllody disease in greenhouses of the Fars and Yazd provinces (Dehghan, Salehi, Khanchezar, & Afshar, 2014).

The phyllody disease in Cucurbitaceae appears to be an emerging and threatening disease in cucumber fields in Yazd province. The infected plants produce no fruits, and the crop loss may reach 100% in case of early infections. Cucumber phyllody was also observed in several other provinces of Iran including Sistan-Baluchistan, Hormozgan, Kerman and Bushehr, however, the phytoplasmas have not been yet characterized (M. Salehi, Siampour, et al., 2015). *O. albicinctus* is also the vector of carrot witches' broom and faba bean phyllody diseases associated with phytoplasmas of 16SrII-D and 16SrII-C subgroups. *O. albicinctus* is very likely to play an important role in the spread of other vegetable diseases associated with 16SrII phytoplasmas in Iran.

Phyllodies of lettuce (*Lactuca sativa* L., Asteraceae) and wild lettuce (*Lactuca serriola* L.) are other important phytoplasma diseases in Iran (M. Salehi, Izadpanah, Nejat, & Siampour, 2007). Lettuce phyllody is observed in several regions of the

Fars province. Wild lettuce phyllody has a wider geographical distribution, which includes Fars, Esfahan, Keman and Yazd provinces, all located in central Iran. Studies have shown that lettuce and wild lettuce phyllodies (Fars strains) share similar molecular and biological characterstics, hence, both diseases are believed to be caused by the same phytoplasma. Molecular analyses have shown that phytoplasmas associated with both diseases are variants of the 16SrIX group, subgroup 16SrIX-D* (16SrIX-D*, Table 3.1). Symptoms include deformed chlorotic and small leaves, proliferation of buds in the crown, stunting and death of the plants (Fig. 3.1). The leafhopper vector of both diseases was identified as *Neoaliturus fenestratus* Herrich-Schäffer (Hemiptera: Cicadellidae) which had the capacity to transmit the diseases to sowthistle and periwinkle, but not to safflower, sunflower, calendula or sesame (M. Salehi, Izadpanah, Nejat, et al., 2007). Interestingly, a 16SrIX-D* phytoplasma strain (GenBank Acc. No. JX483702) was also associated with eggplant big bud in Bushehr province, while phytoplasmas of subgroups 16SrIX-C and 16SrII-D were identified in eggplant in Hormozgan province (Tohidi, Salehi, Ghasemi, Khanchezar, & Shahamiri, 2015).

Iranian cabbage yellows (ICY) associated with a phytoplasma of 16SrVI group ('*Ca.* P. trifolii'), subgroup A, has been found in Zarghan, Fars province. Symptoms were first observed in 2001 with an incidence of 50% in some fields (M. Salehi, Izadpanah, & Siampour, 2007). The main disease symptoms were yellowing, little leaves, stunting, opening of the head and witches' broom. Among several leafhoppers collected from the infected fields, only *Circulifer haematoceps* Mulsant & Rey (Hemiptera: Cicadellidae) was identified as the vector of the ICY phytoplasma in Iran. This vector also totransmited the ICY phytoplasma to cauliflower, rapeseed and periwinkle under the greenhouse conditions (M. Salehi, Izadpanah, & Siampour, 2007).

3.4 Phytoplasma Diseases in Oilseed Crops

Several phytoplasmas have been found in association with diseases of oilseed crops in Iran. Sesame phyllody was the first phytoplasma disease identified and characterized in Iran (M. Salehi & Izadpanah, 1992). Phytoplasma strains of 16SrII, 16SrVI and 16SrIX groups have been associated with seasame phyllodies in Bushehr and Fars provinces in south, Esfahan and Yazd provinces in center and Khorasan Razavi in northeast of Iran (M. Salehi, Esmailzadeh-Hosseini, Salehi, & Bertaccini, 2017) with incidence rates as high as 100%. Phytoplasmas of these groups have been also found causing sesame phyllody in Turkey, Oman, India, Taiwan, Thailand and Pakistan (M. Salehi et al., 2017). The main symptoms of this disease are yellowing, internode shortening, reduced leaf size, virescence, witches' broom and flower sterility.

Phytoplasmas of 16SrII-D (in Fars and Bushehr), 16SrVI-A (in Khorasan-Razavi) and 16SrIX-C (in Bushehr, Khorasan Razavi, Yazd and Esfahan) subgroups were found to be associated with sesame phyllody; however, biological studies showed that phytoplasma strains were differentiated from each other based on the distinct symptoms that each strain developed in periwinkle test plants (M. Salehi et al., 2017). Transmission trials identified *C. haematoceps* as the leafhopper vector of all phytoplasma strains associated with sesame phyllody in Iran (M. Salehi et al., 2017; M. Salehi & Izadpanah, 1992). Experimental transmission assays revealed that *C. haematoceps* could transmit sesame phyllody phytoplasma to sesame and periwinkle plants (M. Salehi & Izadpanah, 1992). *O. albicinctus* was identified as the vector of the sesame phyllody phytoplasma strain of 16SrII-D subgroup in Iran (M. Salehi et al., 2017). However, *O. albicinctus* has been reported as the vector of sesame phyllody phytoplasmas (subgroups 16SrIX-C and group 16SrII) in Turkey and Pakistan (Akhtar et al., 2009; Al-Sakeiti, Al-Subhi, Al-Saady, & Deadman, 2005; Ikten et al., 2014; Nakashima, Hayashi, Chaleeprom, Wongkaew, & Sirithorn, 1995; Rao, Kumar, & Baranwal, 2015; Tseng, Deng, Chang, Huang, & Jan, 2014)0. Both *C. haematoceps* and *O. albicinctus* are abundant leafhopper species in sesame fields in Iran. Moreover, these leafhopper species have wide host range; therefore, the possible role of some other host plants in the epidemiology of sesame phyllody needs to be investigated. *Brassica napus* L. (Brassicaceae) *Lepidium sativum L. (Brassicaceae), Catharanthus roseus (L.) G.Don (Apocynaceae), Lactuca* sp. L. and *Portulaca oleracea* L. (Portulacaceae) were found as the experimental hosts of an unclassified phytoplasma strain of sesame phyllody (M. Salehi & Izadpanah, 1992). Although phytoplasmas of subgroups 16SrII-D, 16SrVI-A and 16SrIX-C have a large host range and widespread geographical distribution in Iran, their relationship to the sesame phyllody phytoplasma has not yet been established.

Rapeseed is an important oilseed crop cultivated in most geographical regions of Iran. Phytoplasma infection of rapseed with 16SrI phytoplasmas has been reported in Europe and Canada (Bertaccini et al., 1998; Olivier, Séguin-Swartz, Hegedus, & Barasubiye, 2006). In Iran, rapeseed phyllody infection was reported from surveys between 2002 and 2004 in Fars, Esfahan, Ghazvin, Kerman and Yazd provinces (M. Salehi, Izadpanah, & Siampour, 2011) with incidence rates as high as 50%. Symptoms included phyllody, proliferation, viresence and floral sterility. Analyses based on the 16S rRNA gene sequence showed that the rapeseed phyllody phytoplasma (Fars strain) is a member of 16SrI-B subgroup (M. Salehi, Izadpanah, et al., 2011). Five leafhopper species were collected from an infected field in Fars province (*C. haematoceps, N. fenestratus, Empoasca decepiens* Paoli (Hemiptera: Cicadellidae), *O. albicinctus* and *Psammotettix striatus* L., Hemiptera: Cicadellidae) but only *C. haematoceps* successfully transmitted the phytoplasma to rapeseed and also to sesame, rocket (*Eruca sativa* Mill., Brassicaceae) periwinkle, stock (*Matthiola sp.* W.T. Alton, Brassicaceae), mustard and radish which developed the phyllody symptoms. Due to the large host range of rapeseed phyllody phytoplasma and its vector, this phytoplasma poses a threat to rapeseed and other economically important crops in Iran.

Rapeseed could be experimentally infected by other phytoplasmas, including sesame phyllody (Fars strain) of 16SrIX and cabbage yellows (Fars strain) of 16Sr VI to develop phyllody symptoms (M. Salehi et al., 2017; M. Salehi & Izadpanah, 1992). Although *C. haematoceps* has been reported as the vector of sesame phyllody and cabbage yellows, the natural infection of rapeseed with these phytoplasmas has not yet been reported in Iran (M. Salehi et al., 2017; M. Salehi, Izadpanah, & Siampour, 2007).

In 2008 and 2009 safflower plants from fields in Fars and Yazd provinces of Iran (M. Salehi, Izadpanah, & Siampour, 2008; M. Salehi, Izadpanah, Siampour, Firouz, & Salehi, 2009) were observed with symptoms of phyllody, virescence, proliferation, floral sterility and little leaf. Grafting and dodder experimental trials coupled with molecular analysis showed that the phytoplasma associated with the disease belonged to 16SrVI-C subgroup (M. Salehi, Izadpanah, Siampour, Firouz, et al., 2009). No Hemiptera vector nor secondary plant reservoirs have been determined for this phytoplasma yet.

Sunflower phyllody is a phytoplasma disease reported from central Iran in the Fars, Yazd, Esfahan and Kerman provinces (M. Salehi, Esmailzadeh-Hosseini, & Salehi, 2015; Tazehkand, Pour, Heydarnejad, Varsani, & Massumi, 2010). Symptoms include phyllody, proliferation, head abnormality and virescence. In all provinces, the disease is associated with a phytoplasma strain belonging to the 16SrII-D subgroup; however, within Kerman province, 16SrII and 16SrVI groups were identified in mixed infections in sunflower plants showing witches' broom symptoms (Tazehkand et al., 2010). The insect vector of this phytoplasma disease has not yet been identified.

In 2011, a phytoplasma of 16SrI group was identified in association with niger seed phyllody disase in Esfahan province, Iran (Vaali, Bahar, & Zirak, 2011). A 16SrI phytoplasma was also previously identified in canola in Kerman province (Tazehkand et al., 2010).

3.5 Phytoplasma Diseases in Alfalfa

Alfalfa (*Medicago sativa* L., Fabaceae) is an economically important forage crop cultivated in most regions of Iran in over 618,000 hectares (Asgharipour, Mousavinik, & Enayat, 2016). Historically, it has been believed that alfalfa was first cultivated in ancient Iran. Alfalfa witches' broom (AlfWB) caused by a diverse group of phytoplasmas is one of the most devastating diseases threatening the production of alfalfa in several countries. Phytoplasma strains belonging to 16SrI, 16SrII, 16SrIII, 16SrVI, 16SrVII and 16SrXII groups have been associated with AlfWB in Argentina, Europe and the Middle East (Conci et al., 2005; Firrao et al., 2004; P. Jones, Arocha, & Plata, 2005; Khan, Botti, Al-Subhi, Gundersen-Rindal, & Bertaccini, 2002; Marcone et al., 1997; Peters et al., 1999). AlfWB symptoms include shoot proliferation, little leaf, virescence, phyllody, flower sterility, dwarfing, major loss of the forage yield, decline and plant death (Esmailzadeh-Hosseini, Salehi, Khodakaramian, Mirchenari, & Bertaccini, 2015; Khan et al., 2002) (Fig. 3.1). AlfWB disease was first observed in 1987 in Kerman province (south of Iran). The AlfWB phytoplasma was first identified in Iran (Fars Province) using Dienes' stain, vector trials and graft transmission in 1993 (M. Salehi & Izadpanah, 1993; M. Salehi et al., 1995). A comprehensive review of the latest status of the disease in Iran has been recently provided (Esmailzadeh-Hosseini, Khodakaramian, Salehi, & Bertaccini, 2016a). AlfWB is distributed in many provinces of Iran and has caused up to 100% of economical losses to the alfalfa production in the country (Table 3.1), particularly in Bushehr and Yazd provinces.

Phytoplasmas of 16SrII and 16SrXII groups are the major causes of AlfWB disease in south and west of Iran, respectively. Phytoplasmas of 16SrII group are the major causes of AlfWB in the Middle East. AlfWB occurs in eight provinces, mainly in southern Iran (Table 3.1). Phytoplasmas associated with AlfWB in Iran have been identified as the members of 16SrII-C and 16SrII-D subgroups (Esmailzadeh-Hosseini, Khodakaramian, et al., 2016a). In Oman and Saudi Arabia, however, only phytoplasmas of the 16SrII-D subgroup were associated with AlfWB (Al-Saleh, Amer, AL-Shahwan, Abdalla, & Damiri, 2014; Khan et al., 2002).

O. albicinctus has been identified as the vector of the AlfWB-16SrII phytoplasma in Fars province. This polyphagous leafhopper has also been found as the vector of other 16SrII phytoplasmas causing carrot witches' broom and faba bean phyllody (M. Salehi, Esmailzadeh-Hosseini, Salehi, & Bertaccini, 2016; M. Salehi, Rasoulpour, & Izadpanah, 2016) in the same regions. The authors stated that alfalfa may serve as a reservoir of faba bean phyllody phytoplasma. Weed species including *Prosopis farcta* (Banks & Sol.) J.F.Macbr. (Fabaceae) and *Cardaria draba* L. (Brassicaceae) have also been found to act as reservoirs of the AlfWB-16SrII phytoplasmas and as preferred hosts for the leafhopper vector in Yazd province (Esmailzadeh-Hosseini, Salehi, & Mirzaie, 2011). Other plant species, tamarisk (*Tamarix aphylla* L. Karst, Tamaricaceae) and saxaul (*Haloxylon aphyllum* Minkw (Amaranthaceae), cultivated nearby the alfalfa fields in Yazd province were found to be a site for survival of *O. albicinctus*; hence, they may play a role in the epidemiology of AlfWB in Iran (Esmailzadeh-Hosseini et al., 2011).

Phytoplasmas of 16SrXII-A subgroup were also associated with AlfWB in Iran. The highest rate of alfalfa infection by the AlfWB phytoplasmas of this subgroup was 14.7% in Chaharmahal-Bakhtiari province in central Iran; however, the major occurrence of this type of AlfWB is in western Iran, which is the major area of alfalfa cultivation (Esmailzadeh-Hosseini, Salehi, Khodakaramian, et al., 2015). No insect vector has been identified for the AlfWB phytoplasmas of 16SrXII-A. The co-occurrence of AlfWB and "bois noir" phytoplasmas (16SrXII-A) in some regions of Iran (eg. in Lorestan province) may have an impact on the epidemiology of these two devastating diseases.

Phytoplasmas of 16SrI and 16Sr VI groups contributed to a minority of AlfWB diseases in Iran (Esmailzadeh-Hosseini, Salehi, Khodakaramian, et al., 2015; Rashidi, Habili, & Ghasemi, 2010).

3.6 Phytoplasma Diseases in Weeds and Ornamental Plants

Phytoplasma infections of weeds and ornamental plants have been reported from diverse geographical regions of Iran. The infection of 10 weed and ornamental plant species by phytoplasmas belonging to 16SrVI (subgroup VI-A) and 16SrI (subgroups 16SrI-A, I-B and I-C) groups was reported in the Markazi province (Babaie et al., 2007). Symptoms included witches' broom, virescence, early decline, phyllody, dwarfing, yellowing, dieback and seed sterility. Phytoplasmas of 16SrVI group were detected in symptomatic milkweed and cockscomb, while those detected in *Tagetes* sp. L. (Asteraceae)—swan plant, periwinkle, *Tanacetum* sp. L. (Asteraceae), prickly lettuce, black-eyed Susan and tickseed—belonged to 16SrI group.

Weeds can serve as reservoirs of phytoplasmas, therefore they play important roles on the survival of phytoplasmas and the emergence and spread of phytoplasma diseases. For example, lettuce and wild lettuce phyllodies in Iran are caused by the same phytoplasma strains (16SrIX-D*) and vectored by the same leafhopper *N. fenestratus* (M. Salehi, Izadpanah, Nejat, et al., 2007). Therefore, wild lettuce is considered an important secondary host for the rapid spread of the phytoplasma to lettuce and other potential crops. As mentioned earlier, weeds may also play a role in the epidemiology of AlfWB in Iran (Esmailzadeh-Hosseini et al., 2011).

Periwinkle is a common ornamental plant cultivated in Iran. It is the experimental host for most of the phytoplasma strains (Marcone, 2014). Natural infection of periwinkle by diverse phytoplasma strains of 16SrI, 16SrVI and 16SrIX groups has been reported in Iran (Table 3.1).

3.7 Phytoplasma Diseases in Pome Fruits

Pear decline (PD), a destructive pear disease caused by '*Ca.* P. pyri' (16SrX-C subgroup) has long been a problem in North America and Europe. A PD-like disease has been also reported from Taiwan (Liu, Chen, & Lin, 2007) where the European psyllid vectors *Cacopsylla pyri* L. (Psyllidae) and *C. pyricola* Foerster are absent (B. Jarausch & Jarausch, 2010; Seemüller & Schneider, 2004). PD is also present in Iran (Hashemi-Tameh, Bahar, & Zirak, 2014b; M. Salehi, Izadpanah, Taghavi, & Rahimian, 2008). It is known that the type of rootstocks plays a role in the severity of symptoms that is possibly related to the overwintering of the phytoplasma in the root system (Schaper & Seemüller, 1982). Depending on the rootstock, PD infected pear trees may develop 'quick decline'; 'slow decline' or 'reddening of the foliage' (Seemüller, 1989). 'Quick decline' is a rapid wilt of the trees while 'slow decline' is a gradual weakening of the pears grown on a less susseptible rootstock. 'Foliage reddening' is a mild slow decline on plants grown on tolerant rootstocks (B. Jarausch & Jarausch, 2010). PD symptoms have been observed in many pear orchards in Fars, Esfahan, Tehran and Khorasan provinces of Iran, which include stunting, abnormal foliage, reddening, yellowing and slow decline (Hashemi-Tameh et al.,

2014b; M. Salehi, Izadpanah, Taghavi, et al., 2008). The phytoplasma associated with PD in Iran (isolate Fars) was a member of 16SrX-C subgroup, related to '*Ca.* P. pyri' (M. Salehi, Izadpanah, Taghavi, et al., 2008). *C.pyricola* is present in Iran, so it is expected to play a significant role in the PD spread in the country (M. Salehi, Izadpanah, Taghavi, et al., 2008).

Recently, it was reported that phytoplasmas related to '*Ca.* P. prunorum' (16SrX-B) and '*Ca.* P. asteris' (16SrI-B) were associated with a pear disease in some orchards in northern and central regions of Iran, respectively (Hashemi-Tameh et al., 2014b). Symptoms included leaf rolling (plants infected by phytoplasmas of 16SrX-B), and yellowing/reddening of leaves and decline (plants infected by phytoplasmas of 16SrI-B). The authors also reported a disease similar to apple proliferation (AP) in apple trees in central Iran (Hashemi-Tameh, Bahar, & Zirak, 2014a). Symptoms included yellowing of leaves, defoliation, proliferation, rosetting and decline. The disease was reported to be associated with phytoplasmas of 16SrI and 16SrII groups, but not with '*Ca.* P. mali'. This was the first report of apple infection by phytoplasmas of 16SrII in the world and the first report of apple infection by phytoplasmas of group 16SrI in Iran (Hashemi-Tameh et al., 2014a).

3.8 Phytoplasma Diseases in Stone Fruits

The genus *Prunus* has around 430 plant species mostly distributed throughout temperate regions of the world. Most of these plant species are economically important as fruit and nut crops. The most important *Prunus* species include peach, almond, nectarine, plum, European plum, Japanese plum, apricot, sweet cherry and sour cherry (Vicente, Manganaris, Cisneros-Zevallos, & Crisosto, 2011). Several phytoplasma diseases from different countries are known as major threats to *Prunus* crop production. Members of 16SrIII, 16SrIX and 16SrX groups are among the most important phytoplasmas causing significant losses to *Prunus* species (Foissac & Wilson, 2010).

Western X-disease (WX) is one of the most important phytoplasma diseases with destructive effects on the production of peach and cherry in North America (Foissac & Wilson, 2010). The causal phytoplasma belongs to 16SrIII group (X-disease group) and is related to '*Ca.* P. pruni'. WX phytoplasma also infects walnut and pecan (Davis et al., 2013; Lee, Davis, & Gundersen-Rindal, 2000). The disease is mainly limited to North America, but it has also been described from Italy (Foissac & Wilson, 2010). Wild chokecherry (*Prunus virginiana* L., Rosaceae) has been identified as the main reservoir plant of the WX phytoplasma (Davis et al., 2013). So far, phytoplasmas of 16SrIII group have not been reported from Iran.

European stone fruit yellows (ESFY) is another important phytoplasma disease of *Prunus* species apparently limited to Europe and Asia Minor (B. Jarausch & Jarausch, 2010). ESFY includes several host-specific diseases such as apricot chlorotic leaf roll, plum leptonecrosis, peach yellowing and decline of plum, peach and almond (Cieślińska, 2011). ESFY is not known to affect cherry and sour cherry

(Foissac & Wilson, 2010). The ESFY phytoplasmas belong to 16SrX-F group, related to '*Ca.* P. prunorum' (Seemüller & Schneider, 2004). They are all vectored by *Cacopsylla pruni* Scopoli (Psyllidae) (Carraro, Osler, Loi, Ermacora, & Refatti, 1998). It is worth noting that other fruit tree phytoplasmas of 16SrX are also transmitted by psyllids in specific manner. For example, apple proliferation ('*Ca.* P. mali', 16SrX-A) is vectored by *C. picta* Foerster (synonym, *Psylla costalis* Flor, Psyllidae) (Frisinghelli, Delaiti, Grando, Forti, & Vindimian, 2000) and *C. melanoneura* Foerster (Tedeschi, Bosco, & Alma, 2002). *C. pyricola* is known to transmit both PD and peach yellow leaf roll (PYLR) phytoplasmas. *C. pruni*, *C. pyricola* and *C. pyri* are among the psyllid fauna of Iran.

Recently, a phytoplasma strain related to '*Ca.* P. prunorum' has been associated with plum yellow leaf stunt disease in northern Iran (Allahverdi, Rahimian, & Babaeizad, 2014a). According to Hashemi-Tameh (2014), '*Ca.* P. prunorum' is also responsible for a PD-like disease in northern Iran. Although Hemiptera vectors are not known for these diseases and considering their vector specificity and the presence of *Cacopsylla* species in Iran, it is expected that *Cacopsylla* plays a role in transmitting such diseases in this country. Therefore, control of psyllids should be considered a routine strategy for the management of phytoplasma fruit tree diseases. Some wild *Prunus* species (*P. spinosa* L. (Rosaceae), *P. cerasifera* Lamotte, *P. amygdalus* (Mill.) D.A. Webb, *P. armeniaca* L., *P. domestica* L.), and *Celtis australis* L. (Cannabaceae), *Fraxinus excelsior* L. (Oleaceae) and *Rosa canina* L. (Rosaceae) are known to host the ESFY phytoplasma around the world (W. Jarausch et al., 2001). The role of bindweed (*Convolvulus arvensis* L., Convolvulaceae) and Bermuda grass in the epidemiology of ESFY has also been suggested (Sanchez-Capuchino, Casanova, & Llácer Inia, 1982).

Almond witches' broom (AlmWB) is the most devastating phytoplasma disease of stone fruits in Iran. According to the Food and Agriculture Organization (FAO), Iran has been among the top three countries for almond production in 2009 (128,464 tons, 8% of the world's production) after the United States and Spain. AlmWB is considered a major threat to almond production in Iran and Lebanon. AlmWB was first reported in Iran and Lebanon in 1995 (Abou-Jawdah, Karakashian, Sobh, Martini, & Lee, 2002; Choueiri et al., 2001; Lova et al., 2011; M. Salehi & Izadpanah, 1995a). The disease was first described in Iran as almond brooming in 1995 (M. Salehi & Izadpanah, 1995a).

Field surveys have estimated a decline of more than 100,000 almond trees affected by AlmWB disease in Lebanon in 2002 (Lova et al., 2011). Investigations have shown that the AlmWB phytoplasma is also associated with similar diseases in peach and nectarine in Iran and Lebanon (Abou-Jawdah, Dakhil, El-Mehtar, & Lee, 2003; M. Salehi, Haghshenas, Khanchezar, & Esmailzadeh-Hosseini, 2011). In Iran, AlmWB was primarily reported from almond groves in Fars province and later found in almond and peach orchards in other provinces (M. Salehi, Haghshenas, et al., 2011; M. Salehi, Izadpanah, & Heydarnejad, 2006). The rapid spread of this devastating disease is alarming not only Iran and Lebanon but also other stone fruit growing countries worldwide.

Sequence analysis of the 16S rRNA gene confirmed that the AlmWB phytoplasma was related to the pigeon pea witches'-broom phytoplasma group, 16SrIX (Abou-Jawdah et al., 2002; Verdin et al., 2003). It was found that two 16SrIX phytoplasma strains were associated with the disease, one from 16SrIX-B subgroup and the other one from 16SrIX-C subgroup (Abou-Jawdah et al., 2002; M. Salehi et al., 2006). An AlmWB phytoplasma strain of the 16SrIX-B was characterized as a novel taxon and named '*Ca.* P. phoenicium' (Firrao et al., 2004; Verdin et al., 2003). All phytoplasma strains characterized in Lebanon belonged to 16SrIX-B subgroup, while those in Iran were phytoplasma strains of both subgroups, 16SrIX-B and 16SrIX-C (Casati et al., 2016; M. Salehi et al., 2006). All phytoplasma strains of 16SrIX-B ('*Ca.* P. phoenicium) have been associaed with stone fruit trees and reported only from Iran and Lebanon. In contrast, phytoplasmas of 16SrIX-C were associated with diseases in several other crops (Casati et al., 2016). Phytoplasmas of the 16SrIX-C were also found in wild plant species in Lebanon, but not in almonds or other stone fruits. Symptoms of AlmWB caused by phytoplasma of both subgroups strains are similar. The most characteristic symptoms on almond trees are shoot proliferation in different parts of the tree canopy resembling witches'broom, development of axillary buds on the branches, general yellowing of leaves, severe dieback and decline of the tree (Fig. 3.1). Considerable loss of production occurs within 1–2 years after the onset of the disease (Lova et al., 2011; M. Salehi et al., 2006). In the case of peach and nectarine trees, the symptoms include early flowering (15 to 20 days earlier than normal) and early emergence of buds on the infected branches. Phyllody and witches' broom symptoms will appear on different parts of the tree while small and yellow leaves are developed on the canopy (Lova et al., 2011; M. Salehi et al., 2006; Verdin et al., 2003).

Almond is so far the only stone fruit tree found to be infected by phytoplasmas of 16SrIX-C in Iran in East Azerbaijan, Fars and Chaharmahal-Bakhtiari provinces (M. Salehi et al., 2006; Zirak, Bahar, & Ahoonmanesh, 2009a). It was shown, however, that some other stone fruits could be infected by this phytoplasma strain under greenhouse conditions (Abbasian, Salehi, & Hasanzadeh, 2009). On the other hand, the natural host range of 16SrIX-B phytoplasma strains ('*Ca.* P. phoenicium') includes almond, wild almond, peach, nectarine, apricot, GF 677, *Anthemis* spp. L. (Asteraceae) and *Smilax aspera* L. (Smilacaceae) (Lova et al., 2011; M. Salehi & Izadpanah, 2001; M. Salehi et al., 2006; M. Salehi, Izadpanah, et al., 2011; Tedeschi et al., 2015). AlmWB caused by phytoplasmas of 16SrIX-B is reported from several provinces of Iran including Fars, Kerman, Kohguiluieh-Boyerahmad and Kurdistan (Pourali & Salehi, 2012; M. Salehi, Salehi, Abbasian, & Izadpanah, 2015).

It has been shown that the polyphagous leafhopper *Asymmetrasca decedens* Paoli, Typhlocibinae, (synonym *Empoasca decedens*) can transmit '*Ca.* P.phoenicium' (16SrIX-B) to almond and peach in Lebanon (Abou-Jawdah et al., 2014). This leafhopper species is also present in almond orchards of Iran (Mozaffarian & Wilson, 2016). The planthoppers *Tachycixius* cf. *cypricus* Dlabola (Hemiptera: Cixiidae) and *Tachycixius viperinus* Dlabola were also reported to transmit '*Ca.* P. phoenicium' to peach in experimental trials in Lebanon (Tedeschi et al., 2015). *Tachycixius* cf. *cypricus* and *T. viperinus* individuals carrying '*Ca.*

P. phoenicium' were collected from *Smilax aspera* and *Anthemis* spp. Interestingly, these two weed species were also suspected as reservoirs of AlmWB phytoplasma in Lebanon (Tedeschi et al., 2015). The vectors of the AlmWB phytoplasma in Iran are still unknown. However, field surveys in almond orchards infected by AlmWB phytoplasma (strain 16SrIX-C) showed that the leafhopper species *Frutioidea bisignata* Mulsant and Rey (Hemiptera: Cicadellidae) and *Zygina flammigera* Fourcroy (Hemiptera: Cicadellidae) are carriers of this phytoplasma strain in Fars province (Siampour, Izadpanah, Afsharifar, Salehi, & Taghizadeh, 2006).

Zirak et al. (2009a) also reported the association of phytoplasmas of 16SrII, 16SrXII and 16SrVI groups with diseased almonds (showing AlmWB-like symptoms) in central Iran. Records also showed that phytoplasmas of 16SrII and 16SrXII groups were associated with a disease in Japanese plum in central Iran (Esfahan, ChaharMahal-Bakhtiari and Tehran) (Zirak, Bahar, & Ahoonmanesh, 2009b). Symptoms included little leaf, yellowing, rosetting and proliferation. Infection of cherry by phytoplasmas of 16SrI and 16SrII groups was also reported in central Iran (Zirak, Bahar, & Ahoonmanesh, 2010).

3.9 Phytoplasma Diseases in Grapevine

Grapevine (*Vitis vinifera* L., Vitaceae) is an important economic plant species cultivated in most regions of Iran. It is a host of several phytoplasmas that cause diseases commonly known as 'grapevine yellows' (GY). GY symptoms include uneven yellowing and reddening of leaves in white and red grapevine varieties, respectively that may turn into necrosis, downward curling of the leaves, development of black pustules on the affected shoots, dieback, shriveling and early drying of the berries (Constable, 2010; E. Salehi, Taghavi, Salehi, & Izadpanah, 2014). Phytoplasmas associated with GY around the world are diverse and belong to the following groups: aster yellows (16SrI), elm yellows (16SrV), Stolbur (16SrXII-A and 16SrXII-B subgroups), X-disease (16SrIII), faba bean phyllody (16SrII), ash yellows (16SrVII) and pigeon pea witches' broom (16SrIX-C subgroup) (Constable, 2010).

Flavescence dorée (FD) is probably the most economically important phytoplasma disease of grapevine and it is caused by phytoplasmas of 16SrV-C and 16SrV-D subgroups. FD was first observed in France in 1957 (Caudwell, 1957) and then introduced and spread in Europe (Boudon-Padieu, 2003). FD has a narrow host range and is vectored almost exclusively by *Scaphoideus titanus* Ball (Hemiptera: Cicadellidae) in a highly specific manner. *S. titanus* exclusively lives on grapes; hence, it plays a crucial role in spreading FD (Constable, 2010). FD has a defined geographical distribution in Europe and to date it has not been detected in vineyards in Iran.

In Australia, GY diseases (known as Australian GY) were associated with phytoplasmas of 16SrI, 16SrII-D, 16SrXII-B and 16SrXXIII (Constable, 2010; Constable, Whiting, Gibb, & Symons, 2002). The insect vector of GY phytoplasmas of 16SrII is still unknown. It has been reported that some GY symptomatic grapevine plants in Iran were infected by phytoplasmas of 16SrII-B subgroup (Ghayeb Zamharir, Paltrinieri, Hajivand, Taheri, & Bertaccini, 2017). In the USA, GY phytoplasmas belong to group 16SrIII (Constable, 2010), which have not been reported in Iran.

"Bois noir" (BN) is the most important and widespread phytoplasma disease on grapevine in Iran. The disease is present in Europe and the Middle East. BN is caused by Stolbur phytoplasma strains (subgroup 16SrXII-A, '*Ca.* P. solani'). The incidence of BN throughout the world has been over 80% in a year in some vineyards (Constable, 2010). In Iran, BN was first reported from the northeast in 2009 (Karimi, Contaldo, Mahmoudi, Duduk, & Bertaccini, 2009). In 2014, BN was found in Lorestan and Fars provinces, in western and southern Iran, respectively (E. Salehi, Taghavi, et al., 2014). BN is now present in the major grapevine-growing provinces (Ghayeb Zamharir et al., 2017; Mirchenari, Massah, & Zirak, 2015) (Table 3.1). BN symptoms were more obvious in late spring in Fars and Lorestan provinces, and mostly limited to one part or one branch of the vine. E. Salehi, Taghavi, et al. (2014) successfully dodder-transmitted the BN phytoplasma from infected grapevine to healthy periwinkle, and found SNPs on the 16S rDNA sequence that distinguished Lorestan and Fars BN strains (16SrXII-A) from a BN strain in Khorasan in northeast of Iran (accession number KJ637210) (Karimi et al., 2009). This suggests that variants of BN phytoplasma strains may be responsible for the occurrence of BN in Iran.

Stolbur phytoplasmas have a wide host range; they have been found in diverse herbaceous or woody plants. These host plants could act as alternate or reservoir host for the BN phytoplasma and also may serve as host for its insect vector (Constable, 2010; Martini, Marcone, Lee, & Firrao, 2014). In Iran, phytoplasmas of 16SrXII-A subgroup have been associated with diseases in several plant species (Table 3.1), but the exact role of these plant species in the occurrence of BN is poorly understood. Maixner (2007) stated that bindweed (*Convolvulus arvensis*) and stinging nettle (*Urtica dioica* L., Urticaceae) play a significant role in the spatio-temporal distribution patterns of BN in Europe, and may affect the population density and movement of its vector, *Hyalesthes obsoletus* Signoret (Hemiptera: Cixiidae). Although the insect vector of the BN phytoplasma is not yet identified in Iran, *H. obsoletus* is present in Iranian vineyards. Grapevine is a dead-end host for the BN phytoplasma (Maixner, Johannesen, Michel, Lux, & Seitz, 2007); therefore the presence of the preferred host plants (bindweed or stinging neetle) in vineyards is a risk factor for the spread of BN. Infection of bindweed by stolbur phytoplasmas was determined in some vineyards of Iran (Salehi et al., unpublished), therefore, its possible role in the epidemiology of BN in Iran needs to be investigated.

Phytoplasmas of the group 16SrVII (ash yellows, '*Ca.* P. fraxini') have been detected in symptomatic grapevines in Qazvin province in central Iran (Ghayeb Zamharir et al., 2017). The association of 16SrVII phytoplasmas with GY has been reported from Chile in South America (Gajardo et al., 2009). Phytoplasmas of 16SrVII have also been detected on date palm in southern Iran, and on olive trees in the north of Iran (Ghayeb Zamharir, Boshehri, & Khajehzadeh, 2016; Ghayeb Zamharir & Razavi, 2016).

3.10 Phytoplasma Diseases in Citrus Crops

Witches' broom disease of lime (WBDL) is probably the most devastating phytoplasma disease of citrus crops so far reported in Iran. WBDL mainly affects small-fruited acid lime trees [Mexican or key lime; *Citrus aurantifolia* (L.) Swingle (Rutaceae)] which is the most popular citrus crop in southern provinces of Iran including Sistan-Baluchestan, Hormozgan, Fars and Kerman (M. Salehi, Heydarnejad, et al., 2005). WBDL was first reported from Oman in 1986 and then found in the United Arab Emirates (UAE) in 1989. In both countries the disease has seriously affected lime plantations (El Shereiqi & Gassouma, 1993). In Iran, WBDL symptoms were first seen in southeast Iran near Nikshahr and Qasr-e-Qand in Sistan-Baluchestan province (approximately 100 km west of the Pakistani border) in 1997. The first surveys identified about 500 severely infected trees, indicating that the WBDL was present in the region for some years (Bove & Garnier, 2000). Within a few years, the disease spread out toward the southwest of the country and severely affected the main lime-growing areas in Hormozgan province. WBDL was observed on lime and bakraee (a *Citrus reticulata* Bianco (Rutaceae) hybrid) groves of Kerman province (M. Salehi, Heydarnejad, et al., 2005). The disease has also reached Fars, another major lime growing region, but quarantine measures prevented the spread of disease across the province. WBDL is easy to diagnose by visual inspection of symptoms that generally involve witches' broom with small, pale and tender leaves (Fig. 3.1), which later dry up and fall, leaving naked branches. The trees decline in few years after the onset of symptoms (Bove & Garnier, 2000). It has been estimated that about 30% of the Mexican lime trees (over half a million trees in 7000 hectares) in Iran have been destroyed by WBDL in about 10 years (Mardi et al., 2011). Due to the high socio-economic impact of WBDL in Iran, government funds were allocated for WBDL research and management in southern Iran (Mardi et al., 2011).

Although lime has been found to be the major plant host of the WBDL phytoplasma, other citrus trees such as grapefruit, citron, limequat and especially bakraee has been also infected in Iran. In the UAE and Oman WBDL was occasionally reported on citron, Palestine sweet lime and sweet limetta (Bagheri, Faghihi, Salehi, & Khanchezar, 2010; Djavaheri & Rahimian, 2004; El Shereiqi & Gassouma, 1993).

'*Ca.* P. aurantifolia' causes WBDL disease in Iran as well as in Oman and the UAE (Zreik, Carle, Bove, & Garnier, 1995). This phytoplasma was assigned to 16SrII group as the only member of the 16SrII-B subgroup (Lee, Gundersen-Rindal, Davis, & Bartoszyk, 1998). Phylogenetic and serological analyses showed that 16SrII-C phytoplasmas, widespread in Iran, were closely related to the WBDL phytoplasma (Siampour, Izadpanah, Galetto, Salehi, & Marzachi, 2013). No serological relationship was found between the WBDL phytoplasma and phytoplasmas of 16SrII-D (Siampour et al., unpublished). However, the WBDL phytoplasma was distinguishable from other phytoplasmas within the group 16SrII (Siampour et al., 2013; Siampour, Izadpanah, Marzachi, & Salehi, 2015).

M. Salehi et al. (2000) showed that '*Ca.* P. aurantifolia' can be transmitted to periwinkle and a number of solanaceous species. Among several citrus species, M. Salehi, Nejat, Tavakkoli, and Izadpanah (2005) found that *Citrus jambhiri* Lush. (Rutaceae) (Rough lemon), *C. limonia* Osbeck (Rangpur lime), *C. macrophylla* Wester (Alemow), *C. medica* L. (Citron) and *C. reticulata* hybrid (Bakraee) could be infected by the WBDL phytoplasma. Persian lime, sweet orange and grapefruit appear to be resistant, although symptoms have been observed in grapefruit, probably due to high WBDL phytoplasma pressure (Bagheri et al., 2010; Bove & Garnier, 2000; Jeger et al., 2017; M. Salehi, Nejat, et al., 2005).

The rapid spread of WBDL suggests involvement of active insect vectors in all the affected areas. Early field surveys showed that the leafhopper *Hishimonus phycitis* Distant (Hemiptera: Cicadellidae) was present on the lime trees in affected areas of Iran and Oman (Bové et al., 1993; Siampour et al., 2006). Serological and DNA hybridization analyses confirmed the presence of WBDL phytoplasma in *H. phycitis* specimens (Bové et al., 1993). In our surveys from lime groves (Siampour et al., unpublished), *H. phycitis* were mainly collected on witches' brooms. Delicate and compact broom tissues appear to favor *H. phycitis* feeding and breeding. The leafhopper was barely collected on other citrus trees (except on infected bakraee). In 2007, M. Salehi, Izadpanah, Siampour, Bagheri, and Faghihi (2007) successfully transmitted the WBDL phytoplasma to bakraee (but not lime) seedlings using *H. phycitis* collected from WBDL phytoplasma infected lime trees. In 2009, a massive inoculation of older lime trees by large numbers of *H. phycitis* leafhoppers collected from WBDL affected lime trees resulted in the transmission of the WBDL phytoplasma and the reproduction of WBDL symptoms (Bagheri, Salehi, Faghihi, Samavi, & Sadeghi, 2009). Quantitative real-time PCR showed that both lime seedlings and older plants may have symptomless infection and that symptom development depends on the phytoplasma concentration (Amiri, Izadpanah, Hamzeh Zarghani, & Salehi, 2016).

3.11 Conclusion

Diverse groups of phytoplasma diseases have been reported from most of the provinces in Iran. Over 70 plant species from 26 botanical families are hosts of phytoplasmas in this country. Fars was the province with the highest diversity of phytoplasma strains and the highest number of plant species found infected by phytoplasmas. Most of the characterized phytoplasmas belonged to 16SrI, 16SrII, 16SrVI, 16SrIX and 16SrXII groups. Phytoplasmas of the groups 16SrX, 16SrXIV, 16SrXXX and 16SrVII are also present in Iran. Although phytoplasmas of the group 16SrII have been classified into 21 subgroups (Yang, Jiang, Che, Cao, & Luo, 2016), only phytoplasma subgroups 16SrII-B, 16SrII-C and 16SrII-D have been identified in Iran, with a more widespread distribution in southern provinces.

WBDL phytoplasma (*'Ca.* P. aurantifolia') is the only phytoplasma strain assigned within 16SrII-B subgroup (Lee et al., 1998; Zreik et al., 1995). It is closely related to phytoplasmas of 16SrII-C group, which possess a wide host range and geographical distribution (Martini et al., 2007; Siampour et al., 2013). WBDL phytoplasma is transmitted by *H. phycitis* (Bagheri et al., 2009; M. Salehi, Izadpanah, Siampour, et al., 2007) and affects the lime industry in southern Iran. The WBDL phytoplasma is not seed-borne, although the germination of seeds obtained from infected trees was shown to be affected (Faghihi et al., 2011). The narrow host range of the WBDL phytoplasma and its insect vector makes it easier to manage the disease using quarantine measures including continuous monitoring, immediate eradication of infected trees, planting of certified seedlings and control of insect populations. Some promising results to control WBDL have been observed after treating infected lime trees with some natural compounds (M. Salehi, Hassanzadeh, Esmailzadeh-Hosseini, Adeli, & Kalantar, 2012). This disease is subject to quarantine for Fars province, an important lime-growing area in Iran.

AlfWB phytoplasma (group 16SrII) is present in some regions of Iran (Esmailzadeh-Hosseini, Khodakaramian, et al., 2016a), and is transmitted specifically by *O. albicinctus* (Esmailzadeh-Hosseini, Khodakaramian, et al., 2016a; M. Salehi et al., 1995). Recently, *O. albicinctus* has been shown to transmit several 16SrII-D phytoplasmas including the cucumber phyllody phytoplasma in southern Iran (M. Salehi, Siampour, et al., 2015).

AlmWB is the most devastating phytoplasma disease of stone fruits in Iran, caused by phytoplasmas of 16SrIX-B and 16SrIX-C (M. Salehi et al., 2006). Phytoplasmas of 16SrIX-C have been also found in black locust, sesame, tomato and lettuce in Iran (Table 3.1). Using certified plants or resistant cultivars could have a significant effect on controlling the spread of AlmWB (EPPO, 2017). AlmWB has been only reported from Iran and Lebanon but due to rapid rate of spread, it poses a potential threat to stone fruits in other nearby countries.

Phytoplasmas of group 16SrXII are widespread throughout Iran causing diseases of high impact on different crops. All strains belong to subgroup 16SrXII-A (related to '*Ca.* P. solani'). Phytoplasma strains of subgroup 16SrXII-A have been identified in tomato, potato, plum, almond, peach, grapes and alfalfa (Table 3.1).

"Bois Noir" (BN), caused by phytoplasmas of 16SrXII-A subgroup, is the most important and widespread disease of grapevine in Iran (Mirchenari et al., 2015; E. Salehi, Taghavi, et al., 2014). Although the Hemiptera vectors of 16SrXII-A phytoplasmas have not been identified in Iran, *H. obsoletus* and its preferred host, bindweed (*C. arvensis*), is present in the country.

Control of insect vectors is one of the major components of the management strategy of phytoplasma diseases. However, the insect vector for the majority of phytoplasma diseases in Iran are yet unknown. It is also known the critical role of weed species in the epidemiology of phytoplasma diseases as reservoirs for the completion of the vector life cycle. As well, the role of secondary plant hosts in the epidemiology of most of phytoplasma diseases in Iran is very poorly understood. Priority must be given to studies focusing on disease epidemiology and plant-host interactions, emphasizing on vector feeding behaviour and migration patterns, and the discrimination of closely related phytoplasma strains. Information from such studies would be important tools to support the effective management of phytoplasma diseases in Iran.

References

Abbasian, M., Salehi, M., & Hasanzadeh, N. (2009). Host range of almond witches' broom phytoplasma among stone fruits. *Iranian Journal of Plant Pathology, 44*(176), 362–363.

Abou-Jawdah, Y., Abdel Sater, A., Jawhari, M., Sobh, H., Abdul-Nour, H., Bianco, P. A., ... Alma, A. (2014). *Asymmetrasca decedens* (Cicadellidae, Typhlocybinae), a natural vector of '*Candidatus* Phytoplasma phoenicium'. *Annals of Applied Biology, 165*(3), 395–403.

Abou-Jawdah, Y., Dakhil, H., El-Mehtar, S., & Lee, I.-M. (2003). Almond witches'-broom phytoplasma: A potential threat to almond, peach, and nectarine. *Canadian Journal of Plant Pathology, 25*(1), 28–32.

Abou-Jawdah, Y., Karakashian, A., Sobh, H., Martini, M., & Lee, I.-M. (2002). An epidemic of almond witches'-broom in Lebanon: Classification and phylogenetic relationships of the associated phytoplasma. *Plant Disease, 86*(5), 477–484.

Akhtar, K. P., Sarwar, G., Dickinson, M., Ahmad, M., Haq, M. A., Hameed, S., & Iqbal, M. J. (2009). Sesame phyllody disease: Its symptomatology, etiology, and transmission in Pakistan. *Turkish Journal of Agriculture and Forestry, 33*(5), 477–486.

Allahverdi, T., Rahimian, H., & Babaeizad, V. (2014a). Association of *Candidatus* Phytoplasma prunorum associated with plum yellow leaf stunt in Iran. *Iranian Journal of Plant Pathology, 50*(2).

Allahverdi, T., Rahimian, H., & Babaeizad, V. (2014b). First report of a *Candidatus* Phytoplasma solani isolate affecting *Sophora alopecuroides* in Iran. *New Disease Reports, 30*(22), 2044-0588.2014.

Al-Sakeiti, M., Al-Subhi, A., Al-Saady, N., & Deadman, M. (2005). First report of witches'-broom disease of sesame (*Sesamum Indicum*) in Oman. *Plant Disease, 89*(5), 530–530.

Al-Saleh, M., Amer, M. A., Al-Shahwan, I. M., Abdalla, O. A., & Damiri, B. V. (2014). Detection and molecular characterization of alfalfa witches' broom phytoplasma and its leafhopper vector in Riyadh Region of Saudi Arabia. *International Journal of Agriculture and Biology, 16*, 300–306.

Amiri, M., Izadpanah, K., Hamzeh Zarghani, H., & Salehi, M. (2016). *Detection of 'Candidatus Phytoplasma aurantifolia' in lime trees during the latent period of the disease.* Paper presented at the 22th Iranian plant protection congress (August 26–30, 2016), College of Agriculture and Natural Resources, University of Tehran, Karaj, Iran, p. 87.

Arocha, Y., Horta, D., Piñol, B., Palenzuela, I., Picornell, S., Almeida, R., & Jones, P. (2005). First report of a phytoplasma associated with Bermuda-grass white leaf disease in Cuba. *Plant Pathology, 54*(2), 233–233.

Arocha, Y., & Jones, P. (2010). Phytoplasma diseases of the Gramineae. In P. G. Weintraub, P. Jones (Eds), *Phytoplasmas: Genomes, plant hosts and vectors* (pp. 170–187). Cambridge, MA: CABI North American Office.

Asgharipour, M., Mousavinik, S. M., & Enayat, F. F. (2016). Evaluation of energy input and greenhouse gases emissions from alfalfa production in the Sistan region, Iran. *Energy Reports, 2*, 135–140.

Azimi, M., Farokhi-Nejad, R., & Mehrabi-Koushki, M. (2017a). First report of a '*Candidatus* Phytoplasma aurantifolia'-related strain associated with leaf roll symptoms on eucalyptus in Iran. *New Disease Reports, 35*, 4.

Azimi, M., Farokhi-Nejad, R., & Mehrabi-Koushki, M. (2017b). First report of *Candidatus* Phytoplasma aurantifolia (16SrII group) associated with *Conocarpus erectus* disease in Iran. *Australasian Plant Disease Notes, 12*(1), 27.

Babaie, G., Khatabi, B., Bayat, H., Rastgou, M., Hosseini, A., & Salekdeh, G. (2007). Detection and characterization of phytoplasmas infecting ornamental and weed plants in Iran. *Journal of Phytopathology, 155*(6), 368–372.

Bagheri, A., Faghihi, M., Khankahdani, H. H., Seyahooei, M. A., Ghanbari, N., & Sarbijan, S. S. (2017). First report of a phytoplasma associated with sapodilla flattened stem disease in Iran. *Australasian Plant Disease Notes, 12*(1), 25.

Bagheri, A., Faghihi, M., Salehi, M., & Khanchezar, A. (2010). *First report of natural infection of grapefruit trees to lime witches broom phytoplasma.* Paper presented at the 19th Iranian plant protection congress (July 31–August 3, 2010), Iranian Research Institute of Plant Protection, Teheran, Iran. Vol. 1, p. 409.

Bagheri, A., Salehi, M., Faghihi, M., Samavi, S., & Sadeghi, A. (2009). Transmission of *Candidatus* Phytoplasma aurantifolia to Mexican lime by the leafhopper *Hishimonus phycitis* in Iran. *Journal of Plant Pathology, 91*(4).

Bahari, A., Alavi, S., Shams-Bakhsh, M., & Saberi, E. (2016). First report of a 16SrII group related phytoplasma associated with witches' broom of *Cupressus sempervirens* var. *horizontalis* in Iran. *New Disease Reports, 34*, 21

Bertaccini, A., Voráckova, Z., Vibio, M., Fránová, J., Navrátil, M., Špak, J., & Nebesárová, J. (1998). Comparison of phytoplasmas infecting winter oilseed rape in the Czech Republic with Italian Brassica phytoplasmas and their relationship to the aster yellows group. *Plant Pathology, 47*(3), 317–324.

Biabani, R., Ghasemi, S., Salehi, M., & Rahimian, H. (2009). Purification and serological study of sugarcane white leaf phytoplasma in Khuzestane province. *Plant Protection Journal, 1*(1).

Boudon-Padieu, E. (2003). *The situation of grapevine yellows and current research directions: Distribution, diversity, vectors, diffusion and control.* Paper presented at the 14th Meeting of ICVG (September 12-17, 2003), Locorotondo (Bari), Italy. *Session, 3*, 47–53.

Bove, J., & Garnier, M. (2000). Witches' broom disease of lime. *Arab Journal of Plant Protection, 18*(2), 148–152.

Bové, J., Zreik, L., Danet, J., Bonfils, J., Mjeni, A., & Garnier, M. (1993). Witches' Broom Disease of Lime Trees: Monoclonal Antibody and DNA Probes for the Detection of the Associated MLO and the Identification of a Possible Vector. In *International Organization of Citrus Virologists Conference Proceedings* (Vol. 12, pp. 342–348): eScholarship, University of California Riverside.

Carraro, L., Osler, R., Loi, N., Ermacora, P., & Refatti, E. (1998). Transmission of European stone fruit yellows phytoplasma by *Cacopsylla pruni*. *Journal of Plant Pathology*, 233–239.

Casati, P., Quaglino, F., Abou-Jawdah, Y., Picciau, L., Cominetti, A., Tedeschi, R., … Lova, M. M. (2016). Wild plants could play a role in the spread of diseases associated with phytoplasmas of pigeon pea witches'-broom group (16SrIX). *Journal of Plant Pathology*, 71-81.

Caudwell, A. (1957). Deux années d'études sur la Flavescence dorée, nouvelle maladie grave de la vigne. *Annales de l'amelioration des plantes, 4*, 359–393.

Choueiri, E., Jreijiri, F., Issa, S., Verdin, E., Bové, J., & Garnier, M. (2001). First report of a phytoplasma disease of almond (*Prunus amygdalus*) in Lebanon. *Plant Disease, 85*(7), 802–802.

Cieślińska, M. (2011). European stone fruit yellows disease and its causal agent '*Candidatus* Phytoplasma prunorum'. *Journal of Plant Protection Research, 51*(4), 441–447.

Conci, L., Meneguzzi, N., Galdeano, E., Torres, L., Nome, C., & Nome, S. (2005). Detection and molecular characterisation of an alfalfa phytoplasma in Argentina that represents a new subgroup in the 16S rDNA ash yellows group ('*Candidatus* Phytoplasma fraxini'). *European Journal of Plant Pathology, 113*(3), 255–265.

Constable, F. E. (2010). Phytoplasma epidemiology: Grapevines as a model. In P. G. Weintraub & P. Jones (Eds.), *Phytoplasmas: Genomes, plant hosts and vectors* (Vol. 1, pp. 188–212). Preston, UK: CABI.

Constable, F. E., Whiting, J., Gibb, K. S., & Symons, R. (2002). A new grapevine yellows phytoplasma from the Buckland Valley of Victoria, Australia. *VITIS, 41*(3), 147–153.

Davis, R. E., Zhao, Y., Dally, E. L., Lee, I.-M., Jomantiene, R., & Douglas, S. M. (2013). '*Candidatus* Phytoplasma pruni', a novel taxon associated with X-disease of stone fruits, *Prunus* spp.: Multilocus characterization based on 16S rRNA, secY, and ribosomal protein genes. *International Journal of Systematic and Evolutionary Microbiology, 63*(2), 766–776.

Dehghan, H., Salehi, M., Khanchezar, A., & Afshar, H. (2014). Biological and molecular characterization of a phytoplasma associated with greenhouse cucumber phyllody in Fars province. *Iranian Journal of Plant Pathology, 50*(4).

Djavaheri, M., & Rahimian, H. (2004). Witches'-broom of bakraee (*Citrus reticulata* hybrid) in Iran. *Plant Disease, 88*(6), 683–683.

El Shereiqi, R., & Gassouma, S. (1993). Witches' broom disease of lime in the United Arab Emirates. In *International Organization of Citrus Virologists Conference Proceedings* (Vol. 12, pp. 453-454): eScholarship University of California Riverside.

EPPO. (2017). Mini data sheet on '*Candidatus* Phytoplasma phoenicium' Retrieved May 16, 2018, from EPPO https://gd.eppo.int

Esmailzadeh-Hosseini, S., Khodakaramian, G., Salehi, M., & Bertaccini, A. (2016a). Characterization of 16SrII group phytoplasmas associated with alfalfa (*Medicago sativa*) witches' broom disease in diverse areas of Iran. *Journal of Crop Protection, 5*(4), 581–590.

Esmailzadeh-Hosseini, S., Khodakaramian, G., Salehi, M., & Bertaccini, A. (2016b). Molecular identification and phylogenetic analysis of phytoplasmas associated with alfalfa witches' broom diseases in the western areas of Iran. *Phytopathogenic Mollicutes, 6*(1), 16–22.

Esmailzadeh-Hosseini, S., Salehi, M., Khodakaramian, G., Mirchenari, S. M., & Bertaccini, A. (2015). An up to date status of alfalfa witches' broom disease in Iran. *Phytopathogenic Mollicutes, 5*(1), 9–18.

Esmailzadeh-Hosseini, S., Salehi, M., Mirchenari, S., & Bertaccini, A. (2016). First report of a 16SrII-D phytoplasma associated with *Calendula officinalis* phyllody in Iran. *New Disease Reports, 34*, 22.

Esmailzadeh-Hosseini, S., Salehi, M., Mirchenari, S., Tarizeh, D., & Gholampoor, H. (2015). First report of a 16SrVI group related phytoplasma associated with cucumber phyllody in a greenhouse in Iran. *New Disease Reports, 32*, 5.

Esmailzadeh-Hosseini, S., Salehi, M., & Mirzaie, A. (2011). Alternate hosts of alfalfa witches' broom phytoplasma and winter hosts of its vector *Orosius albicinctus* in Yazd-Iran. *Bulletin of Insectology, 64*, S247–S248.

Faghihi, M., Bagheri, A., Askari Seyahooei, M., Pezhman, A., & Faraji, G. (2017). First report of a '*Candidatus* Phytoplasma aurantifolia'-related strain associated with witches'-broom disease of limequat in Iran. *New Disease Reports, 35*, 24–24.

Faghihi, M., Bagheri, A., Bahrami, H., Hasanzadeh, H., Rezazadeh, R., Siampour, M., … Izadpanah, K. (2011). Witches'-broom disease of lime affects seed germination and seedling growth but is not seed transmissible. *Plant Disease, 95*(4), 419–422.

Faghihi, M., Siampour, M., Zaeifi, M., Bagheri, A., Salehi, M., & Samavi, S. (2010). First report of a phytoplasma associated with *Periploca aphylla* witches' broom in Iran. *Plant Pathology, 59*(2), 400.

Faghihi, M., Taghavi, S., Safaei, A., Siampour, M., & Najafabadi, S. (2016). First report of a phytoplasma associated with bell pepper big bud disease in Iran. *New Disease Reports, 33*, 15.

Faghihi, M., Taghavi, S., Salehi, M., Sadeghi, M., Samavi, S., & Siampour, M. (2014). Characterisation of a phytoplasma associated with Petunia witches' broom disease in Iran. *New Disease Reports, 30*, 21.

Fattahi, M., Salehi, M., Sharzehi, A., & Esmailzadeh-Hosseini, S. (2016). Partial biological and molecular characteristics of a phytoplasma associated with Behshahr (Mazandaran) periwinkle phyllody. *Iranian Journal of Plant Pathology, 52*(1).

Firrao, G., Andersen, M., Bertaccini, A., Boudon, E., Bove, J., Daire, X., … Gibb, K. S. (2004). '*Candidatus* Phytoplasma', a taxon for the wall-less, non-helical prokaryotes that colonize plant phloem and insects. *International Journal of Systematic and Evolutionary Microbiology, 54*(4), 1243–1255.

Foissac, X., & Wilson, M. R. (2010). Current and possible future distributions of phytoplasma diseases and their vectors. In P. G. Weintraub & P. Jones (Eds.), *Phytoplasmas: Genomes, Plant Hosts and Vectors; Weintraub, PG, Jones, P., Eds* (pp. 309–324). Oxfordshire, UK: CABI.

Frisinghelli, C., Delaiti, L., Grando, M., Forti, D., & Vindimian, M. (2000). *Cacopsylla costalis* (Flor 1861), as a vector of apple proliferation in Trentino. *Journal of Phytopathology, 148*(7-8), 425–431.

Gajardo, A., Fiore, N., Prodan, S., Paltrinieri, S., Botti, S., Pino, A., … Bertaccini, A. (2009). Phytoplasmas associated with grapevine yellows disease in Chile. *Plant Disease, 93*(8), 789–796.

Ghayeb Zamharir, M. (2014). Molecular study of a new disease of peach in Iran associated with a phytoplasma. *Advances in Microbiology, 4*(1), 20.

Ghayeb Zamharir, M., Boshehri, S. M. S., & Khajehzadeh, Y. (2016). *Candidatus* phytoplasma fraxini related (16S rRNAVII) strain associated with date yellows disease in Iran. *Australasian Plant Disease Notes, 11*(1), 31.

Ghayeb Zamharir, M., Paltrinieri, S., Hajivand, S., Taheri, M., & Bertaccini, A. (2017). Molecular identification of diverse '*Candidatus* Phytoplasma'species associated with grapevine decline in Iran. *Journal of Phytopathology, 165*(7-8), 407–413.

Ghayeb Zamharir, M., & Razavi, R. (2016). First finding of a '*Candidatus* Phytoplasma fraxini'-related strain associated with disease of olive in Iran. *New Disease Reports, 34*, 10.

Hanboonsong, Y., Ritthison, W., Choosai, C., & Sirithorn, P. (2006). Transmission of sugarcane white leaf phytoplasma by *Yamatotettix flavovittatus*, a new leafhopper vector. *Journal of economic entomology, 99*(5), 1531–1537.

Hashemi-Tameh, M., Bahar, M., & Zirak, L. (2014a). '*Candidatus* Phytoplasma asteris' and '*Candidatus* Phytoplasma aurantifolia', New Phytoplasma Species Infecting Apple Trees in Iran. *Journal of Phytopathology, 162*(7-8), 472–480.

Hashemi-Tameh, M., Bahar, M., & Zirak, L. (2014b). Molecular characterization of phytoplasmas related to apple proliferation and aster yellows groups associated with pear decline disease in Iran. *Journal of Phytopathology, 162*(10), 660–669.

Heshmati, G. A. (2013). Indigenous plant species from the drylands of Iran, distribution and potential for habitat maintenance and repair. In *Combating Desertification in Asia, Africa and the Middle East* (pp. 355-375). Dordrecht, The Netherlands: Springer.

Hogenhout, S. A., Oshima, K., Ammar, E. D., Kakizawa, S., Kingdom, H. N., & Namba, S. (2008). Phytoplasmas: Bacteria that manipulate plants and insects. *Molecular plant pathology, 9*(4), 403–423.

Hosseini, P., Bahar, M., Madani, G., & Zirak, L. (2011). Molecular characterization of phytoplasmas associated with potato purple top disease in Iran. *Journal of Phytopathology, 159*(4), 241–246.

Hosseini, S., Bahar, M., & Zirak, L. (2013a). Characterization of phytoplasmas related to peanut witches'-broom and stolbur groups associated with alfalfa diseases in Iran. *Journal of Plant Diseases and Protection, 120*(2), 70–76.

Hosseini, S., Bahar, M., & Zirak, L. (2013b). Detection and Identification of a 16 S r II Group Phytoplasma Causing Clover Little Leaf Disease in I ran. *Journal of Phytopathology, 161*(4), 295–297.

Ikten, C., Catal, M., Yol, E., Ustun, R., Furat, S., Toker, C., & Uzun, B. (2014). Molecular identification, characterization and transmission of phytoplasmas associated with sesame phyllody in Turkey. *European Journal of Plant Pathology, 139*(1), 217–229.

Jamshidi, E., Jafarpour, B., Rouhani, H., & Salehi, M. (2014). Association of members of clover proliferation (16SrVI) and pigeon pea witches' broom (16SrIX) phytoplasma groups with tomato big bud disease in Iran. *Iranian Journal of Plant Pathology, 50*(2), 77–89.

Jarausch, B., & Jarausch, W. (2010). Psyllid vectors and their control. In P. Weintraub & P. Jones (Eds.), *Phytoplasmas: Genomes, plant hosts and vectors* (pp. 250–271). Oxford, UK: CAB International.

Jarausch, W., Jarausch, B., Danet, J., Broquaire, J., Dosba, F., Saillard, C., & Garnier, M. (2001). Detection and indentification of European stone fruit yellows and other phytoplasmas in wild plants in the surroundings of apricot chlorotic leaf roll-affected orchards in southern France. *European Journal of Plant Pathology, 107*(2), 209–217.

Jeger, M., Bragard, C., Candresse, T., Chatzivassiliou, E., Dehnen-Schmutz, K., Gilioli, G., … MacLeod, A. (2017). Pest categorisation of Witches' broom disease of lime (*Citrus aurantifolia*) phytoplasma. *EFSA Journal, 15*(10), 5027–5049.

Jones, P., Arocha, Y., & Plata, G. (2005). First report of a 'Candidatus Phytoplasma asteris' isolate associated with a witches' broom disease of alfalfa in Bolivia. *Plant Pathology, 54*(4), 559–559.

Jones, R., & Barbetti, M. J. (2012). Influence of climate change on plant disease infections and epidemics caused by viruses and bacteria. *Plant Sciences Reviews, 22*, 1–31.

Karimi, M., Contaldo, N., Mahmoudi, B., Duduk, B., & Bertaccini, A. (2009). *Identification of stolbur-related phytoplasmas in grapevine showing decline symptoms in Iran*. Le Progrès agricole et viticole, Hors Série, Extended abstract of the 16th Meeting ICVG, Dijion, France, August 31–September 4., *31*, 208–209.

Karimzade, M., Siampour, M., Zakiaghl, M., & Mehrvar, M. (2018). Identification and characterization of a phytoplasma associated with black locust yellow disease in two provinces of Iran. *Crop Protection, 110*, 261–268.

Khan, A., Botti, S., Al-Subhi, A., Gundersen-Rindal, D., & Bertaccini, A. (2002). Molecular identification of a new phytoplasma associated with alfalfa witches'-broom in Oman. *Phytopathology, 92*(10), 1038–1047.

Lee, I.-M., Bottner-Parker, K., Zhao, Y., Davis, R. E., & Harrison, N. (2010). Phylogenetic analysis and delineation of phytoplasmas based on secY gene sequences. *International Journal of Systematic and Evolutionary Microbiology, 60*(12), 2887–2897.

Lee, I.-M., Davis, R. E., & Gundersen-Rindal, D. E. (2000). Phytoplasma: Phytopathogenic mollicutes. *Annual Reviews in Microbiology, 54*(1), 221–255.

Lee, I.-M., Gundersen-Rindal, D. E., Davis, R. E., & Bartoszyk, I. M. (1998). Revised classification scheme of phytoplasmas based on RFLP analyses of 16S rRNA and ribosomal protein gene sequences. *International Journal of Systematic and Evolutionary Microbiology, 48*(4), 1153–1169.

Liu, H.-L., Chen, C.-C., & Lin, C.-P. (2007). Detection and identification of the phytoplasma associated with pear decline in Taiwan. *European Journal of Plant Pathology, 117*(3), 281–291.

Lova, M. M., Quaglino, F., Abou-Jawdah, Y., Choueiri, E., Sobh, H., Casati, P., … Bianco, P. A. (2011). Identification of new 16SrIX subgroups,-F and-G, among '*Candidatus* Phytoplasma phoenicium'strains infecting almond, peach and nectarine in Lebanon. *Phytopathologia mediterranea, 50*(2), 273–282.

Maixner, M. (2007). Biology of *Hyalesthes obsoletus* and approaches to control this soilborne vector of Bois noir disease. *IOBC WPRS Bulletin, 30*(7), 3–11.

Maixner, M., Johannesen, J., Michel, K., Lux, B., & Seitz, A. (2007). Host plant specificity of *Hyalesthes obsoletus* and consequences for "bois noir" epidemiology. *Bulletin of Insectology, 60*(2), 399.

Marcone, C. (2014). Molecular biology and pathogenicity of phytoplasmas. *Annals of Applied Biology, 165*(2), 199–221.

Marcone, C., Ragozzino, A., & Seemüller, E. (1997). Detection of Bermuda grass white leaf disease in Italy and characterization of the associated phytoplasma by RFLP analysis. *Plant Disease, 81*(8), 862–866.

Marcone, C., Schneider, B., & Seemüller, E. (2004). '*Candidatus* Phytoplasma cynodontis', the phytoplasma associated with Bermuda grass white leaf disease. *International Journal of Systematic and Evolutionary Microbiology, 54*(4), 1077–1082.

Mardi, M., Khayam Nekouei, S., Farsad, L. K., Ehya, F., Shabani, M., Shafiee, M., … Hosseini Salekdeh, G. (2011). Witches' broom disease of Mexican lime trees: Disaster to be addressed before it will be too late. *Bulletin of Insectology, 64 (Suplement)*, S205-S206.

Martini, M., Lee, I.-M., Bottner, K., Zhao, Y., Botti, S., Bertaccini, A., … Khan, A. (2007). Ribosomal protein gene-based phylogeny for finer differentiation and classification of phytoplasmas. *International Journal of Systematic and Evolutionary Microbiology, 57*(9), 2037–2051.

Martini, M., Marcone, C., Lee, I.-M., & Firrao, G. (2014). The Family Acholeplasmataceae (Including Phytoplasmas). In E. Rosenberg, E. F. DeLong, S. Lory, E. Stackebrandt, & F. Thompson (Eds.), *The Prokaryotes: Firmicutes and Tenericutes* (pp. 469–504). Berlin/Heidelberg, Germany: Springer.

Mirchenari, S. M., Massah, A., & Zirak, L. (2015). 'Bois noir': New phytoplasma disease of grapevine in Iran. *Journal of Plant Protection Research, 55*(1), 88–93.

Mirzaie, A., Esmailzadeh-Hosseini, S., Jafari-Nodoshan, A., & Rahimian, H. (2007). Molecular characterization and potential insect vector of a phytoplasma associated with garden beet witches' broom in Yazd, Iran. *Journal of Phytopathology, 155*(4), 198–203.

Miyazaki, A., Shigaki, T., Koinuma, H., Iwabuchi, N., Rauka, G. B., Kembu, A., … Maejima, K. (2017). '*Candidatus* Phytoplasma noviguineense', a novel taxon associated with Bogia coconut syndrome and banana wilt disease on the island of New Guinea. *International Journal of Systematic and Evolutionary Microbiology, 68*(1), 170–175.

Mohammadi Farsani, T., Etemadi, N., Sayed-Tabatabaei, B. E., & Talebi, M. (2011). Assessment of genetic diversity of bermudagrass (*Cynodon dactylon*) using ISSR markers. *International journal of molecular sciences, 13*(1), 383–392.

Mozaffarian, F., & Wilson, M. R. (2016). A checklist of the leafhoppers of Iran (Hemiptera: Auchenorrhyncha: Cicadellidae). *Zootaxa, 4062*(1), 1–63.

Nakashima, K., Hayashi, T., Chaleeprom, W., Wongkaew, P., & Sirithorn, P. (1995). Detection of DNA of phytoplasmas associated with phyllody disease of sesame in Thailand. *Japanese Journal of Phytopathology, 61*(6), 519–528.

Olivier, C., Séguin-Swartz, G., Hegedus, D., & Barasubiye, T. (2006). First Report of "*Candidatus* Phytoplasma asteris"-Related Strains in *Brassica rapa* in Saskatchewan, Canada. *Plant Disease, 90*(6), 832–832.

Omar, A. F., & Foissac, X. (2012). Occurrence and incidence of phytoplasmas of the 16SrII-D subgroup on solanaceous and cucurbit crops in Egypt. *European Journal of Plant Pathology, 133*(2), 353–360.

Pearce, T., Scott, J., & Pethybridge, S. (2011). First Report of Witch's Broom Phytoplasma (16SrII-D Group) in Purple Coneflower in Australia. *Plant Disease: An International Journal of Applied Plant Pathology, 95*(6), 773.

Peters, R., Lee, M., Grau, C., Driscoll, S., Winberg, R., Kurtzweil, N., … Lee, I.-M. (1999). First report of aster yellows phytoplasma in alfalfa. *Plant Disease, 83*(5), 488–488.

Pourali, H., & Salehi, M. (2012). Genetic diversity of the phytoplasma isolates associated with almond witches' broom in Iran. *Iranian Journal of Plant Pathology, 48*(3), 61–74.

Rao, G., Kumar, A., & Baranwal, V. (2015). Classification of the sesame phytoplasma strains in India at the 16Sr subgroup level. *Journal of Plant Pathology, 97*(3).

Rao, G., Srivastava, S., Gupta, P., Sharma, S., Singh, A., Singh, S., … Marcone, C. (2008). Detection of sugarcane grassy shoot phytoplasma infecting sugarcane in India and its phylogenetic relationships to closely related phytoplasmas. *Sugar Tech, 10*(1), 74–80.

Rashidi, M., Habili, N., & Ghasemi, A. (2010). First report of a stolbur phytoplasma associated with witches' broom of Japanese spindle (*Euonymus japonicus*). *Plant Pathology, 59*(4), 796–796.

Salehi, E., Salehi, M., & Masoumi, M. (2016). Biological and molecular characterization of the phytoplasma associated with tomato big bud disease in Zanjan province, Iran. *Iranian Journal of Plant Pathology, 52*(3).

Salehi, E., Salehi, M., Taghavi, S., & Izadpanah, K. (2014). A 16SrII-D Phytoplasma strain associated with Tomato Witches'-Broom in Bushehr province, Iran. *Journal of Crop Protection, 3*(3), 377–388.

Salehi, E., Salehi, M., Taghavi, S., & Izadpanah, K. (2016). First report of a 16SrIX group (pigeon pea witches'-broom) phytoplasma associated with grapevine yellows in Iran. *Journal of Plant Pathology, 98*(2).

Salehi, E., Taghavi, S., Salehi, M., & Izadpanah, K. (2014). Partial biological and molecular characterization of phytoplasmas associated with grapevine yellows in Fars and Lorestan provinces of Iran. *Iranian Journal of Plant Pathology, 50*(2).

Salehi, M., & Esmailzadeh-Hosseini, S. (2016). The first report of a 16SrXII-A phytoplasma associated with tomato big bud disease in Iran. *Journal of Plant Pathology, 98*(3).

Salehi, M., Esmailzadeh-Hosseini, S., Rasoulpour, R., Salehi, E., & Bertaccini, A. (2016). Identification of a phytoplasma associated with pomegranate little leaf disease in Iran. *Crop Protection, 87*, 50–54.

Salehi, M., Esmailzadeh-Hosseini, S., & Salehi, E. (2015). Characterisation of a phytoplasma associated with sunflower phyllody in Fars, Isfahan and Yazd provinces of Iran. *New Disease Reports, 31*, 6–6.

Salehi, M., Esmailzadeh-Hosseini, S., & Salehi, E. (2016). First report of a '*Candidatus* phytoplasma asteris' related phytoplasma associated with eucalyptus little leaf disease in Iran. *Journal of Plant Pathology, 98*(1), 175.

Salehi, M., Esmailzadeh-Hosseini, S., Salehi, E., & Bertaccini, A. (2016). Molecular and biological characterisation of a 16SrII phytoplasma associated with carrot witches' broom in Iran. *Journal of Plant Pathology*, 83-90.

Salehi, M., Esmailzadeh-Hosseini, S., Salehi, E., & Bertaccini, A. (2017). Genetic diversity and vector transmission of phytoplasmas associated with sesame phyllody in Iran. *Folia microbiologica, 62*(2), 99–109.

Salehi, M., Ghasemi, S., Taker-Khani, K., & Izadpanah, K. (2001). Phytoplasmal white leaf disease of sugarcane in Iran. *Iranian Journal of Plant Pathology, 37*(3/4), 98–99.

Salehi, M., Haghshenas, F., Khanchezar, A., & Esmailzadeh-Hosseini, S. (2011). Association of '*Candidatus* Phytoplasma phoenicium'with GF-677 witches' broom in Iran. *Bulletin of Insectology, 64*, 113–114.

Salehi, M., Hassanzadeh, H., Esmailzadeh-Hosseini, S., Adeli, S., & Kalantar, F. (2012). *Effect of Azino-Acphon compound on control of witches' broom of lime in field.* Paper presented at the 20th Iranian Plant Protection Congress, Shiraz, Iran, Vol 1, p. 497.

Salehi, M., Heydarnejad, J., & Izadpanah, K. (2005). Molecular characterization and grouping of 35 phytoplasmas from central and southern provinces in Iran. *Iranian Journal of Plant Pathology, 41*(1), 62–65.

Salehi, M., Hosseini, S. A. E., Salehi, E., & Bertaccini, A. (2016). Occurrence and Characterization of a 16Sr II-D Subgroup Phytoplasma Associated with Parsley Witches' Broom Disease in Iran. *Journal of Phytopathology, 164*(11-12), 996–1002.

Salehi, M., & Izadpanah, K. (1992). Etiology and transmission of sesame phyllody in Iran. *Journal of Phytopathology, 135*(1), 37–47.

Salehi, M., & Izadpanah, K. (1993). A comparative study of transmission, host range and symptomatology of sesame phyllody and alfalfa witches broom. *Iranian Journal of Plant Pathology, 29*, 157–158.

Salehi, M., & Izadpanah, K. (1995a). *Almond Brooming.* Paper presented at the 12th Iranian Plant Protection Congress (September 2-7, 1995), Univeristy of Teheran, Karaj, Iran. p238.

Salehi, M., & Izadpanah, K. (1995b). *Big bud of tomato and eggplant in Fars.* Paper presented at the 12th Iranian Plant Protection Congress (September 2-7, 1995), University of Teheran, Karaj, Iran. p 239.

Salehi, M., & Izadpanah, K. (2001). Peach witches' broom disease in Fars province. *Iranian Journal of Plant Pathology, 37*(3/4).

Salehi, M., & Izadpanah, K. (2005). Partial characterization of a phytoplasma associated with witches' broom disease of sugar beet in Iran. *Iranian Journal of Plant Pathology, 41*(1).
Salehi, M., Izadpanah, K., & Behjatnia, A. (2000). Detection of phytoplasmas from Iran using polymerase chain reaction. *Iranian Journal of Plant Pathology, 36*(3/4).
Salehi, M., Izadpanah, K., & Ebrahim-Nesbat, F. (1995). Etiology, Transmission and host range of Alfalfa witches, broom in southern Iran. *Iranian Journal of Plant Pathology, 31*, 1–9.
Salehi, M., Izadpanah, K., & Heydarnejad, J. (2006). Characterization of a new almond witches' broom phytoplasma in Iran. *Journal of Phytopathology, 154*(7--8), 386–391.
Salehi, M., Izadpanah, K., Nejat, N., & Siampour, M. (2007). Partial characterization of phytoplasmas associated with lettuce and wild lettuce phyllodies in Iran. *Plant Pathology, 56*(4), 669–676.
Salehi, M., Izadpanah, K., & Rahimian, H. (1997). Witches'-broom disease of lime in Sistan, Baluchistan. *Iran. J. Plant Pathology, 33*, 3–4.
Salehi, M., Izadpanah, K., & Siampour, M. (2007). Characterization of a phytoplasma associated with cabbage yellows in Iran. *Plant Disease, 91*(5), 625–630.
Salehi, M., Izadpanah, K., & Siampour, M. (2008). First report of '*Candidatus* Phytoplasma trifolii'-related strain associated with safflower phyllody disease in Iran. *Plant Disease, 92*(4), 649–649.
Salehi, M., Izadpanah, K., & Siampour, M. (2011). Occurrence, molecular characterization and vector transmission of a phytoplasma associated with rapeseed phyllody in Iran. *Journal of Phytopathology, 159*(2), 100–105.
Salehi, M., Izadpanah, K., Siampour, M., Bagheri, A., & Faghihi, S. (2007). Transmission of '*Candidatus* Phytoplasma aurantifolia'to Bakraee (*Citrus reticulata* hybrid) by feral *Hishimonus phycitis* leafhoppers in Iran. *Plant Disease, 91*(4), 466–466.
Salehi, M., Izadpanah, K., Siampour, M., Firouz, R., & Salehi, E. (2009). Molecular characterization and transmission of safflower phyllody phytoplasma in Iran. *Journal of Plant Pathology*, 453–458.
Salehi, M., Izadpanah, K., Siampour, M., & Taghizadeh, M. (2009). Molecular characterization and transmission of Bermuda grass white leaf phytoplasma in Iran. *Journal of Plant Pathology*, 655–661.
Salehi, M., Izadpanah, K., Taghavi, S., & Rahimian, H. (2008). Characterization of a phytoplasma associated with pear decline in Iran. *Journal of Phytopathology, 156*(7-8), 493–495.
Salehi, M., Nejat, N., Tavakkoli, A., & Izadpanah, K. (2005). Reaction of *Citrus* cultivars to '*Candidatus* phytoplasma aurantifolia' in Iran. *Iranian Journal of Plant Pathology, 41*(3), 363–376.
Salehi, M., Rasoulpour, R., & Izadpanah, K. (2016). Molecular characterization, vector identification and partial host range determination of phytoplasmas associated with faba bean phyllody in Iran. *Crop Protection, 89*, 12–20.
Salehi, M., Salehi, E., Abbasian, M., & Izadpanah, K. (2015). Wild almond (*Prunus scoparia*), a potential source of almond witches' broom phytoplasma in Iran. *Journal of Plant Pathology, 97*(2), 377–381.
Salehi, M., Siampour, M., Esmailzadeh-Hosseini, S., & Bertaccini, A. (2015). Characterization and vector identification of phytoplasmas associated with cucumber and squash phyllody in Iran. *Bull Insectol, 68*, 311–319.
Samavi, S., Faghihi, M., Hasanzadeh, H., Bagheri, A., Salehi, M., & Sotoudehnia, P. (2012). First report of the natural occurrence of group 16SrII '*Candidatus* Phytoplasma aurantifolia'in two *Solanum* species in Iran. *New Disease Reports, 26*, 23.
Sanchez-Capuchino, J., Casanova, R., & Llácer Inia, G. (1982). An approach to control apricot chlorotic leaf roll and epidemiological observations. *Acta Horticulturae, 121*, 435–438.
Schaper, U., & Seemüller, E. (1982). Condition of the phloem and the persistence of mycoplasmalike organisms associated with apple proliferation and pear decline. *Phytopathology, 72*, 736–742.

Seemüller, E. (1989). Pear decline. In P. Fridlund (Ed.), *Virus and viruslike diseases of pome fruits and simulating noninfectious disorders* (pp. 188–201). Washington State University, Pullman, Washington: Cooperative Extension College of Agriculture and Home Economics.

Seemüller, E., & Schneider, B. (2004). '*Candidatus* Phytoplasma mali','*Candidatus* Phytoplasma pyri'and '*Candidatus* Phytoplasma prunorum', the causal agents of apple proliferation, pear decline and European stone fruit yellows, respectively. *International Journal of Systematic and Evolutionary Microbiology, 54*(4), 1217-1226.

Seyahooei, M. A., Hemmati, C., Faghihi, M. M., & Bagheri, A. (2017). First report of a '*Candidatus* Phytoplasma trifolii'-related strain associated with *Suaeda aegyptiaca* and its potential vector in Iran. *Australasian Plant Disease Notes, 12*(1), 24.

Siampour, M., Izadpanah, K., Afsharifar, A., Salehi, M., & Taghizadeh, M. (2006). Detection of phytoplasma in insects collected in witches' broom affected lime groves. *Iranian Journal of Plant Pathology, 42*(1).

Siampour, M., Izadpanah, K., Galetto, L., Salehi, M., & Marzachi, C. (2013). Molecular characterization, phylogenetic comparison and serological relationship of the Imp protein of several '*Candidatus* Phytoplasma aurantifolia'strains. *Plant Pathology, 62*(2), 452–459.

Siampour, M., Izadpanah, K., Marzachi, C., & Salehi, M. (2015). Identification and characterization of conserved and variable regions of lime witches' broom phytoplasma genome. *Microbiology, 161*(9), 1741–1751.

Sichani, F. V., Bahar, M., & Zirak, L. (2011). Characterization of stolbur (16SrXII) group phytoplasmas associated with *Cannabis sativa* witches'-broom disease in Iran. *Plant Pathology Journal, 10*(4), 161–167.

Sichani, F. V., Bahar, M., & Zirak, L. (2014). Characterization of phytoplasmas related to aster yellows group infecting annual plants in Iran, based on the studies of 16S rRNA and rp genes. *Journal of Plant Protection Research, 54*(1), 1–8.

Singh, J., Rani, A., Kumar, P., Baranwal, V., Saroj, P., & Sirohi, A. (2012). First report of a 16SrII-D phytoplasma '*Candidatus* Phytoplasma australasia'associated with a tomato disease in India. *New Disease Reports, 26*, 14.

Snehi, S., Khan, M., Raj, S., Mall, S., Singh, M., & Rao, G. (2008). Molecular identification of '*Candidatus* Phytoplasma cynodontis' associated with Bermuda grass white leaf disease in India. *Plant Pathology, 57*(4), 770–770.

Soltani, M., Laux, P., Kunstmann, H., Stan, K., Sohrabi, M., Molanejad, M., … Rousta, I. (2016). Assessment of climate variations in temperature and precipitation extreme events over Iran. *Theoretical and Applied Climatology, 126*(3-4), 775–795.

Strauss, E. (2009). Phytoplasma research begins to bloom. *Science, 325*(5939), 388–390.

Tazehkand, S. A., Pour, A. H., Heydarnejad, J., Varsani, A., & Massumi, H. (2010). Identification of phytoplasmas associated with cultivated and ornamental plants in Kerman province, Iran. *Journal of Phytopathology, 158*(11-12), 713–720.

Tedeschi, R., Bosco, D., & Alma, A. (2002). Population dynamics of *Cacopsylla melanoneura* (Homoptera: Psyllidae), a vector of apple proliferation phytoplasma in northwestern Italy. *Journal of economic entomology, 95*(3), 544–551.

Tedeschi, R., Picciau, L., Quaglino, F., Abou-Jawdah, Y., Molino Lova, M., Jawhari, M., … Abdul-Nour, H. (2015). A cixiid survey for natural potential vectors of '*Candidatus* Phytoplasma phoenicium'in Lebanon and preliminary transmission trials. *Annals of Applied Biology, 166*(3), 372–388.

Tohidi, Z., Salehi, M., Ghasemi, S., Khanchezar, A., & Shahamiri, S. M. (2015). Association of a 16SrIX-C phytoplasma with eggplant phyllody in Iran. *Journal of Crop Protection, 4*(2), 247–256.

Tseng, Y.-W., Deng, W.-L., Chang, C.-J., Huang, J.-W., & Jan, F.-J. (2014). First report on the association of a 16SrII-A phytoplasma with sesame (*Sesamum indicum*) exhibiting abnormal stem curling and phyllody in Taiwan. *Plant Disease, 98*(7), 990–990.

Vaali, F., Bahar, M., & Zirak, L. (2011). Niger seed (*Guizotia abyssinica*), a new host of '*Candidatus* Phytoplasma asteris' in Iran. *Journal of Phytopathology, 159*(4), 321–323.

Verdin, E., Salar, P., Danet, J.-L., Choueiri, E., Jreijiri, F., El Zammar, S., … Garnier, M. (2003). '*Candidatus* Phytoplasma phoenicium'sp. nov., a novel phytoplasma associated with an emerging lethal disease of almond trees in Lebanon and Iran. *International Journal of Systematic and Evolutionary Microbiology, 53*(3), 833–838.

Vicente, A. R., Manganaris, G. A., Cisneros-Zevallos, L., & Crisosto, C. H. (2011). 13 Prunus. In L. A. Terry (Ed.), *Health-promoting properties of fruits and vegetables* (pp. 238–259). Oxford, UK: CAB International.

Weintraub, P. G. (2007). Insect vectors of phytoplasmas and their control-an update. *Bulletin of Insectology, 60*(2), 169.

Yang, Y., Jiang, L., Che, H., Cao, X., & Luo, D. (2016). Identification of a novel subgroup 16SrII-U phytoplasma associated with papaya little leaf disease. *International Journal of Systematic and Evolutionary Microbiology, 66*(9), 3485–3491.

Zhao, Y., & Davis, R. E. (2016). Criteria for phytoplasma 16Sr group/subgroup delineation and the need of a platform for proper registration of new groups and subgroups. *International Journal of Systematic and Evolutionary Microbiology, 66*(5), 2121–2123.

Zibadoost, S., & Rastgou, M. (2016). Molecular identification of phytoplasmas associated with some weeds in West Azarbaijan province, Iran. *Acta Agriculturae Slovenica, 107*(1), 129–136.

Zibadoost, S., Rastgou, M., & Tazehkand, S. A. (2016). Detection and molecular identification of '*Candidatus* phytoplasma trifoli'infecting some cultivated crops and vegetables in West Azarbaijan province, Iran. *Australasian Plant Disease Notes, 11*(1), 3.

Zirak, L., Bahar, M., & Ahoonmanesh, A. (2009a). Characterization of phytoplasmas associated with almond diseases in Iran. *Journal of Phytopathology, 157*(11-12), 736–741.

Zirak, L., Bahar, M., & Ahoonmanesh, A. (2009b). Molecular characterization of phytoplasmas related to peanut witches' broom and stolbur group infecting plum in Iran. *Journal of Plant Pathology, 91*(3), 713–716.

Zirak, L., Bahar, M., & Ahoonmanesh, A. (2010). Characterization of phytoplasmas related to '*Candidatus* Phytoplasma asteris' and peanut WB group associated with sweet cherry diseases in Iran. *Journal of Phytopathology, 158*(1), 63–65.

Zreik, L., Carle, P., Bove, J. M., & Garnier, M. (1995). Characterization of the mycoplasmalike organism associated with witches'-broom disease of lime and proposition of a *Candidatus* taxon for the organism,"*Candidatus* phytoplasma aurantifolia". *International Journal of Systematic and Evolutionary Microbiology, 45*(3), 449–453.

Chapter 4
Diversity of Phytoplasmas in Cuba, Their Geographic Isolation and Potential Development of Management Strategies

Karel Acosta, Madelaine Quiñones Pantoja, and Edel Pérez-López

Abstract Phytoplasmas ('*Candidatus* Phytoplasma') are plant pathogens associated with important economic losses in crop production around the world. Similarly to most tropical countries, Cuba is seriously affected by plant diseases associated with phytoplasmas and few management strategies are available to growers to control the pathogens. In this chapter, we describe phytoplasma diseases caused by infection of single or multiple strains of phytoplasmas and affecting three economically important crop plants of Cuba, sapodilla (*Manilkara zapota* L. (Sapotaceae), soybean (*Glycine max* (L.) Merr., Fabaceae) and papaya (*Carica papaya* L., Caricaeae). Geographical distribution and high diversity of phytoplasma strains and their potential insect vectors are discussed. Based on available data, research avenues in epidemiology and management of phytoplasma diseases affecting Cuban crops are suggested.

Keywords Soybean · Sapodilla · Papaya · Tropical · Island · Caribbean

K. Acosta
Universidad de Las Tunas (ULT), Las Tunas, Cuba
e-mail: karelap@ult.edu.cu

M. Q. Pantoja
Centro Nacional de Sanidad Agropecuaria (CENSA),
San José de Las Lajas, Mayabeque, Cuba
e-mail: madeqp@censa.edu.cu

E. Pérez-López (✉)
Department of Biology, University of Saskatchewan, Saskatoon, SK, Canada
e-mail: epl733@mail.usask.ca

C. Y. Olivier et al. (eds.), *Sustainable Management of Phytoplasma Diseases in Crops Grown in the Tropical Belt*, Sustainability in Plant and Crop Protection 12,
https://doi.org/10.1007/978-3-030-29650-6_4

4.1 Introduction

Phytoplasmas (class: Mollicutes) are wall-less plant pathogens that were first reported in Japan by Doi, Teranaka, Yora, and Asuyama (1967) and have, since then, been associated with hundreds of plant diseases in various economically important crops around the world (I.-M. Lee, Davis, & Gundersen-Rindal, 2000; Maejima, Oshima, & Namba, 2014; Marcone, 2015). Phytoplasmas are transmitted from plant to plant by species of phloem feeder insects and are obligate parasites of their plant hosts and their insect vectors (Rojas-Martínez, 2009; Weintraub & Beanland, 2006).

Phytoplasmas are well adapted to their parasitic lifestyle, as they shift gene expression patterns when switching hosts and possess genes encoding for transporters, allowing them to import hosts metabolites and nutrients for their own benefits (Makarova, MacLean, & Nicolaisen, 2015; Oshima et al., 2011). Phytoplasmas also produce virulence factors that can modify the gene expressions of their hosts, hence altering their physiology (Sugio & Hogenhout, 2012; Zhao, Davis, Wei, Shao, & Jomantiene, 2014). Virulence factors and nutrients intakes by the pathogen induce plant malformations and severe perturbations in metabolism and hormonal balances, leading to phytoplasma disease-like symptoms such as flower malformations (virescence and phyllody), growth aberrations (witches'-broom), yellowing, stunting and decline. While most phytoplasma diseases are caused by single strain infection, double phytoplasma infection of a host is common and have been observed in potato, apple trees and papaya among many other crops (Acosta et al., 2013; Leyva-López, Ochoa-Sánchez, Leal-Klevezas, & Martínez-Soriano, 2002; Seemüller, Kiss, Sule, & Schneider, 2010).

Because phytoplasmas have not been isolated in axenic cultures, they have been grouped under a provisional genus, '*Candidatus* Phytoplasma' and are classified based on comparison of the highly conserved 16S rRNA gene sequences, phylogenetic analysis and RFLP patterns (Firrao et al., 2004; I.-M. Lee, Gundersen-Rindal, Davis, & Bartoszyk, 1998). Other less conserved genes are also being used for a finer classification (Hodgetts & Dickinson, 2009; Pérez-López, Olivier, Luna-Rodríguez, & Dumonceaux, 2016).

Cuba, a country located in the tropical belt, has undergone tremendous agricultural changes since the 1990s, to achieve food sovereignty and to generate cash via exportation of high value crop products (Palma, Toral, Vázquez, Fuentes, & Hernández, 2015). Crops for local uses include rice, soybean and vegetables and are produced on agricultural and urban lands using mostly organic and agroecological diversification methods (Altieri, Funes-Monzote, & Petersen, 2012; Reardon & Pérez, 2010). Cash crops include high value tropical fruits such as grapefruits and processed products such as sugar, tobacco or coffee (Clouse, 2014; Palma et al., 2015). However, with $526 million per year during 2012–2014, Cuba's agricultural exports are limited and new cash crops such as papaya or sapodilla are being developed (McMinimy, 2016).

Cuba has been identified as one of the top five countries in Latin America with the highest number of phytoplasmas reported in the literature (Pérez-López, Olivier,

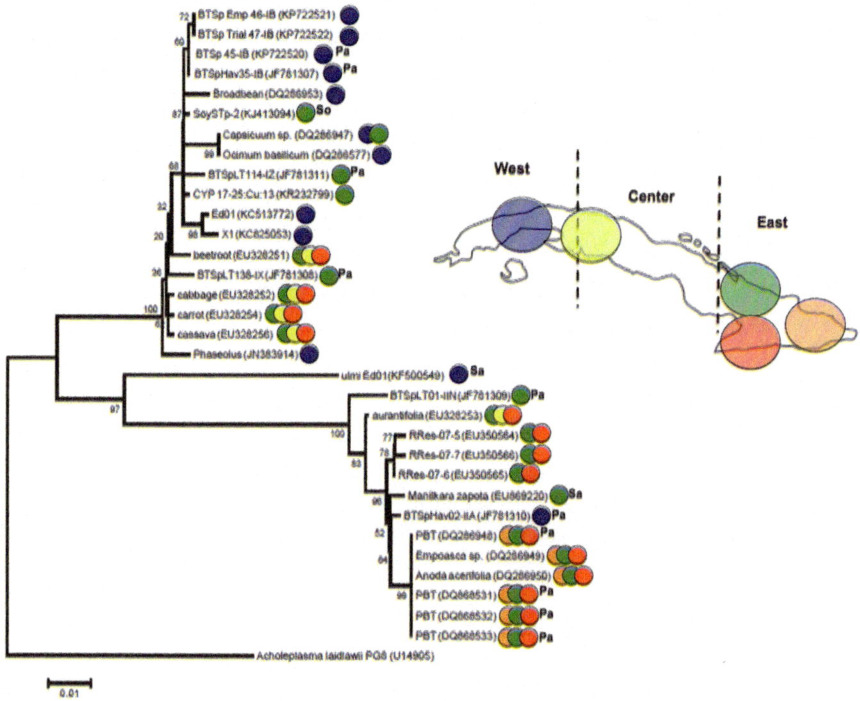

Fig. 4.1 Phylogenetic tree using minimum evolution method of the 16S rDNA sequences reported in Cuba (longer than 1kb), and geographic distribution of the strains. Colors in the circles correspond to the colors in the map. Sa (Saposilla), So (soybean), Pa (papaya)

et al., 2016). Therefore, phytoplasma diseases are a threat to the existing local crops and the development of new cash crops. Most of phytoplasma-infected plants have been reported in the West and East of Cuba, with a low number of reports in the Center of the country (Fig. 4.1). This discrepancy is not related to the intensity of agricultural activity, but rather to the presence of research institutions working on plant pathology.

Leafhoppers, planthoppers and psyllids are known to be insect vectors of phytoplasma diseases (Weintraub & Wilson, 2010). As a neotropical country with various climatic regions, Cuba has a high diversity of Hemiptera insects (Hidalgo-Gato & Rodríguez-León, 2010; Rojas, 2014). Recently, a few leafhopper vectors of phytoplasmas and several leafhopper species known to transmit phytoplasmas in other countries have been identified in Cuba (Acosta, Zamora, et al., 2017; Arocha, Piñol, López, et al., 2007; Pérez et al., 2010; Sánchez Castro, Quiñones Pantoja, Piñol, & Fernández Argudín, 2015). However, the majority of the phytoplasma vectors in Cuban crops have not been identified.

Knowledge of the epidemiology of phytoplasma diseases is paramount for the development of integrated management. In Cuba, data and information on phytoplasma diseases and their vectors are still incomplete and not yet conducive to the development of disease management techniques. In this chapter, we will describe the occurrence of diverse phytoplasma diseases in three economically important crop plants, sapodilla, soybean and papaya. Geographical distribution, isolation and high diversity of phytoplasma strains in Cuba will then be discussed and insects species reported as potential vectors will be presented. Based on the data available, research and potential management techniques will be suggested.

4.2 Economically Important Plants Infected by Phytoplasmas in Cuba

4.2.1 Sapodilla

Sapodilla (*Manilkara zapota* L.) is a fruit tree belonging to the family Sapotaceae. Sapodilla has been identified as a potential export crop because of its nutritional values and delicious flavor (Acosta, Piñol, Acosta, Countín, & Arocha, 2009). Pilot cultivation was established in eastern and western Cuba and in both regions, symptoms typical of phytoplasma diseases, such as leaf yellowing, witches' broom, were reported (Fig. 4.2) (Acosta et al., 2009; Pérez-López, Hernández-Rodríguez, Pantoja, & Zaldívar, 2013). In western Cuba, the phytoplasma affecting sapodilla was identified as a '*Candidatus* Phytoplasma ulmi'-related strain (16SrV), while in eastern Cuba, the phytoplasma was identified as a '*Ca.* P. aurantifolia'-related strain (16SrII) (Fig. 4.1). Strains related to '*Ca.* P. aurantifolia' have been identified in Cuba in other plants such as papaya (Acosta et al., 2009; Arocha, Piñol, Acosta, et al., 2009). However, the detection of '*Ca.* P. ulmi'-related strain in Sapodilla trees in Cuba is the first report of the presence of a group 16SrV phytoplasma in Latin America and the first report of the presence of a group 16SrV phytoplasma in sapodilla tree (Pérez-López, Luna-Rodríguez, Olivier, & Dumonceaux, 2016).

Pilot cultivation plants have been set up in various regions of Cuba by importation of selected fruit trees and shrubs from other tropical and subtropical countries

Fig. 4.2 Symptoms reported for Sapodilla (**a**), soybean (**b**), and papaya (**c**), affected by phytoplasmas in Cuba

to study the possibilities of cultivating edible tree/shrub fruits for local needs and as cash crops for export markets (Moreno-Casasola, Guevara, Vázquez-González, & Eusebio, 2016). As '*Ca*. P. ulmi was found in one of the cultivation trials, it can be hypothesized that the '*Ca*. P. ulmi'-related strain detected in Cuba might be the result of an introduction of infected plant material. Phytoplasmas belonging to the elm yellows phytoplasma group are widely spread in Europe, and have been listed by the European and Mediterranean Plant Protection Organization (EPPO) as an A1 quarantine pest (Hodgetts, Flint, & Fox, 2015).

4.2.2 Soybean

Soybean (*Glycine max* (L.) Merr., Fabaceae) is one of the most widely marketed crops in the world, mostly due to the high value of the seeds as animal and human food and for its oil (Soto, Ferreira, Delgado, & Enríquez, 2013). In Cuba, soybean is used to elaborate a broad spectrum of products with the main product being a base for poultry and porcine feeding (Romero, Ruz, & González, 2013). In 2013, soybean plants showing stunting, chlorosis, crinkle and aborted seed pods (Fig. 4.2), were observed in Eastern Cuba (Acosta et al., 2015). Symptomatic plants were associated with phytoplasma strains related to '*Ca*. P. asteris' (Acosta et al., 2015; I.-M. Lee et al., 2004), and identified as members of three different subgroups, 16SrI-B, 16SrI-W, and 16SrI-Y (Fig. 4.1). Aster yellow (AY) phytoplasmas (16SrI) have been detected in soybean plants in many countries including the US (M. Lee, Grau, Lukaesko, & Lee, 2002) (Fig. 4.3).

4.2.3 Papaya

Papaya (*Carica papaya* L., Caricaeae) is a fruit tree of high economic relevance in Cuba with an annual production of over 55,000 tons and a strong potential for development as an export crop (Acosta et al., 2013; Arocha, Piñol, Acosta, et al., 2009). However, papaya plants are highly susceptible to various pathogens such as viruses, bacteria, Rickettsia-related bacteria and phytoplasmas (Arocha, Piñol, Acosta, et al., 2009; Luis-Pantoja et al., 2015; Pérez-López & Luis-Pantoja, 2012). Papaya bunchy top (PBT) and Bunchy top symptom (BTS) are important diseases affecting papaya plants in Cuba. PBT symptoms include leaf chlorosis, leaf margin necrosis, reduced growth of leaf and petiole, as well as shortened internodes and the presence of oily spots on stems and petioles. BTS symptoms includes leaf yellowing, small or no flowers and fruit production, and shortened internodes of the crown giving the typical "bunchy top" appearance of the crown (Acosta et al., 2013). BTS has been associated with phytoplasmas of the group 16SrII-P and 16SrII-Q (related to '*Ca*. P. aurantifolia') and group 16SrI-B, while PBT is associated with Rickettsia-related bacteria (Fig. 4.2) (Arocha, Piñol, Acosta, et al., 2009; Luis-Pantoja et al., 2015).

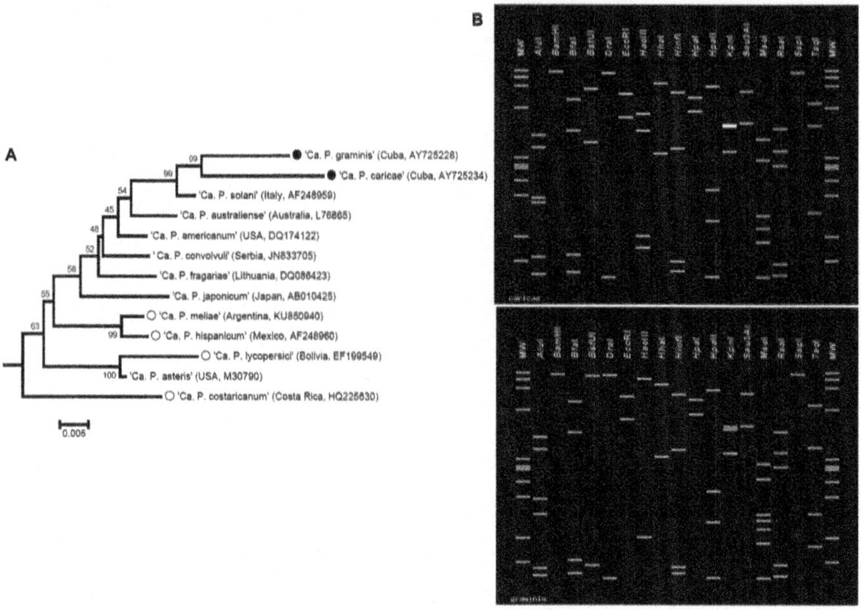

Fig. 4.3 Characterization of '*Ca.* Phytoplasma' species described in Cuba. (**a**) Phylogenetic subtree using minimum evolution method of the 16S rDNA sequences of the '*Ca.* Phytoplasma' species branched with the Cuban species (black circle). (**b**) Distinctive RFLP pattern obtained with *i*Phyclassifier from *in silico* digestion of 16S rRNA gene (R16F2n/R2 fragment) from '*Ca.* P. caricae' and (**c**) '*Ca.* P. graminis'. Lanes labelled MW represents φX174DNA-HaeIII digestion

BTS-associated phytoplasma group 16SrII have been detected only in papayas grown in eastern region of Cuba, while BTS-associated phytoplasma group 16SrI-B was identified in papayas grown both in western and eastern Cuba, with the higher diversity of phytoplasmas located in the east (Fig. 4.1).

Mixed infection of PBT-associated rickettsia and BTS-associated phytoplasma were found in papaya trees exhibiting symptoms of BTS and PBT (Acosta et al., 2013). Similarly, the presence of papaya ringspot virus (PRSV) and BTS-associated phytoplasma was observed in papaya trees exhibiting BTS and PRSV symptoms (Arocha, Piñol, Acosta, et al., 2009).

Other phytoplasma strains have been detected in symptomatic papaya plants in Cuba. For example, 16SrI phytoplasmas were identified in papaya exhibiting symptoms of mosaic, yellow crinkling and apical necrosis (Rojas-Martínez, Zavaleta-Mejía, & Rivas-Valencia, 2011). Subgroups 16SrI-B, -L, -X, -Z were identified in papaya, with 16SrI-X and 16SrI-Z reported only in this country (Acosta et al., 2011; Pérez-López, Luna-Rodríguez, et al., 2016). 16SrX phytoplasmas were detected in papaya exhibiting symptoms of yellowing, wrinkling, necrosis, phyllody, virescence and stunted crown leaves (Arocha, Horta, Peralta, & Jones, 2003). Subgroups 16SrII-P and 16SrII-Q, associated with BTS disease, have also been reported only in Cuba (Pérez-López, Luna-Rodríguez, et al., 2016). In brief, phytoplasmas affecting papaya plants in Cuba are very diverse with the identification of strains belonging to the groups 16SrI, 16SrII, 16SrX, and 16SrXVII (Fig. 4.1) (Acosta et al., 2009; Acosta et al., 2013; Acosta, Zamora, et al., 2017; Arocha et al., 2003; Arocha, López, Piñol, et al., 2005).

4.3 Diversity of '*Candidatus* Phytoplasma' Species in Cuba

Two of the 42 '*Ca.* Phytoplasma' species described so far (Davis, Zhao, Wei, Dally, & Lee, 2017), were identified in Cuba in association with diseased sugarcane ('*Ca.* P. graminis') and papaya plants ('*Ca.* P. caricae') (Arocha, López, Piñol, et al., 2005). '*Ca.* P. graminis' and '*Ca.* P. caricae' are the reference strains for the groups 16SrXVI and 16SrXVII, respectively. Moreover, phytoplasmas belonging to group 16SrI, 16SrII, 16SrV, 16SrX, 16SrXIV and 16SrXVI have also been detected in plants and insects in Cuba.

Based on 16S rRNA gene sequence analysis, Zhao, Wei, Davis, and Lee (2010) divided the phytoplasma clade into three distinct subclades: two major subclades I and II containing most of the phytoplasma strains and a smaller subclade III. Based on this subclade classification, phylogenetic analysis using 16S rRNA gene sequences showed that Cuban phytoplasma strains are very diverse as they are grouped into the two major subclades I and II (Fig. 4.1) (Arocha, López, Piñol, et al., 2005; Pérez-López, Luna-Rodríguez, et al., 2016). Subclades I and II contain the majority of the '*Ca.* Phytoplasma' species detected and described in Latin America such as '*Ca.* P. maliae' from Argentina (Fernández, Galdeano, Kornowski, Arneodo, & Conci, 2016), '*Ca.* P. hispanicum' from Mexico (Davis, Harrison, Zhao, Wei, & Dally, 2016), '*Ca.* P. costaricanum' (I.-M. Lee, Bottner-Parker, Zhao, Villalobos, & Moreira, 2011), and '*Ca.* P. lycopersici' (Arocha, Piñol, López, et al., 2007). Interestingly, phytoplasmas of the group 16SrXIII, such as those related to '*Ca.* P. hispanicum', are widely spread in Latin America, but have never been reported in Cuba, probably due to the geographic isolation of Cuba (Pérez-López, Luna-Rodríguez, et al., 2016). Details of natural hosts, accession numbers and 16Sr group/subgroup classification of phytoplasma strains identified in Cuba are indicated in Table 4.1.

4.4 Insect Vectors Transmitting Phytoplasmas in Cuba

Phytoplasma are transmitted by sap-feeders, mostly belonging to the hemipteran group of Cicadellidea (leafhoppers) which contains more than 75% of confirmed vectors. Insect vectors of phytoplasmas have also been found in four families of planthoppers (Fulgoroidea), two genera of psyllid and four species of true bugs (Heteroptera) (Weintraub & Beanland, 2006).

Table 4.1 Phytoplasma strains detected in Cuba, based on peer reviewed reports published in national and international journals

GeneBank Accession Number	Host	Group/ Subgroup	[a]References
DQ286577	Basil (*Ocimum basilicum*)	16SrI-AD[b]	Arocha, Piñol, Picornell, Almeida, Jones, et al., (2006) and Pérez-López, Luna-Rodríguez, et al. (2016)
DQ286953	Broadbean (*Vicia faba*)	16SrI-AE[b]	Arocha, Piñol, Picornell, Almeida, and Jones (2007) and Pérez-López, Luna-Rodríguez et al. (2016)
DQ286947	Sweet pepper (*Capsicum annuum*)	16SrI-AD[b]	Arocha, Piñol, Picornell, et al. (2007) and Pérez-López, Luna-Rodríguez, et al. (2016)
KP722520, KP722521, KP722522, KJ413095	Papaya (*Carica papaya*), *Empoasca papayae*, soybean (*Glycine max*)	16SrI-B	Acosta et al. (2015) and Acosta, Zamora et al. (2017)
EU328254, EU328252, EU328251, EU328256	Carrot (*Daucus carota*), cabbage (*Brassica Oleracea*), beetroot (*Beta vulgaris*), cassava (*Manihot esculenta*)	16SrI-nc	Arocha, Piñol, Almeida et al. (2009)
JN383913 JN383914	Tomato (*Solanum Lycopersicum* Common bean (*Phaseolus vulgaris*)	16SrI-nc 16SrI-nc	Zamora, Acosta, and Martínez (2012) and Zamora, Acosta, Pinol, Quinones, and Bertaccini (2014)
KC825053	Strawberry (*Fragaria x ananassa*)	16SrI-nc	Ferriol-Marchena et al. (2013)
KC513772	Macadamia nut (*Macadamia integrifolia*)	16SrI-nc	Pérez-López, Hernández-Rodríguez, et al. (2013)
JF781311	Papaya (*Caricapapaya*)	16SrI-Z[b]	Acosta, Piñol, et al. (2017) and Pérez-López, Luna-Rodríguez et al. (2016)
JF781307	Papaya (*Carica papaya*)	16SrI-L[b]	Acosta et al. (2013) and Pérez-López, Luna-Rodríguez et al. (2016)
JF781308	Papaya (*Carica papaya*)	16SrI-X[b]	Acosta et al. (2013)
KJ413093	Soybean (*Glycine max*)	16SrI-W[b]	Acosta et al. (2015)
KJ413094	Soybean (*Glycine max*)	16SrI-Y[b]	Acosta et al. (2015)
EU328253	Radish (*Raphanus sativus*)	16SrII-nc	Arocha, Piñol, Almeida et al. (2009)
EU869220	Sapodilla (*Manilkara zapota*)	16SrII-nc	Acosta et al. (2009)

(continued)

Table 4.1 (continued)

GeneBank Accession Number	Host	Group/ Subgroup	[a]References
EU350565, EU350564, EU350566	Papaya (*Carica papaya*)	16SrII-nc	Arocha, Piñol, Acosta, et al. (2009)
FJ649200, FJ649198, FJ649199	*Empoasca papayae*	16SrII-nc	Pérez et al. (2010)
DQ286948, DQ286949, DQ868531, DQ868532, DQ286950, DQ286952, DQ286951, DQ868533	Papaya (*Carica papaya*), *Empoasca spp, Anoda acerifolia, Euphorbia heterophylla, Malvastrum coromandelianum, Rhynchosia minima*	16SrII-P[b]	Arocha, Piñol, López, et al. (2007), Arocha, Piñol, Picornell, Almeida, and Jones (2006), and Pérez-López, Luna-Rodríguez, et al. (2016)
JF781310	Papaya (*Carica papaya*)	16SrII-Q[b]	Acosta et al. (2013) and Pérez-López, Luna-Rodríguez et al. (2016)
JF781309	Papaya *Carica papaya*)	16SrII-N[b]	Acosta et al. (2013)
KF500549	Sapodilla (*Manilkara zapota*)	16SrV-nc	Pérez-López, Pantoja, Hernández-Rodríguez, and Bárzaga (2013)
AY725238	Bermuda-grass (*Cynodon dactylon*)	16SrXIV-nc	Arocha, López, Piñol, et al. (2005)
AY257547	Papaya (*Carica papaya*)	16SrX-nc	Arocha et al. (2003)
AY725237, AY257548	Planthopper (*Saccharosydne saccharivora)* sugarcane (*Saccharum ssp.*)	16SrXVI-A[c]	Arocha, López, Fernández, et al. (2005)
AY725234	Papaya (*Carica papaya*)	16SrXVII-A[c]	Arocha, López, Piñol, et al. (2005)

[a]Classification based on the use of *i*PhyClassifier software
[b]Subgroups only reported in Cuba to date
[c]Groups only reported in Cuba to date
nc Not classified mainly because the sequence reported is shorter than F2nR2 sequence

In Cuba, two insects have been identified as capable of transmitting phytoplasmas, the planthopper *Saccharosydne saccharivora* Westwood (Hemiptera: Delphacidae) and the leafhopper *Empoasca papayae* Oman (Hemiptera: Cicadellidae) (Acosta, Zamora, et al., 2017; Arocha, López, Fernández, et al., 2005; Pérez et al., 2010). *S. saccharivora* is able to transmit sugarcane yellow leaf phytoplasma (SCYLP), a phytoplasma related to '*Ca.* P. graminis' specie (Arocha, López, Fernández, et al., 2005; Arocha, López, Piñol, et al., 2005), and considered to be a serious pest for many countries producing sugarcane (CABI, 2018). *E. papayae* feeds on papaya plants, and is able to transmit strains belonging to groups 16SrI and 16SrII which are associated with BTS disease (Arocha, Piñol, Acosta, et al., 2009). *E. papayae* has also been identified as vector of Rickettsia-like bacteria, pathogens associated with PBT disease in Cuba (Acosta, Zamora, et al., 2017).

Recently, several insect species sampled in economically important crops were found positive for the presence of phytoplasma. For example, planthopper *Cedusa* spp. Fowler (Hemiptera: Derbidae) sampled in sugarcane fields and leafhopper *Empoasca canavalia* DeLong, sampled in Chaya fields tested positive for the presence of phytoplasma, making them potential phytoplasma vectors of SCYLP and PBT phytoplasma, respectively (Arocha, López, Piñol, et al., 2005; Sánchez Castro et al., 2015).

Cuba hosts many insect species known to be phytoplasma vectors in other countries. For example *Dalbulus maidis* DeLong & Wolcott (Hemiptera: Cicadellidae) and *D. elimatus* Ball. are vectors of maize bushy stunt phytoplasma in Mexico ('*Ca.* P. asteris', group 16SrI-B) (Nault, 1990; Weintraub & Beanland, 2006). These two leafhopper species, as well as the AY phytoplasma strain they transmit, are present in Cuba (Acosta et al., 2015; Acosta, Zamora, et al., 2017; Zanol, 2006). The leafhopper species *Scaphytopius fuliginosus* Osborn (Hempitera: Cicadellidae), and *Acinopterus angulatus* Lawson (Hemiptera: Cicadellidae), both present in Cuba, were reported to be vector of the Machismo disease in soybean in Mexico and AY diseases in various vegetables and fruit trees in California, respectively (Weintraub & Beanland, 2006).

Moreover, climatic changes inducing stronger storms or winds, human travels and commercial transportations increase risks of pathogens export/importation towards Cuba, hence increasing risks of phytoplasma diseases occurrence in Cuban agriculture. For example, Citrus is produced throughout Cuba and is one of Cuba's top exports as well as an important source of fresh fruits for local consumers. Cuba's citrus industry has not been heavily challenged by insect pests and diseases, but its proximity to Florida and Caribbean island nations, where the citrus greening disease attacking citrus trees is prevalent, is a risk for the Cuban citrus industry (Office of Global Analysis, 2008).

Knowledge of phytoplasma epidemiology and identification of confirmed and potential insect vector and plant hosts are a necessity to forecast phytopathological issues and to plan and develop sustainable management techniques.

4.5 Existing and Potential Management Techniques

Being phloem limited, phytoplasma are difficult to reach and control. Phytoplasma can be killed with antibiotics, an expensive management method rarely allowed in the fields. Moreover, antibiotics do not provide long-term protection and do not always give 100% elimination of phytoplasma (Bertaccini & Duduk, 2010; Kawabe et al., 2011). Elimination of phytoplasma via hot water treatment or cryotherapy, sometimes associated with *in vitro* cultures and/or antibiotic treatments has been demonstrated to be efficient method to obtain phytoplasma-free plant materials for planting (Bertaccini & Duduk, 2010). Of interest for Cuba, thermotherapy (50 °C for 2 h) of sugarcane stalks eliminate phytoplasma in sugarcanes and thermotherapy associated with *in vitro* shoot tip cultures and antibiotic treatments provide phytoplasma-free cassava plants (Carvalho et al., 2017; Soufi & Komor, 2014).

Genetic resistance in plants is the best method to manage phytoplasma diseases (Jarausch & Torres, 2014). Although much progress have been done in understanding resistance in plants, genetic resistance is rare and few resistant varieties of commercially important plant species have been produced (Jarausch & Torres, 2014). In Cuba, coconut palm cultivars resistant to Lethal yellowing (LY) phytoplasma were developed in the 1960s when an epidemic of LY spread in the country. Those cultivars are performing very well in mixed cultures in areas where LY is present (Seemüller & Harries, 2010). However, for other economically important crops, the production of resistant plant species is still a long way off, but there are strong potential for improvement. For example, differences in susceptibility to phytoplasma diseases among sugarcane cultivars have been observed in Cuba and could be used to develop resistant varieties (Arocha, López, Fernández, et al., 2005). The lower susceptibility of sugarcane has been associated with a low phytoplasma titer in the plants and might be correlated with the recovery phenomenon observed in phytoplasma-infected apple trees and grapevine (Soufi & Komor, 2014).

Vector control is the most common technique used for managing phytoplasma diseases. Insecticides are commonly used when the identity of the vector and the epidemiology of the diseases are known (Weintraub, 2007). However, insecticides are expensive and not environmentally friendly, and in many cases, cannot be implemented when the identity and biology of the vectors are not known. Moreover, Cuba moved in the early 1980s, from an agriculture highly dependent on external inputs to a semi-organic agriculture using IPM, local resources and low external inputs (Nicholls, Pérez, Vasquez, & Altieri, 2002). For example, Cuba is mass-producing its own biological control agents for several pests, as well as producing biopesticide using neem and paradise plant extracts as well as extracts made from *Lantana camara* plants. Cuban made biopesticides are being used successfully on various phloem feeders, including leafhoppers (Nicholls et al., 2002; Orsini, Kahane, Nono-Womdim, & Gianquinto, 2013; Pérez-López & Luis-Pantoja, 2012).

Insect vectors can also be controlled by netting. For example, Australia is controlling phytoplasma diseases in papaya using insect-proof screening (Walsh, Guthrie, & White, 2006). Netting blocks many insect pests and does not seem to change the sugar content in the fruits. However, this method is expensive and reduces fruit weight and production due to lower pollination levels (Walsh et al., 2006). Overall, this method is cost effective for cash-crops, organically grown fruits or when a premium price can be obtained for the fruits. Other management techniques in papaya crops, that help to lower the phytoplasma inoculum load, consist of rouging and ratooning (removing symptomatic shoots and allowing lateral shoot development) diseased trees (Guthrie, White, Walsh, & Scott, 1998). Hydroponic cultures for high value crops are being used and might help reduce the incidence of phytoplasma diseases. However, phytoplasma infections have been detected in Cuban hydroponic cultures of carrots, beetroots, radishes and seem to be spread by insect vectors (Arocha & Jones, 2009). Other management include push-pull strategies using semiochemicals to repel insects from important crop plants toward a trap crop (La Grange, Schröder, Glinwood, Ignell, & Krüger, 2017).

Weeds are known to play a key role in the epidemiology of phytoplasma diseases, since they can be an alternate host and reservoir for phytoplasma and/or leafhopper vectors, and can therefore influence the population density of the vectors and act as a source of phytoplasma inoculum (Marzachi et al., 2009; Olivier, Séguin-Swartz, Galka, & Olfert, 2011). In Cuba, several species of weeds (*Cynodon dactylon, Conyza canadensis, Sorghum halepense* and *Macroptilium lathyroides*) have been found to be infected with phytoplasma and might play a role as inoculum/vector reservoir (Arocha, López, Piñol, et al., 2005). Weed removal are essential in phytoplasma disease management to reduce the level of inoculum (Rao, Mall, & Marcone, 2012; Weintraub & Wilson, 2010).

4.6 Conclusion

Cuba is an archipelago located in the Caribbean, and its geographic isolation along with plant importation, have played a role in the particular diversity of phytoplasmas and the presence and diversity of insect vectors in the country. The local diversity and the high amount of unique strains detected in Cuba, are key factors to take into account for future development of disease management strategies in economically important crops. The very specific agroecosystems in Cuba, mostly composed of organic urban and rural farming but also of large-scale rural farming, will need to be taken into account for management development. This will require extensive epidemiological studies on phytoplasma diseases, as well as extension work to increase grower's knowledge in phloem feeder's transmitted diseases (Pérez-López, Wist, Rodríguez, Luna-Rodríguez, & Olivier, 2018).

References

Acosta, K. I., Piñol, B., Acosta, E., Countín, P., & Arocha, Y. (2009). First report on detection of '*Candidatus* Phytoplasma aurantifolia'(group 16SrII) affecting sapodilla in eastern Cuba. *Plant Pathology, 58*(2), 391–391.

Acosta, K. I., Piñol, B., Zamora-Gutierrez, L., Quiñones-Pantoja, M. L., Miranda-Cabrera, I., Leyva-López, N. E., & Arocha, Y. (2017). A phytoplasma representative of a new subgroup 16SrI-Z associated with Bunchy Top Symptoms (BTS) on papaya in Cuba. *Revista de Protección Vegetal, 32*(1), 52–59.

Acosta, K. I., Silva, F., Zamora, L., Quiñones, M., Martínez, R., Piñol, B., … Zerbini, F. (2015). Detection and molecular characterization of an aster yellows group phytoplasma associated with soybean and evidence of two new 16SrI subgroups in Cuba. *Journal of Plant Pathology, 97*(2), 339–344.

Acosta, K. I., Zamora, L., Fernández, A., Arocha, Y., Martínez, Y., Santos, M., … Leyva, N. (2011). First report of '*Candidatus* Phytoplasma asteris' (group 16SrI) affecting papaya in Cuba. *New Disease Reports, 24*, 29.

Acosta, K. I., Zamora, L., Piñol, B., Fernández, A., Chávez, A., Flores, G., … Arocha, Y. (2013). Identification and molecular characterization of phytoplasmas and rickettsia pathogens associated with 'Bunchy Top Symptom' (BTS) and 'Papaya Bunchy Top'(PBT) of papaya in Cuba. *Crop Protection, 45*, 49–56.

Acosta, K. I., Zamora, L., Piñol, B., Quiñones, M. L., Ramos, P. L., Luis, M., … Arocha, Y. (2017). Empoasca papayae Oman, 1937 (Hemiptera: Cicadellidae) the simultaneous vector of phytoplasmas and rickettsia associated with "Bunchy Top Symptom" in Cuba. *Anales de Biología, 39*, 35–42.

Altieri, M. A., Funes-Monzote, F. R., & Petersen, P. (2012). Agroecologically efficient agricultural systems for smallholder farmers: Contributions to food sovereignty. *Agronomy for Sustainable Development, 32*(1), 1–13.

Arocha, Y., Horta, D., Peralta, E., & Jones, P. (2003). First report on molecular detection of phytoplasmas in papaya in Cuba. *Plant Disease, 87*(9), 1148–1148.

Arocha, Y., & Jones, P. (2009). Phytoplasma diseases of the Gramineae. In P. G. Weintraub & P. Jones (Eds.), *Phytoplasmas: Genomes, plant hosts and vectors* (pp. 170–187). Preston, UK: MPG Books Group.

Arocha, Y., López, M., Fernández, M., Piñol, B., Horta, D., Peralta, E., … Wilson, M. R. (2005). Transmission of a sugarcane yellow leaf phytoplasma by the delphacid planthopper *Saccharosydne saccharivora*, a new vector of sugarcane yellow leaf syndrome. *Plant Pathology, 54*(5), 634–642.

Arocha, Y., López, M., Piñol, B., Fernandez, M., Picornell, B., Almeida, R., … Jones, P. (2005). '*Candidatus* Phytoplasma graminis' and '*Candidatus* Phytoplasma caricae', two novel phytoplasmas associated with diseases of sugarcane, weeds and papaya in Cuba. *International Journal of Systematic and Evolutionary Microbiology, 55*(6), 2451–2463.

Arocha, Y., Piñol, B., Acosta, K. I., Almeida, R., Devonshire, J., Van de Meene, A., … Lucas, J. (2009). Detection of phytoplasma and potyvirus pathogens in papaya (*Carica papaya* L.) affected with 'Bunchy Top Symptom'(BTS) in eastern Cuba. *Crop Protection, 28*(8), 640–646.

Arocha, Y., Piñol, B., Almeida, R., Acosta, K. I., Quiñones, M. L., Zayas, T., … Lucas, J. (2009). First report of phytoplasmas affecting organoponic crops in central and eastern Cuba. *Plant Pathology, 58*(4), 793–793.

Arocha, Y., Piñol, B., López, M., Miranda, I., Almeida, R., Wilson, M. R., & Jones, P. (2007). Bunchy top symptom' of papaya in Cuba: New insights. *Bulletin of Insectology, 60*(2), 393.

Arocha, Y., Piñol, B., Picornell, B., Almeida, R., & Jones, P. (2006). First report of a 16SrII ('*Candidatus* Phytoplasma aurantifolia') group phytoplasma associated with a bunchy-top disease of papaya in Cuba. *Plant Pathology, 55*(6), 821.

Arocha, Y., Piñol, B., Picornell, B., Almeida, R., & Jones, P. (2007). Broad bean and sweet pepper: Two new hosts associated with *Candidatus* phytoplasma asteris (16SrI phytoplasma group) in Cuba. *Plant Pathology, 56*(2), 345.

Arocha, Y., Piñol, B., Picornell, B., Almeida, R., Jones, P., & Boa, E. (2006). Basil little leaf: A new disease associated with a phytoplasma of the 16SrI (Aster Yellows) group in Cuba. *Plant Pathology, 55*(6), 822.

Bertaccini, A., & Duduk, B. (2010). Phytoplasma and phytoplasma diseases: A review of recent research. *Phytopathologia Mediterranea, 48*(3), 355–378.

CABI. (2018). *Saccharosydne saccharivora* (west indian canefly). Retrieved March 15, 2019, from CAB International https://www.cabi.org/cpc/datasheet/48877

Carvalho, M., Oliveira, E., Souza, A., Pereira, J., Diamantino, M., & Oliveira, S. (2017). Cleaning cassava genotypes infected with cassava frogskin disease via *in vitro* shoot tip culture. *Genetics and Molecular Research, 16*(2).

Clouse, C. (2014). *Farming Cuba: Urban agriculture from the ground up* (M. Baber, Ed.). New York, NY: Princetown Architectural Press.

Davis, R. E., Harrison, N. A., Zhao, Y., Wei, W., & Dally, E. L. (2016). '*Candidatus* Phytoplasma hispanicum', a novel taxon associated with Mexican periwinkle virescence disease of

Catharanthus roseus. *International Journal of Systematic and Evolutionary Microbiology, 66*(9), 3463–3467.

Davis, R. E., Zhao, Y., Wei, W., Dally, E. L., & Lee, I.-M. (2017). '*Candidatus* Phytoplasma luffae', a novel taxon associated with witches' broom disease of loofah, *Luffa aegyptica* Mill. *International Journal of Systematic and Evolutionary Microbiology, 67*(8), 3127–3133.

Doi, Y., Teranaka, M., Yora, K., & Asuyama, H. (1967). Mycoplasma-or PLT group-like microorganisms found in the phloem elements of plants infected with mulberry dwarf, potato witches' broom, aster yellows, or paulownia witches' broom. *Japanese Journal of Phytopathology, 33*(4), 259–266.

Fernández, F. D., Galdeano, E., Kornowski, M. V., Arneodo, J. D., & Conci, L. R. (2016). Description of '*Candidatus* Phytoplasma meliae', a phytoplasma associated with chinaberry (*Melia azedarach* L.) yellowing in South America. *International Journal of Systematic and Evolutionary Microbiology, 66*(12), 5244–5251.

Ferriol-Marchena, X., Hernández-Rodríguez, L., Luis-Pantoja, M., García-García, G., Banguela-Castillo, A., & Pérez, J. (2013). First report of a '*Candidatus* Phytoplasma asteris' isolate affecting a strawberry nursery in Cuba. *New Disease Reports, 28*, 19.

Firrao, G., Andersen, M., Bertaccini, A., Boudon, E., Bove, J., Daire, X., … Gibb, K. S. (2004). '*Candidatus* Phytoplasma', a taxon for the wall-less, non-helical prokaryotes that colonize plant phloem and insects. *International Journal of Systematic and Evolutionary Microbiology, 54*(4), 1243–1255.

Guthrie, J. N., White, D., Walsh, K. B., & Scott, P. (1998). Epidemiology of phytoplasma-associated papaya diseases in Queensland, Australia. *Plant Disease, 82*(10), 1107–1111.

Hidalgo-Gato, M. M., & Rodríguez-León, R. (2010). Auchenorrhyncha (Insecta: Hemiptera) del area protegida Mil Cumbres (Pinar Del Rio, Cuba). *Boletín de la SEA* (46), 491–497.

Hodgetts, J., & Dickinson, M. (2009). Phytoplasma phylogeny and detection based on genes other than 16S rRNA. In P. G. Weintraub & P. Jones (Eds.), *Genomes, plant hosts and vectors* (pp. 93–112). Oxfordshire, UK: CAB International.

Hodgetts, J., Flint, L., & Fox, A. (2015). First report of '*Candidatus* phytoplasma ulmi' (16SrV-A) associated with *Ulmus* cultivar Morfeo (elm) in the United Kingdom. *New Disease Reports, 32*, 26–26.

Jarausch, B., & Torres, E. (2014). Management of phytoplasma-associated diseases. In A. Bertaccini (Ed.), *Phytoplasmas and phytoplasma disease management: How to reduce their economic impact* (pp. 199–208). Bologna, Italy: IPWG – International Phytoplasmologist Working Group.

Kawabe, Y., Tsuda, J., Matsuura, K., Ogawa, S., Usami, Y., & Kusunoki, M. (2011). Analysis of oxytetracycline and detection of phytoplasma after the trunk injection of oxytetracycline formulation in *Elaeocarpus sylvestris* var. *ellipticus* infected with Elaeocarpus yellows. *Journal of Tree Health, 15*, 97–101.

La Grange, R., Schröder, M. L., Glinwood, R., Ignell, R., & Krüger, K. (2017). Leafhopper interactions with host plants - a role for volatile cues? *IOBC-WPRS Bulletin, 126*, 22–26.

Lee, I.-M., Bottner-Parker, K. D., Zhao, Y., Villalobos, W., & Moreira, L. (2011). '*Candidatus* Phytoplasma costaricanum'a novel phytoplasma associated with an emerging disease in soybean (*Glycine max*). *International Journal of Systematic and Evolutionary Microbiology, 61*(12), 2822–2826.

Lee, I.-M., Davis, R. E., & Gundersen-Rindal, D. E. (2000). Phytoplasma: Phytopathogenic mollicutes. *Annual Reviews in Microbiology, 54*(1), 221–255.

Lee, I.-M., Gundersen-Rindal, D. E., Davis, R. E., & Bartoszyk, I. M. (1998). Revised classification scheme of phytoplasmas based on RFLP analyses of 16S rRNA and ribosomal protein gene sequences. *International Journal of Systematic and Evolutionary Microbiology, 48*(4), 1153–1169.

Lee, I.-M., Gundersen-Rindal, D. E., Davis, R. E., Bottner, K., Marcone, C., & Seemüller, E. (2004). 'Candidatus Phytoplasma asteris', a novel phytoplasma taxon associated with aster yellows and related diseases. *International Journal of Systematic and Evolutionary Microbiology, 54*(4), 1037–1048.

Lee, M., Grau, C., Lukaesko, L., & Lee, I.-M. (2002). Identification of aster yellows phytoplasmas in soybean in Wisconsin based on RFLP analysis of PCR-amplified products (16S rDNAs). *Canadian Journal of Plant Pathology, 24*(2), 125–130.

Leyva-López, N. E., Ochoa-Sánchez, J. C., Leal-Klevezas, D. S., & Martínez-Soriano, J. P. (2002). Multiple phytoplasmas associated with potato diseases in Mexico. *Canadian Journal of Microbiology, 48*(12), 1062–1068.

Luis-Pantoja, M., Ramos-González, P., Naranjo, M., Hernández-Rodríguez, L., Rodríguez, J., & Pérez-López, E. (2015). Rickettsia-related bacteria associated with papaya plants showing bunchy top disease in Cuba. *Journal of General Plant Pathology, 81*(2), 166–168.

Maejima, K., Oshima, K., & Namba, S. (2014). Exploring the phytoplasmas, plant pathogenic bacteria. *Journal of General Plant Pathology, 80*(3), 210–221.

Makarova, O., MacLean, A. M., & Nicolaisen, M. (2015). Phytoplasma adapt to the diverse environments of their plant and insect hosts by altering gene expression. *Physiological and Molecular Plant Pathology, 91*, 81–87.

Marcone, C. (2015). Current status of phytoplasma diseases of forest and landscape trees and shrubs. *Journal of Plant Pathology, 97*(1), 9–36.

Marzachi, C., Coulibaly, A., Coulibaly, N., Sangare, A., Diarra, M., De Gregorio, T., & Bosco, D. (2009). Cotton virescence phytoplasma and its weed reservoir in Mali. *Journal of Plant Pathology, 91*(3), 717–721.

McMinimy, M. A. (2016). *US Agricultural Trade with Cuba: Current Limitations and Future Prospects*. Retrieved from https://fas.org/sgp/crs/row/R44119.pdf

Moreno-Casasola, P., Guevara, M. M., Vázquez-González, C., & Eusebio, A. J. (2016). La conservación de los servicios ecosistémicos y el cambio climático: algunas alternativas. In P. Moreno-Casasola (Ed.), *Servicios ecosistémicos de las selvas y bosques costeros de Veracruz* (pp. 312–358). Xalapa, Mexico: D.R.

Nault, L. (1990). Evolution of an insect pest: Maize and the corn leafhopper, a case study. *Maydica, 35*(2), 165–175.

Nicholls, C. I., Pérez, N., Vasquez, L., & Altieri, M. A. (2002). The development and status of biologically based integrated pest management in Cuba. *Integrated Pest Management Reviews, 7*(1), 1–16.

Office of Global Analysis. (2008). *Cuba's Food & Agriculture Situation Report.* Retrieved from https://thecubaneconomy.com/articles/2012/02/us-department-of-agriculture-cubas-food-agriculture-situation-report-2008/

Olivier, C. Y., Séguin-Swartz, G., Galka, B., & Olfert, O. (2011). Aster yellows in leafhoppers and field crops in Saskatchewan, Canada, 2001–2008. *The Americas Journal of PLant Science and Biotechnology, 141*, 425–462.

Orsini, F., Kahane, R., Nono-Womdim, R., & Gianquinto, G. (2013). Urban agriculture in the developing world: A review. *Agronomy for Sustainable Development, 33*(4), 695–720.

Oshima, K., Ishii, Y., Kakizawa, S., Sugawara, K., Neriya, Y., Himeno, M., ... Yamaji, Y. (2011). Dramatic transcriptional changes in an intracellular parasite enable host switching between plant and insect. *PLoS One, 6*(8), e23242.

Palma, I. P., Toral, J. N., Vázquez, M. R. P., Fuentes, N. F., & Hernández, F. G. (2015). Historical changes in the process of agricultural development in Cuba. *Journal of Cleaner Production, 96*, 77–84.

Pérez, K. A., Piñol, B., Arocha, Y., Wilson, M., Boa, E., & Lucas, J. (2010). Transmission of the phytoplasma associated with bunchy top symptom of papaya by *Empoasca papayae* Oman. *Journal of Phytopathology, 158*(3), 194–196.

Pérez-López, E., Hernández-Rodríguez, L., Pantoja, M., & Zaldívar, M. H. (2013). First report of a '*Candidatus* Phytoplasma asteris' isolate affecting macadamia nut trees in Cuba. *New Disease Reports, 28*, 1.

Pérez-López, E., & Luis-Pantoja, M. (2012). Principales enfermedades causadas por bacterias en los cultivos frutícolas de papaya (*Carica papaya* L.), piña (*Ananas comosus* (L.) Merr.) y mango (*Mangifera indica* L.). *CITRIFRUT, 26*, 28–34.

Pérez-López, E., Luna-Rodríguez, M., Olivier, C. Y., & Dumonceaux, T. J. (2016). The underestimated diversity of phytoplasmas in Latin America. *International Journal of Systematic and Evolutionary Microbiology, 66*(1), 492–513.

Pérez-López, E., Olivier, C. Y., Luna-Rodríguez, M., & Dumonceaux, T. J. (2016). Phytoplasma classification and phylogeny based on in silico and in vitro RFLP analysis of cpn60 universal target sequences. *International Journal of Systematic and Evolutionary Microbiology, 66*(12), 5600–5613.

Pérez-López, E., Pantoja, M., Hernández-Rodríguez, L., & Bárzaga, I. (2013). First report of a '*Candidatus* Phytoplasma ulmi'isolate affecting sapodilla trees in western Cuba. *New Dis Rep, 28*, 18.

Pérez-López, E., Wist, T., Rodríguez, Y., Luna-Rodríguez, M., & Olivier, C. Y. (2018). Maize bushy stunt in native corn: Implications for Mexican "subsistence farmers". *Environment, Development and Sustainability, 20*, 1797–1805.

Rao, G. P., Mall, S., & Marcone, C. (2012). Recent biotechnological approaches in diagnosis and management of sugarcane phytoplasma diseases. *Functional Plant Science and Biotechnoly, 6*, 19–29.

Reardon, J. A. S., & Pérez, R. A. (2010). Agroecology and the development of indicators of food sovereignty in Cuban food systems. *Journal of Sustainable Agriculture, 34*(8), 907–922.

Rojas, M. L. (2014). Fauna de insectos en cayos del golfo de Ana María, Cuba/Insects from the gulf of Ana María keys, Cuba. *Revista de Investigaciones Marinas, 32*(2), 15–19.

Rojas-Martínez, R. I. (2009). Insect vectors of phytoplasmas. In K. Del Claro, P. S. Oliveira, & V. Rico-Gray (Eds.), *Tropical biology and conservation management-volume 7: Phytopathology and entomology* (Vol. VII, pp. 46–59). Oxford, UK: Eolss Publishers.

Rojas-Martínez, R. I., Zavaleta-Mejía, E., & Rivas-Valencia, P. (2011). Presencia de fitoplasmas en papayo (*Carica papaya*) en México. *Revista Chapingo. Serie horticultura, 17*(1), 47–50.

Romero, A., Ruz, R., & González, M. (2013). Evaluación de siete cultivares de soya (*Glycine max*) en las condiciones edafoclimáticas del municipio Majibacoa, Las Tunas. *Pastos y Forrajes, 36*(4), 459–463.

Sánchez Castro, A., Quiñones Pantoja, M. L., Piñol, B., & Fernández Argudín, B. M. (2015). Primer informe de Typhlocybinae como vectores potenciales de fitoplasmas en *Cnidoscolus chayamansa* (Miller) IM Johnst.(chaya) en Cuba. *Revista de Protección Vegetal, 30*(2), 148–157.

Seemüller, E., & Harries, H. (2010). Plant resistance. In P. G. Weintraub & P. Jones (Eds.), *Phytoplasmas: Genomes, plant hosts and vectors. CAB International, Oxfordshire, UK* (pp. 147–169). Preston, UK: CABI.

Seemüller, E., Kiss, E., Sule, S., & Schneider, B. (2010). Multiple infection of apple trees by distinct strains of '*Candidatus* Phytoplasma Mali' and its pathological relevance. *Phytopathology, 100*(9), 863–870.

Soto, N., Ferreira, A., Delgado, C., & Enríquez, G. A. (2013). *In vitro* regeneration of soybean plants of the Cuban Incasoy-36 variety. *Biotecnología Aplicada, 30*(1), 29–33.

Soufi, Z., & Komor, E. (2014). Latent infection of asymptomatic Hawaiian sugarcane cultivars with 16SrI and 16SrXI phytoplasmas. *Journal of General Plant Pathology, 80*(3), 255–263.

Sugio, A., & Hogenhout, S. A. (2012). The genome biology of phytoplasma: Modulators of plants and insects. *Current Opinion in Microbiology, 15*(3), 247–254.

Walsh, K. B., Guthrie, J. N., & White, D. (2006). Control of phytoplasma diseases of papaya in Australia using netting. *Australasian Plant Pathology, 35*(1), 49–54.

Weintraub, P. G. (2007). Insect vectors of phytoplasmas and their control-an update. *Bulletin of Insectology, 60*(2), 169.

Weintraub, P. G., & Beanland, L. (2006). Insect vectors of phytoplasmas. *Annual Review of Entomology, 51*, 91–111.

Weintraub, P. G., & Wilson, M. R. (2010). Control of phytoplasma diseases and vectors. In P. G. Weintraub & P. Jones (Eds.), *Phytoplasmas: Genomes, plant hosts and vectors* (pp. 233–249). Oxfordshire, UK: CAB International.

Zamora, L., Acosta, K. I., & Martínez, Y. (2012). First report of '*Candidatus* Phytoplasma asteris' (16SrI group) affecting common bean in Cuba. *New Disease Reports, 25*, 4.

Zamora, L., Acosta, K. I., Pinol, B., Quinones, M., & Bertaccini, A. (2014). First report of '*Candidatus* Phytoplasma asteris' (16SrI group) causing stunt of tomato in Cuba. *New Disease Reports, 30*, 10.

Zanol, K. M. (2006). Catálogo dos Deltocephalinae (Hemiptera: Cicadellidae) neotropicais (incluíndo o norte do México). Parte I—Athysanini e Deltocephalini excluídos. *Acta Biológica Paranaense, 35*.

Zhao, Y., Davis, R. E., Wei, W., Shao, J., & Jomantiene, R. (2014). Phytoplasma genomes: Evolution through mutually complementary mechanisms, gene loss and horizontal acquisition. In D. C. Gross, A. Lichens-Park, & C. Kole (Eds.), *Genomics of plant-associated Bacteria* (pp. 235–271). Berlin/Heidelberg, Germany: Springer Berlin Heidelberg.

Zhao, Y., Wei, W., Davis, R. E., & Lee, I.-M. (2010). Recent advances in 16S rRNA gene-based phytoplasma differentiation, classification and taxonomy. In P. G. Weintraub & P. Jones (Eds.), *Phytoplasmas: Genomes, plant hosts and vectors* (pp. 64–92). Oxfordshire, UK: CABI.

Chapter 5
Integrated Management of Napier Grass Stunt Disease in East Africa

George O. Asudi, Francis N. Muyekho, Charles A. O. Midega, and Zeyaur R. Khan

Abstract Napier grass stunt (NGS) disease is a serious disease of the most important fodder crop in East Africa. The disease is caused by a phytoplasma, '*Candidatus* Phytoplasma oryzae', a cell wall-less bacterium belonging to the 16SrXI group, in the class Mollicutes. The disease affects Napier grass causing up to 100% crop loss in infested farms. Despite the absence of a commercially available, effective control method, several techniques have been attempted by farmers to support the integrated management of NGS disease in the affected regions. This chapter reviews various methods used by farmers to control the disease in Kenya, Tanzania and Uganda. We also propose suggestions that could be used to curb the spread of the disease to regions not yet reached by the pathogen.

Keywords Napier grass stunt disease · Phytoplasma · Insect vector · Transmission and control

---

G. O. Asudi (✉)
Biochemistry, Microbiology and Biotechnology Department, Kenyatta University, Nairobi, Kenya

Matthias-Schleiden-Institute for Genetics, Bioinformatics and Molecular Botany – Department of Plant Physiology, Friedrich-Schiller-University, Jena, Germany
e-mail: gasudii@gmail.com

F. N. Muyekho
Masinde Muliro University of Science and Technology, Kakamega, Kenya
e-mail: fnmuyekho57@gmail.com

C. A. O. Midega · Z. R. Khan
International Centre of Insect Physiology and Ecology (icipe), Nairobi, Kenya
e-mail: cmidega@icipe.org; zkhan@icipe.org

C. Y. Olivier et al. (eds.), *Sustainable Management of Phytoplasma Diseases in Crops Grown in the Tropical Belt*, Sustainability in Plant and Crop Protection 12,
https://doi.org/10.1007/978-3-030-29650-6_5

5.1 Introduction

Napier grass, *Pennisetum purpureum* Schumach (Poaceae), also known as elephant grass (Fig. 5.1a) is a perennial feed crop with a vigorous root system, sometimes stoloniferous with a creeping rhizome. The grass is indigenous to the Zambezi valley and is currently grown in many countries (Farrell, Simons, & Hillocks, 2002). It grows naturally in damp grassland, forest margins or riverbeds, reaching about 10 m high and producing up to 29 tonnes/ha of dry matter. The grass is mainly propagated from cuttings of a node or root crown of 1–2 splits and is highly heterozygous, hence giving rise to a very heterogeneous population of seedlings not "true to type" (Orodho, 2006). The grass can withstand repeated harvests with 4–6 cuts per year producing 50–150 tonnes/ha of green matter. If regularly fertilized and irrigated, it can also withstand heavy grazing and can provide a considerable bulk of feed to the livestock, as it exhibits rapid regrowth and produces a high palatable leafy biomass (Orodho, 2006). The crop is best replanted every 5 or 6 years (Farrell et al., 2002). This high performance has led to its widespread use as a fodder crop in the eastern Africa.

In East Africa, Napier grass is generally grown and managed as a pure stand but also in mixtures with other legumes planted as intercrops, within either the same row or alternate rows. Legume species include *Desmodium uncinatum* (Jacq.) DC. (Fabaceae), *D. intortum* (Mill.) Urb, *Pueraria phaseoloides* (Roxb.) Benth (Fabaceae), *Centrosema pubescens* Benth. (Fabaceae), *Neonotonia wightii* (Wight & Arn.) J.A. Lackey (Fabaceae) and *Stylosanthes guianensis* (Aublet) Sw. (Fabaceae) (Cook, Khan, & Pickett, 2007; Orodho, 2006). When intercropped with herbaceous legumes, cutting or grazing management of Napier grass is usually adjusted to favour the legumes in order to maintain a satisfactory mixed turf. The fodder grass can also be farmed as an alley crop with other fodder legumes such as *Leucaena leucocephala* (Lam.) de Wit (Fabaceae), *Calliandra calothyrsus* Meisn. (Fabaceae), *Sesbania sesban* (L.) Merr. (Fabaceae) and *Gliricidia sepium* (Jacq.)

Fig. 5.1 (**a**) Healthy Napier grass, (**b**) dairy cows being fed on the crop

Kunth ex Walp. (Fabaceae). Legumes content improve the quality of Napier grass-based feed because of the high crude protein content increasing the overall yield of the crop as a result of nitrogen fixation.

5.2 Economic Importance of Napier Grass

Napier grass constitutes 40–80% of forage in East Africa and is used mostly by smallholder dairy farmers practicing intensive (zero grazing) and semi-intensive dairy cattle production (Orodho, 2006). In East Africa, small-scale dairy farming are on the increase because of the market liberalization in the dairy sub-sector (Bingi & Tondel, 2015). Most farmers are also turning to dairying as an important alternative income earner, as other cash crops are becoming less profitable. These dairy farming are in densely populated areas and consist mostly of small size farms, which restrain natural grazing. Therefore, cattle are fed mainly on crop residues, protein concentrates and planted fodder, mostly Napier grass (Fig. 5.1b) (Orodho, 2006; Staal et al., 1998; Staal et al., 1999). The crop is also the most promising and high yielding fodder and gives dry matter yields that surpass most popular tropical pasture grasses including Rhodes grass *Chloris gayana* Kunth, (Poaceae), Setaria (*Setaria sphacelata*, (Schum) Stapf & Hubb, (Poaceae)) and Kikuyu grass (*Pennisetum clandestinum* Hochst ex Chiov.) (Anindo & Potter, 1994; Humphreys, 1994; Skerman & Riveros, 1989). Though not practiced in Kenya, Napier grass can be converted into hay or silage of high quality with or without additives and be stored for use during the seasonal scarcity of feed, in dry seasons. In Taiwan, the grass is used to produce dehydrated grass pellets, which are used as supplementary stock feed (Orodho, 2006). Additionally, Napier grass is widely used for soil and water conservation on hilly slope areas and as building material or mulch in other farming systems (Orodho, 2006). The crop has also been exploited in several countries such as South Africa, India, Brazil and Thailand for renewable energy production such as biogas, ethanol and fuel.

Napier grass is used in East Africa to control cereal stem borers *Chilo partellus* Swinhoe (Lepidoptera: Crambidae) and *Busseola fusca* Fuller (Lepidoptera: Noctuidae), the most deleterious insect pests of cereals and the main target pests in the 'push-pull' strategy (PPS). The latter involves intercropping cereals with *Desmodium* spp. Desv (the push crop) and planting Napier grass (the pull crop) around as a trap plant, as it attracts more stem borers than the main cereal crops (Cook et al., 2007; Pickett, Woodcock, Midega, & Khan, 2014). This strategy leads to a decrease in stem borer population and an increase in cereal grain yield (Khan, Midega, Bruce, Hooper, & Pickett, 2010; Midega, Khan, Amudavi, Pittchar, & Pickett, 2010). This PPS strategy also helps farmers producing fodder in the cereal-livestock mixed farming systems practiced in the East African regions. Farmers who either keep no livestock or have excess fodder can sell bundles of Napier grass to other growers in need of fodder, making the crop a valuable source of income (Khan et al., 2014).

5.3 Napier Grass Stunt Disease Outbreak in East Africa

Napier grass stunt (NGS) disease was first reported in the Bungoma district of western Kenya in 1997. It was later described in Uganda in 2001 and in Tanzania in 2008 (Alicai, Kabirizi, Byenkya, Kayiwa, & Ebong, 2004; Jones, Devonshire, Holman, & Ajanga, 2004; Nielsen, Ebong, Kabirizi, & Nicolaisen, 2007; Orodho, 2006). In the field, NGS disease is expressed in the re-growths of Napier grass after several cuttings or grazing by animals. The disease symptoms include foliar yellowing, small leaves, bushy appearance, yellow to purple streaking, proliferation of tillers and shortening of internodes to the extent that clumps appear severely stunted, with a low biomass yield. Affected shoots become pale yellow-green in colour and seriously dwarfed (Fig. 5.2a, b). Often the whole stool is affected, produces no yield, and eventually dies. NGS disease has spread to several locations in Kenya, Uganda and Tanzania causing serious economic losses in the smallholder dairy industry and posing a significant threat to cereal production in the region (Asudi et al., 2015; Kabirizi, Nielsen, Nicolaisen, Byenkya, & Alacai, 2007; Kawube et al., 2015; Khan et al., 2014; Orodho, 2006). Many smallholder farmers have lost their entire Napier grass crops and have been forced to reduce their number of animals and/or purchase fodder from the local market (Arocha & Jones, 2009).

Napier grass stunt disease is caused by phytoplasma, a non-culturable, wall-less, obligate bacterium in the class Mollicutes. Globally, phytoplasmas are associated with numerous plant diseases of crops, vegetables, fruits, grasses and ornamentals resulting into difficult phytosanitary conditions and serious economic losses worldwide (IRPCM Phytoplasma/Spiroplasma Working Team–Phytoplasma Taxonomy Group, 2004; Lee, Davis, & Gundersen-Rindal, 2000). Based on the 16S rDNA sequences, phytoplasmas associated with NGS diseases in Tanzania, Kenya and Uganda belong to '*Ca.* Phytoplasma oryzae' or rice yellow dwarf (RYD) phytoplasma and are taxonomically classified as 16SrXI group members (Asudi, Van den Berg, Midega, Schneider, et al., 2016; Jones et al., 2004; Nielsen et al., 2007).

Fig. 5.2 (**a**) Severely stunted, bushy Napier grass with foliar yellowing and reduced biomass surrounded by healthy looking plants and (**b**) the spread of Napier grass stunt disease in an affected farm

5.4 Napier Grass Stunt Phytoplasma Transmission

Phytoplasma diseases are primarily spread by sap-sucking insect vectors belonging to the families Cicadellidae (leafhoppers), Fulgoridae (planthoppers) and Psyllidae (psyllids) (Weintraub & Beanland, 2006). Spread can also occur through vegetative propagation of infected plant materials and through vascular connections made by parasitic plants such as dodder (*Cuscuta* spp. L., Convolvulaceae) between infected and uninfected host plants (Eden-Green, 2010). The hemipteran insect vectors feed on phloem tissues, where they acquire phytoplasmas and transmit them from plant to plant during feeding (Lee et al., 2000; Obura et al., 2009; Weintraub & Beanland, 2006). Artificial inoculation of monocots such as Napier grass with phytoplasmas through graft transmission is not possible. Therefore, NGS is spread primarily through the introduction of infected cuttings by farmers and/or insect vectors carrying the phytoplasma (Koji et al., 2012; Orodho, 2006). In Kenya, the leafhopper *Maiestas banda* Kramer (Hemiptera: Cicadellidae) (Fig. 5.3), was identified as the vector of NGS disease (Obura et al., 2009). It occurs in large numbers on Napier grass, suggesting a major role in the disease epidemics in the region (Koji et al., 2012). This small hemipteran insect with a triangular vertex belongs to the Deltocephalini (subfamily: Deltocephalinae) tribe, which comprises insects that are mostly phytophagous on plant species belonging to the family Poaceae (Satoshi, 1999).

Fig. 5.3 Dorsal view of *Maiestas banda* (Hemiptera: Cicadellidae), a vector of Napier grass stunt phytoplasma in Kenya (not shown to scale) (Obura et al. 2009)

5.5 Management of Napier Grass Stunt Disease in East Africa

Different measures are used by farmers to manage NGS diseases in East Africa. The most commonly used methods are cultural techniques that include roguing, replanting with healthy planting materials, burning of infected plants and cutting (Asudi et al., 2015).

5.5.1 Roguing

Sanitation is one of the many tactics needed for an effective disease management strategy in greenhouses and fields. It includes any practice that aims at preventing the spread of pathogens by removing diseased and asymptomatically infected tissue (Weintraub & Beanland, 2006). Roguing of symptomatic plants is effective in most Napier grass farms and involves the removal of all the plants exhibiting NGS symptoms. Collected symptomatic plants should be burned or buried to avoid further insect vector feeding (Asudi et al., 2015). As the mode of spread of Napier grass is often through vegetative propagation (Orodho, 2006), removal of infected plants reduces phytoplasma inoculum that could be spread to other tillers and lowers the risks of inadvertently used infected grasses for transplants. The removal of infected plants also reduces insect pressure by lowering the density of vector populations (Weintraub & Beanland, 2006). However, the development of visible symptoms on phytoplasma-infected Napier grass plants takes time and symptoms are expressed only after several cuts or grazing by animals, therefore allowing the disease to spread further or to be unintentionally distributed to growers (Asudi et al., 2015; Kabirizi et al., 2007; Orodho, 2006).

Another example of roguing that successfully controls phytoplasma-disease is the removal of phytoplasma reservoir hosts. This process has been used to control X-disease by removal of chokecherry trees (*Prunus virginiana* L. ,Rosaceae) that are reservoir hosts for X-disease, from areas within 150 m from peach (*Prunus persica* L., Rosaceae) orchards (Parker, Palmiter, Gilmer, & Hickey, 1963). Recently, we have found several wild grass species including *Brachiaria brizantha* Hochst ex A. Rich (Poaceae), *Chloris gayana, Coix lacryma-jobi* L. (Poaceae), *Cynodon dactylon* (L.) Pers. (Poaceae), *Digitaria scalarum* Chiov. (Poaceae), *Eleusine indica* (L.) Gaerth (Poaceae), *Enteropogon macrostachyus* (Hochst.) Munro (Poaceae), *Hyparrhenia cymbaria* (L.) Stapf (Poaceae), *H. rufa* (Nees) Stapf, *Sporobolus pyramidalis* Beauv (Poaceae) and *Sorghum versicolor* Andersson (Poaceae) that are natural hosts of NGS phytoplasma (Asudi, Van den Berg, Midega, Schneider, et al., 2016; Obura, Masiga, Midega, et al., 2011). Most of these phytoplasma-infected wild grass hosts show no symptoms and are indistinguishable from healthy conspecific grasses. Therefore, we have recommended their removal in and around Napier grass farms.

The vegetation composition surrounding a field has a fundamental effect on the presence and dispersal of phytoplasma vectors, as insect vectors might use the surrounding plants as alternate hosts, and can move from the wild plants to the cultivated vegetation (Weintraub & Beanland, 2006). Removing potential sources of inoculum in the surrounding areas of crops is an efficient technique for reducing the spread of phytoplasma diseases (Bertaccini & Duduk, 2010). However, cultivated crops such as sorghum (*Sorghum bicolor* (L.) Moench, Poaceae), sugarcane (*Saccharum officinarum* L., Poaceae), rice (*Oryza sativa* L., Poaceae), finger millet (*Eleusine coracana* Gaertn) and maize (*Zea mays* L. Poaceae) grown by the majority of farmers as staple foods or cash crops in the region, are also known to host NGS phytoplasma, without symptom development and may complicate roguing as a method of NGS control (Asudi, Van den Berg, Midega, Pickett, & Khan, 2016).

Generally, smallholder farmers in East Africa practice crop-livestock mixed farming where Poaceae crops are grown together with planted fodder grasses, and wild grasses surrounding the crops are left for livestock (Asudi et al., 2015). Therefore, effective control of NGS by roguing methods should be implemented with the integration of disease-resistant/tolerant Napier grass cultivars and/or with alternative fodder plants.

5.5.2 Use of Napier Grass Stunt Disease Tolerant Cultivars

Many efforts have been made by national and international research institutes to select tolerant or resistant Napier grass cultivars. Mulaa et al. (2010) identified NGS disease tolerant varieties among 200 Napier grass clones collected from farmers' fields in western Kenya and reported tolerance variations in response to the pathogen. Most of the tolerant clones selected by Mulaa et al. (2010) were characterised by poor agronomic values but were used by farmers to control the disease. In Uganda, screening trials by Kawube et al. (2014) resulted into mixed outcomes with most Napier grass clones, except one, being highly susceptible. However, Kawube et al. (2014) recommended the use of one of their moderately susceptible clones (16837) since it had suitable agronomic values consisting of high harvestable biomass and relatively high leaf-to-stem ratio. Wamalwa et al. (2017) screened 65 Napier grass accessions collected worldwide including the tolerant accessions identified by Mulaa et al. (2010) using the NGS phytoplasma transmitting vector *M. banda* (Obura et al., 2009). Wamalwa et al. (2017) reported that most of accessions were highly susceptible. A few were tolerant (accessions 16822, 16817 and 16817) and only one was resistant (accession 16807). Unlike random screening methods of Napier grass varieties in the field employed by Mulaa et al. (2010) and Kawube et al. (2014), the Wamalwa et al. (2017) study was carried out in a controlled environment with netted insect cages placed in a screen house, ensuring effective insect transmission only from infected to healthy plants.

Other Napier grass cultivars such as 'South Africa' and 'Ouma 2' (Fig. 5.4), which appear more tolerant to NGS pathogen were also selected and are currently

Fig. 5.4 Tolerant Napier grass cultivars selected and currently being used to manage Napier grass stunt disease (**a**) 'Ouma 2' cultivar and (**b**) 'South Africa' cultivar (ICIPE 2014)

used in NGS integrated management. These cultivars are currently multiplied and supplied to the farmers for management of NGS in the region (ICIPE, 2014). The tolerant or resistant Napier grass accessions identified by Wamalwa et al. (2017) are likely to stand for even longer period in the farmers' fields. However, most selected Napier grass clones have recently been observed to have lost resistance (Francis Muyekho, personal observation), possibly due to more aggressive strains of the pathogen. Therefore, cultivars identified by Wamalwa et al. (2017) will require further multi-locational trials and evaluation before their resistance is validated.

5.5.3 Quality Seed Production and Distribution Systems

5.5.3.1 Introduction

For many years, Napier grass in East Africa has been grown from cuttings and root splits (Orodho, 2006) without serious challenges from pests, leading to low enforcement of phytosanitary standards on seed production and distribution by government agencies (Ajanga et al., 2012). The emergence of NGS and smut diseases, which are spread through vegetative planting materials have posed a major threat to forage production (Alicai et al., 2004; Asudi et al., 2015; Farrell et al., 2002; Jones et al., 2004; Koji et al., 2012). Vector control using systemic pesticide in fodder crops is very expensive, environmentally unfriendly and also poses health risks through accumulation of hazardous residues in the food chain (Iftikhar, Siddiqui, & Rehman, 2014; Khaniki, 2007; Pandit & Sahu, 2002). Though advances have been made in selecting resistant or tolerant Napier grass cultivars to stunt and smut diseases (sect. 1.5.2, this chapter), availability of disease-free, pest-free and true-to-type planting material is an important prerequisite for achieving the desired yield. Extension agents and even Napier grass experts find it difficult to differentiate Napier grass clones phenotypically (Jorge et al., 2014), which further complicates data observation and records. High-quality vegetatively propagated planting material must not only be free of pathogens, but should also be traceable (Fajardo et al., 2010). Therefore, there is a need to develop a set of universal quality standards for cuttings and root split seed multiplication and distribution systems of planting materials. Growers also need to be made aware of the characteristics and benefits of high quality planting material, a knowledge currently lacking in East Africa.

5.5.3.2 Micropropagation

One way of ensuring quality seed production could be through micropropagation. Using the appropriate growth media and conditions, explants of vegetatively propagated plants can be induced to rapidly produce multiple shoots and new roots. Recently, a tissue culture protocol for micropropagation of Napier grass has been developed (Mwisa, Muyekho, Mulaa, Ndege, & Gitonga, 2013) and this could be used to mass propagate resistant Napier grass cultivars. However, this potential has not been fully exploited—especially in regard to the development of pathogen free-Napier grass plants.

5.5.3.3 Hot Water Treatment

Hot water treatment (HWT) has been used successfully to sanitize phytoplasma and other pathogens-infected plant materials including seeds and storage organs since many decades (Baker, 1962; Caudwell, Larrue, Boudon-Padieu, & McLean, 1997). In grapevines, effective HWT treatments for Flavescence dorée (FD) ranged from 10 h at 40 °C to 10 min at 55 °C. The eggs of the leafhopper vector of FD, *Scaphoideus titanus* Ball (Hemiptera: Cicadellidae), were also killed by immersion at 50 °C for 45 min (Caudwell et al., 1997). Chalak et al. (2013) tested six tissue culture protocols on Chardonnay grapevine infected with Bois Noir phytoplasma and showed that heat or HWT of stem cuttings led to a full eradication of the pathogens from the shoots. In sugarcane, Kaewmanee and Hanboonsong (2011) showed that a combination of HWT and antibiotics can reduce the percentage of sugarcanes infected with sugarcane white leaf (SCWL) phytoplasma. In Kenya, research has shown that incubating root split or cane cuttings in hot water (52 °C) for 2 h followed by withdrawal to cool temperatures, can reduce NGS disease incidences by up to 90% in plants for up to 1 year after replanting (Francis Muyekho, Personal Communication). However, overexposure in hot water leads to bud death, resulting in poor or no germination (Francis Muyekho, personal communication). This result suggests that heat treatment has the ability to kill the phytoplasma and that quality seed could be maintained by HWT or heat treatment. However, the method needs improvement and has not been widely used in the management of NGS disease in the region.

5.5.3.4 Seed Multiplication Protocols and Standards for Napier Grass

Seed multiplication and certification protocols and standards for Napier grass have recently been developed by researchers in Kenya (Ajanga et al., 2012). The aim is to maintain and provide to the public high quality propagating materials that are genetically pure and free from pests, and to ensure that farmers have timely access to quality planting material for higher production and better disease and pest management. Seed certification also helps minimizing the possible outbreaks of new pests.

The full protocol comprises three stages and starts with a primary seed multiplication (Fajardo et al., 2010). The source of seeds for multiplication is scientifically confirmed as tolerant/resistant clones, tissue culture plantlets or hot water treated materials that are completely free from NGS disease. This seed crop is multiplied on supervised farms, usually on research station or well-maintained nurseries and is regularly inspected by experts for vigour, off types, diseases and insect pests until the optimal stage for seed harvesting is reached (16 to 26 weeks after planting). The maximum permissible contamination at the time of harvesting should be less than 0.05% of plants infected with pathogens and less than 5% of plants infested with borers. Subject to good inspection report, a first ratoon crop can be used as seed.

Secondary seed multiplication is usually sourced from the primary seed multiplication. This is normally done on large farms or Agricultural Training Centres. The seed crop is regularly inspected by experts to maintain vigour, varietal purity and to eliminate diseases and insect pests. The maximum permissible infected plants is less than 0.5% for diseases and 5% for borers at the time of seed harvesting. The purpose is to generate as much planting material as possible for sale or supply to farmer groups.

Tertiary seeds are mainly produced on farms and may involve interested farmer groups. The production at this stage may be for seed and feed. The sources of planting material are usually from the secondary multiplication sites. Farmers are trained on seed production and disease management by experts. Where possible, an expert should certify the material before it is given to other farmers. Fields should be established in areas free from Napier re-growth. Inspection and roguing should be performed during the growing season to minimize insect pests, diseases and off type variety. Regular training of stakeholders and monitoring are critical at this stage.

5.5.4 Effects of Defoliation (Cutting) on Management of Napier Grass Stunt Disease

In East Africa, dairy farming is practiced by 80% of smallholder farmers who raise their cattle in a cut-and-carry system, requiring consistent supply of roughage feed that accounts for 70% of the feeding costs (Staal et al., 1998). Forage supply, both in quantity and quality, requires management technologies that can enhance production for faster re-growth, increased tillering and leaf-to-stem ratio. Defoliation (basal cutting height and frequency) is one of the strategies (Jørgensen et al., 2010; Munyasi, Auma, Ngode, & Muyekho, 2015; Tessema, Mihret, & Solomon, 2010).

Most studies show that lenient basal cutting at a height of 5–15 cm with harvesting intervals of 4–10 weeks (depending on the clone or variety) increases dry matter yield and quality of Napier grass. These management practices were highly successful in its production prior to the emergence of NGS diseases (Bayble, Melaku, & Prasad, 2007; Jørgensen et al., 2010; Manyawu, Chakoma, Sibanda, Mutisi, & Chakoma, 2003; Munyasi et al., 2015; Orodho, 2006; Rengsirikul et al., 2011; Tessema et al., 2010). However, since phytoplasma infection interferes with nutrient uptake is and translocation to the shoots (IRPCM Phytoplasma/Spiroplasma Working Team–Phytoplasma Taxonomy Group, 2004; Lee et al., 2000), the interaction between cutting management practices and disease severity need to be assessed.

Many of the studies reported on farmer's perceptions and field observations suggest that severe cutting height combined with very frequent harvesting or grazing, increases disease severity (Asudi et al., 2015; Jorge et al., 2014; Kabirizi et al., 2007; Kawube et al., 2015). Physiologically, it is generally accepted that biomass regeneration after cutting is influenced by certain factors: the amount of residual photosynthetic tissue left after cutting, carbohydrates and other reserves in the remaining stubble stem and root, nutrient and water uptake, and the quantity and activity of remaining meristems (Harris, 1978; Wadi, Ishii, & Idota, 2004). The lower rate of re-growth after cutting observed in the fields suggests that both cutting and NGS disease interferes with nutrient and water uptake and translocation of nutrients to the shoot. For example, in Thailand it has been demonstrated that dry matter yield and tiller numbers increased with increasing cutting height up to a maximum of 20 cm from the base when grass was harvested at monthly intervals (Jørgensen et al., 2010; Lounglawan, Lounglawan, & Suksombat, 2013; Wadi et al., 2004). In Kenya, Munyasi et al. (2015) evaluated the effects at three harvest intervals of 4, 8 and 12 weeks alongside three basal heights of defoliations of 5, 10 and 15 cm in two NGS disease tolerant grasses, Guinea grass (*Panicum maximum* Jacq, Poaceae) and Guatemala grass (*Tripsacum laxum* Nash, Poaceae) (Fig. 5.5), and a new NGS disease tolerant Napier grass cv 'Ouma 3'. Munyasi et al. (2015) showed that a 4-week interval of harvesting with defoliation heights of 10 and 15 cm significantly increased biomass yield in Napier cv 'Ouma 3' as compared to Guatemala and Guinea grasses. It was suggested that the biomass yield increase could be due to increased leafing and tillering of the next re-growths, as observed in other studies (Jørgensen et al., 2010; Lounglawan et al., 2013; Wadi et al., 2004; Wangchuk, Rai, Nirola, Dendup, & Mongar, 2015). However, frequent and intense cuttings decrease yield and persistence of the Napier grass sward, probably due to a reduced carbohydrate level in the stem, hence affecting the persistence of the crop in subsequent harvests, especially during the dry season (Wadi et al., 2004). An extended harvesting interval beyond 10 weeks reduces biomass yield and quality, due to reduced re-growth, leafiness and tillering—mostly attributed to aging of leaves and senescence (Munyasi et al., 2015; Ruiz, Febles, & Diaz, 2012).

Fig. 5.5 Guatemala grass (*Tripsacum laxum*) grown in a farmer's plot in Uganda

Whereas the previous studies showed that lenient-frequent cutting is best in long-term production of high-quality feed of many tropical species, it is generally recognized that the optimal management of pasture grasses differs with plant species (growth habit), location (soil), year (climate), and cultural management practices (Onyeonagu & Asiegbu, 2013; Orodho, 2006). Therefore, further research on defoliation and re-growth for new NGS disease resistant or tolerant clones in combination with other cultural management practices is required for sustainable increase in yields of Napier grass, in different African agro-ecological zones.

5.5.5 Control of the Insect Vector

Even though the insect vector of NGS phytoplasma in Napier grass is known (Obura et al., 2009), the use of environmentally friendly insecticides or pesticides to control the vector is currently not available. The use of conventional insecticides targeting only the vector population might aid in management, therefore reducing intra-crop transmission and disease incidence (Pilkington et al., 2004). However, conventional insecticides do not always control the disease because pathogen transmission might occur faster than the insecticides actions, vectors can fly away from the sprayed fields, and new vectors might come from surrounding habitats (Wally et al., 2004).

Currently there is no report about the use of insecticides for control of *M. banda* as most of the farmers are unaware of the vector presence (Asudi et al., 2015). One control strategies will be to identify insecticides that are environment and human health friendly. As *M. banda* can feed on and transmit NGS phytoplasma to other cultivated crops such as rice, sorghum, maize, finger millet and sugarcane, spraying Napier grass crops could push the leafhopper to change host plants, therefore rendering the use of insecticides very unpredictable and hindering the identification of proper targets for control strategies (Asudi, Van den Berg, Midega, Pickett, et al., 2016).

A range of predators including Miridae (Hemiptera) and spiders prey on adults and nymphs of leafhoppers and planthoppers, and species of parasitoids have been associated with Auchenorrhyncha (Waloff & Jervis, 1987). Spiders and ants are major predators of both adults and nymphs of *M. banda*, especially in wooden insect cages in screenhouses (Asudi, personal observation). However, further research both in the fields and screenhouses need to be carried out to determine the effects of predators on population of *M. banda* and prevalence of NGS disease.

5.5.6 Use of Healthy Napier Grass Plants and Quarantine

The high incidences and severities of NGS disease in many areas occurred because of the introduction of the phytoplasma to farmers' fields through infested planting materials (Koji et al., 2012; Orodho, 2006). Therefore, use of healthy plant materials and quarantine measures could help reduce the phytoplasma spread in the region. Due to its perennial nature and poor maintenance by dairy farmers, Napier grass is harvested and allowed to regrow for several years before replantation, thus contributing to the spread of the disease (Asudi et al., 2015). In addition, the fodder crop is vegetatively propagated through cuttings and exchange of cuttings between farmers, hence greatly contributing to the spread of the disease within and between farms, when diseased but asymptomatic cuttings are used (Koji et al., 2012; Orodho, 2006). Whereas quarantine would be an easier method to minimise disease spread, such measures will require the collaborative development of policies and guidelines in each country in order to be effective.

5.5.7 Use of Alternative Grasses

Besides Napier grass, the use of other fodder and ley grasses such as Guatemala grass, Guinea grass, giant setaria (*Seteria sphacealata* var. *splendida*), fodder sorghums, Rhodes grass (*Chloris gayana*), signal grass (*Bricharia brizantha*) and *Brachiaria* hybrid cultivar (cv) Mulato II, is an alternative in farms that are severely infested by NGS disease. Some of these grasses give comparatively good yields and can be used to replace Napier grass where farm sizes are not a major limiting factor (Asudi et al., 2015; Khan et al., 2014; Mutimura, Ebong, Rao, & Nsahlai, 2016). For instance, Guinea and Guatemala grasses harvested according to a 35-day interval at a height of 10 cm from the ground, cumulatively yields dry matter comparable to

Fig. 5.6 Climate-smart push-pull technology: green leaf *Desmodium* is planted between the rows of sorghum, while *Brachiaria* cultivar Mulato II, instead of Napier grass, is planted around this intercrop (Murage et al. 2015)

other Napier cultivars such as 'Ouma 3' (Munyasi et al., 2015). Therefore can be alternatives to Napier grass in severely NGS disease infested areas.

The NGS tolerant *Brachiaria* hybrid cv Mulato II is used as a drought-tolerant trap crop in climate-smart PPS (push-pull system) and has been used to replace Napier grasses (Fig. 5.6). Mulato II is also better than Napier grass as a potential source of protein and energy. Because of its leafiness and thinner stems the animals tend to eat more and Mulato II is reported to yield much more dry matter than Napier grass (Maass et al., 2015; Murage, Midega, Pittchar, Pickett, & Khan, 2015; Mutimura et al., 2016).

Alternative fodder grasses have been found to produce better yields and have been introduced in some countries such as Tanzania and Uganda to alleviate dairy feed shortages and to improve dairy production (Asudi et al., 2015; Pallangyo, Maeda, & Mkonyi, 2008). However, considering that NGS-tolerant Napier grass may not yield as highly as alternative grasses, there is a need for vigorous breeding and integrated pest management strategies for Napier grass to sustainably increase forage yields on farms.

5.6 Recommendations and Future Research

Napier grass remains the most significant fodder plant of choice for most smallholder dairy farmers in East Africa because of: (i) its persistence after harvest and ability to produce high yields and fast re-growths, (ii) its wider adaptability to different agro-ecological zones and (iii) its potential for integration into the smallholder cereal-dairy production system through the PPS (Maass et al., 2015; Murage et al., 2015; Orodho, 2006). Through years of simple selection and breeding, a number of high yielding and resilient cultivars have been made available for production in different regions of East Africa (Negawo, Teshome, Kumar, Hanson, & Jones, 2017; Orodho, 2006). However, despite the efforts made so far, the emergence of NGS and smut diseases in East Africa coupled with other factors pose serious challenges to its sustained production as a major roughage feed, for the smallholder dairy industry in the region. Therefore, the need for the development of an integrated disease management is urgent. Good agricultural practices, implementation of management strategies and farmer's education through extension packages will help improve NGS disease management.

Current molecular diagnostic tools including use of nested and real-time PCR and LAMP have in the recent past aided detection, identification and characterization of the NGS phytoplasma (Asudi, Van den Berg, Midega, Schneider, et al., 2016; Obura, Masiga, Wachira, Gurja, & Khan, 2011; Wamalwa et al., 2017). In addition, recent development of more efficient, easy-to-use, rapid and sensitive field diagnostic tool of NGS disease will assist in the early detection, monitor and diagnosis of the disease and its spread (Wambua et al., 2017). Seasonal monitoring of the *M. banda* populations and movement from Napier grass to other plant hosts could also provide information on the disease spread, which should facilitate the prediction of future NGS disease outbreaks. Other management strategies could include the use of molecular assisted breeding for the development of resistance in Napier grass (Negawo et al., 2017), or the introduction or editing of genes (Nejat, Rookes, Mantri, & Cahill, 2017). For example, transgenic lines with resistance to different groups of insect pests can be generated using genes from various origins as well as approaches based on RNA interference (RNAi) technology, plant-derived genes for proteinase inhibitors, amylase inhibitors or lectins (Perilla-Henao & Casteel, 2016). Overall, further studies are required to elucidate the important factors involved in the NGS phytoplasma / insect vector / plant host three-way interactions, in order to develop asustainable disease management.

References

Ajanga, S., Muyekho, F., Mulaa, M., Mwenge, S., Kariuki, I., Masibili, M., … Sudi, G. (2012). *Guidelines for production of disease free Napier grass seed (East African agricultural productivity project). 8p*. Nairobi, Kenya: Kenya Agricultural Research Institute.

Alicai, T., Kabirizi, J., Byenkya, S., Kayiwa, S., & Ebong, C. (2004). Assessment of the magnitude and farmers' management practices of the elephant grass stunting disorder in Masaka District. Namulonge Agricultural and Animal Production Research Institute, Kampala, Uganda.

Anindo, D., & Potter, H. (1994). Seasonal variation in productivity and nutritive value of Napier grass at Muguga, Kenya. *East African Agricultural and Forestry Journal, 59*(3), 177–185.

Arocha, Y., & Jones, P. (2009). Phytoplasma diseases of the Gramineae. In P. G. Weintraub & P. Jones (Eds.), *Phytoplasmas: Genomes, plant hosts and vectors* (pp. 170–187). Preston, UK: MPG Books Group.

Asudi, G. O., Van den Berg, J., Midega, C. A., Pickett, J. A., & Khan, Z. R. (2016). The significance of Napier grass stunt phytoplasma and its transmission to cereals and sugarcane. *Journal of Phytopathology, 164*(6), 378–385.

Asudi, G. O., Van den Berg, J., Midega, C. A., Pittchar, J., Pickett, J. A., & Khan, Z. R. (2015). Napier grass stunt disease in East Africa: Farmers' perspectives on disease management. *Crop Protection, 71*, 116–124.

Asudi, G. O., Van den Berg, J., Midega, C. A., Schneider, B., Seemüller, E., Pickett, J. A., … Khan, Z. R. (2016). Detection, identification, and significance of phytoplasmas in wild grasses in East Africa. *Plant Disease, 100*(1), 108–115.

Baker, R. (1962). Thermotherapy of planting material. *Phytopathology, 52*, 1244–1255.

Bayble, T., Melaku, S., & Prasad, N. (2007). Effects of cutting dates on nutritive value of Napier (*Pennisetum purpureum*) grass planted sole and in association with Desmodium (*Desmodium intortum*) or Lablab (*Lablab purpureus*). *Livestock Research for Rural Development, 19*(1), 120–136.

Bertaccini, A., & Duduk, B. (2010). Phytoplasma and phytoplasma diseases: A review of recent research. *Phytopathologia Mediterranea, 48*(3), 355–378.

Bingi, S., & Tondel, F. (2015). Recent developments in the dairy sector in Eastern Africa. *European Centre for Development Policy Management, Briefing Note* (78-September 2015), 19p.

Caudwell, A., Larrue, J., Boudon-Padieu, E., & McLean, G. (1997). Flavescence dorée elimination from dormant wood of grapevines by hot-water treatment. *Australian Journal of Grape and Wine Research, 3*(1), 21–25.

Chalak, L., Dekwaneh, B., Elbitar, A., Mourad, N., Mortada, C., & Choueiri, E. (2013). Elimination of grapevine Bois Noir Phytoplasma by tissue culture coupled or not with heat therapy or hot water treatment. *Advances in Crop Science and Technology, 1*(2), 107.

Cook, S. M., Khan, Z. R., & Pickett, J. A. (2007). The use of push-pull strategies in integrated pest management. *Annual Review of Entomology, 52*, 375–400.

Eden-Green, S. (2010). Phytoplasmas: Genomes, plant hosts and vectors. *Plant Pathology, 59*(6), 1177–1178.

Fajardo, J., Lutaladio, N., Larinde, M., Rosell, C., Roca, W., & Chujoy, E. (2010). *Quality declared planting material: protocols and standards for vegetatively propagated crops* (Technical paper 195). Retrieved from FAO, Rome (Italia). www.fao.org/3/a-i1195e.pdf

Farrell, G., Simons, S., & Hillocks, R. (2002). Pests, diseases and weeds of Napier grass, Pennisetum purpureum: A review. *International Journal of Pest Management, 48*(1), 39–48.

Harris, W. (1978). Defoliation as a determinant of the growth, persistence and composition of pasture. In J. Wilson (Ed.), *Plant relations in pastures* (pp. 67–85). Brisbane, Australia: Commonwealth Scientific and Industrial Research Organization.

Humphreys, L. R. (1994). *Tropical forages: Their role in sustainable agriculture*. Harlow, UK: Longman Scientific and Technical.

ICIPE. (2014). Solving Napier stunt disease to save the smallholder dairy sector in East Africa – A success story. http://www.push-pull.net/napier_stunt_brochure.pdf: ICIPE, Nairobi, Kenya.

Iftikhar, B., Siddiqui, S., & Rehman, S. (2014). Assessment of the dietary transfer of pesticides to dairy milk and its effect on human health. *African Journal of Biotechnology, 13*(3).

IRPCM Phytoplasma/Spiroplasma Working Team–Phytoplasma Taxonomy Group. (2004). 'Candidatus Phytoplasma', a taxon for the wall-less, non-helical prokaryotes that colonize plant phloem and insects. *International Journal of Systematic and Evolutionary Microbiology, 54*(4), 1243–1255.

Jones, P., Devonshire, B., Holman, T., & Ajanga, S. (2004). Napier grass stunt: A new disease associated with a 16SrXI group phytoplasma in Kenya. *Plant Pathology, 53*(4), 519–519.

Jorge, A., Lukuyu, B., Marita, C., Mwangi, D., Kinuthia, E., Baltenweck, I., …Poole, J. (2014). *Assessing the uptake and disease impact of Napier grass in Kenya* (D. Odongo Ed.). Addis Abeba, Ethiopia: International Livestock Research Institute.

Jørgensen, S., Pookpakdi, A., Tudsri, S., Stölen, O., Ortiz, R., & Christiansen, J. L. (2010). Cultivar-by-cutting height interactions in Napier grass (*Pennisetum purpureum* Schumach) grown in a tropical rain-fed environment. *Acta Agriculturae Scandinavica Section B–Soil and Plant Science, 60*(3), 199–210.

Kabirizi, J., Nielsen, S., Nicolaisen, M., Byenkya, S., & Alacai, T. (2007, October 27–31). *Napier stunt disease in Uganda: farmers' perceptions and impact on fodder production.* Paper presented at the 8th African Crop Science Society Conference, El-Minia, Egypt.

Kaewmanee, C., & Hanboonsong, Y. (2011). Evaluation of the efficiency of various treatments used for sugarcane white leaf phytoplasma control. *Bulletin of Insectology, 64*(Suppl 1), S197–S198.

Kawube, G., Alicai, T., Otim, M., Mukwaya, A., Kabirizi, J., & Talwana, H. (2014). Resistance of Napier grass clones to Napier grass stunt disease. *African Crop Science Journal, 22*(3), 229–236.

Kawube, G., Talwana, H., Nicolaisen, M., Alicai, T., Otim, M., Kabirizi, J., … Nielsen, S. L. (2015). Napier grass stunt disease prevalence, incidence, severity and genetic variability of the associated phytoplasma in Uganda. *Crop Protection, 75*, 63–69.

Khan, Z. R., Midega, C. A., Bruce, T. J., Hooper, A. M., & Pickett, J. A. (2010). Exploiting phytochemicals for developing a 'push–pull' crop protection strategy for cereal farmers in Africa. *Journal of Experimental Botany, 61*(15), 4185–4196.

Khan, Z. R., Midega, C. A., Nyang'au, I., Murage, A., Pittchar, J., Agutu, L., … Pickett, J. A. (2014). Farmers' knowledge and perceptions of the stunting disease of Napier grass in Western Kenya. *Plant Pathology, 63*(6), 1426–1435.

Khaniki, G. (2007). Chemical contaminants in milk and public health concerns: A review. *International Journal of Dairy Science, 2*(2), 104–115.

Koji, S., Fujinuma, S., Midega, C. A., Mohamed, H. M., Ishikawa, T., Wilson, M. R., … Pickett, J. A. (2012). Seasonal abundance of Maiestas banda (Hemiptera: Cicadellidae), a vector of phytoplasma, and other leafhoppers and planthoppers (Hemiptera: Delphacidae) associated with Napier grass (*Pennisetum purpureum*) in Kenya. *Journal of Pest Science, 85*(1), 37–46.

Lee, I.-M., Davis, R. E., & Gundersen-Rindal, D. E. (2000). Phytoplasma: Phytopathogenic mollicutes. *Annual Reviews in Microbiology, 54*(1), 221–255.

Lounglawan, P., Lounglawan, W., & Suksombat, W. (2013, November 23–24). Effect of cutting interval and cutting height on yield and chemical composition of King Napier grass (*Pennisetum purpureum* x *Pennisetum americanum*). Paper presented at the 4th international conference on agriculture and animal science (CAAS 2013), Phuket, Thailand.

Maass, B. L., Midega, C. A., Mutimura, M., Rahetlah, V. B., Salgado, P., Kabirizi, J. M., … Rao, I. M. (2015). Homecoming of Brachiaria: Improved hybrids prove useful for African animal agriculture. *East African Agricultural and Forestry Journal, 81*(1), 71–78.

Manyawu, G., Chakoma, C., Sibanda, S., Mutisi, C., & Chakoma, I. (2003). The effect of harvesting interval on herbage yield and nutritive value of Napier grass and hybrid Pennisetums. *Asian Australasian Journal of Animal Sciences, 16*(7), 996–1002.

Midega, C. A., Khan, Z. R., Amudavi, D. M., Pittchar, J., & Pickett, J. A. (2010). Integrated management of Striga hermonthica and cereal stemborers in finger millet (*Eleusine coracana* (L.) Gaertn.) through intercropping with *Desmodium intortum. International Journal of Pest Management, 56*(2), 145–151.

Mulaa, M., Awalla, B., Hanson, J., Proud, J., Cherunya, A., Wanyama, J., … Muyekho, F. (2010, November 8–12). *Stunting disease incidence and impact on Napier grass (Pennisetum purpureum Schumach) in western Kenya.* Paper presented at the 12th KARI biennial conference, Nairobi, Kenya.

Munyasi, J., Auma, E., Ngode, L., & Muyekho, F. (2015). Evaluation of biomass yield of selected fodder species in relation to frequency of harvest alongside defoliation heights in Western Kenya. *Journal of Agricultural Extension and Rural Development, 7*(8), 257–262.

Murage, A., Midega, C. A., Pittchar, J., Pickett, J. A., & Khan, Z. R. (2015). Determinants of adoption of climate-smart push-pull technology for enhanced food security through integrated pest management in eastern Africa. *Food Security, 7*(3), 709–724.

Mutimura, M., Ebong, C., Rao, I. M., & Nsahlai, I. V. (2016). Change in growth performance of crossbred (Ankolex Jersey) dairy heifers fed on forage grass diets supplemented with commercial concentrates. *Tropical Animal Health and Production, 48*(4), 741–746.

Mwisa, P., Muyekho, F., Mulaa, M., Ndege, D., & Gitonga, R. (2013, November 12–13). *Effects of Napier accessions in the validation of tissue culture protocol for mass production of disease free planting materials*. Paper presented at the EAAPP Mini-Conference and stakeholders' open day, RDCoE/Morendat Training Centre, Naivasha, Kenya.
Negawo, A. T., Teshome, A., Kumar, A., Hanson, J., & Jones, C. S. (2017). Opportunities for Napier grass (*Pennisetum purpureum*) improvement using molecular genetics. *Agronomy, 7*(2), 28.
Nejat, N., Rookes, J., Mantri, N. L., & Cahill, D. M. (2017). Plant–pathogen interactions: Toward development of next-generation disease-resistant plants. *Critical Reviews in Biotechnology, 37*(2), 229–237.
Nielsen, S. L., Ebong, C., Kabirizi, J., & Nicolaisen, M. (2007). First report of a 16SrXI group phytoplasma (Candidatus Phytoplasma oryzae) associated with Napier grass stunt disease in Uganda. *Plant Pathology, 56*(6), 1039–1039.
Obura, E., Masiga, D., Midega, C. A., Otim, M., Wachira, F., Pickett, J. A., … Khan, Z. R. (2011). Hyparrhenia grass white leaf disease, associated with a 16SrXI phytoplasma, newly reported in Kenya. *New Disease Reports, 24*, 17.
Obura, E., Masiga, D., Wachira, F., Gurja, B., & Khan, Z. R. (2011). Detection of phytoplasma by loop-mediated isothermal amplification of DNA (LAMP). *Journal of Microbiological Methods, 84*(2), 312–316.
Obura, E., Midega, C. A., Masiga, D., Pickett, J. A., Hassan, M., Koji, S., … Khan, Z. R. (2009). Recilia banda Kramer (Hemiptera: Cicadellidae), a vector of Napier stunt phytoplasma in Kenya. *Naturwissenschaften, 96*(10), 1169–1176.
Onyeonagu, C., & Asiegbu, J. (2013). Harvest frequency effect on plant height, grass tiller production, plant cover and percentage dry matter production of some forage grasses and legumes in the derived savannah, Nigeria. *African Journal of Agricultural Research, 8*(7), 608–618.
Orodho, A. (2006). *The role and importance of Napier grass in the smallholder dairy industry in Kenya*. Retrieved from http://www.fao.org/ag/agp/agpc/doc/newpub/napier/napier_kenya.htm
Pallangyo, B., Maeda, C., & Mkonyi, S. (2008). *Napier grass (Pennisetum purpurium Schum.) diversity, uses and diseases in Eastern, Northern and lake zones of Tanzania*. Paper presented at the 1st National Plant Protection Advisory Committee ad hoc meeting, sugarcane research institute, Kibaha Tanzania, September 19, 2008.
Pandit, G., & Sahu, S. (2002). Assessment of risk to public health posed by persistent organochlorine pesticide residues in milk and milk products in Mumbai, India. *Journal of Environmental Monitoring, 4*(1), 182–185.
Parker, K., Palmiter, D., Gilmer, R., & Hickey, K. (1963). X-disease of peach and cherry trees and its control. *Cornell Extension Bulletin, 1100*, 1–12.
Perilla-Henao, L. M., & Casteel, C. L. (2016). Vector-borne bacterial plant pathogens: Interactions with hemipteran insects and plants. *Frontiers in Plant Science, 7*, 1163.
Pickett, J. A., Woodcock, C. M., Midega, C. A., & Khan, Z. R. (2014). Push–pull farming systems. *Current Opinion in Biotechnology, 26*, 125–132.
Pilkington, L., Gurr, G., Fletcher, M., Elliott, E., Nikandrow, A., & Nicol, H. (2004). Reducing the immigration of suspected leafhopper vectors and severity of Australian lucerne yellows disease. *Australian Journal of Experimental Agriculture, 44*(10), 983–992.
Rengsirikul, K., Ishii, Y., Kangvansaichol, K., Pripanapong, P., Sripichitt, P., Punsuvon, V., … Tudsri, S. (2011). Effects of inter-cutting interval on biomass yield, growth components and chemical composition of napiergrass (Pennisetum purpureum Schumach) cultivars as bioenergy crops in Thailand. *Grassland Science, 57*(3), 135–141.
Ruiz, T., Febles, G., & Diaz, H. (2012). Plantation distance, frequency and cutting height on the biomass production of *Tithonia diversifolia* collection 10 during the year. *Cuban Journal of Agricultural Science, 46*(4), 423–426.
Satoshi, K. (1999). The phylogeny of the genera in the tribes Deltocephalini, Paralimnini, and their allies (Homoptera, Cicadellidae, Deltocephalinae). *Esakia: Occasional Papers of the Hikosan Biological Laboratory in Entomology, 39*, 65–108.

Skerman, P., & Riveros, F. (1989). *Tropical grasses*. Rome, Italy: FAO.

Staal, S., Chege, L., Kenyanjui, M., Kimari, A., Lukuyu, B., Njubi, D., … Wambugu, M. (1998). A cross-sectional survey of Kiambu district for the identification of target groups of smallholder dairy producers. In *KARI/ILRI collaborative project research report*. Nairobi, Kenya: Kenyan Agricultural Research Institute.

Staal, S., Kruska, R., Baltenweck, I., Kenyanjui, M., Wokabi, A., Njubi, D., … Thorpe, W. (1999). *Combined household and GIS analysis for smallholder production systems: An application to intensifying smallholder dairy systems in central Kenya.* Paper presented at the Third International Symposium on Systems Approaches for Agricultural Development, (SAAD-III), Lima, Peru, November 8–10, 1999.

Tessema, Z., Mihret, J., & Solomon, M. (2010). Effect of defoliation frequency and cutting height on growth, dry-matter yield and nutritive value of Napier grass (*Pennisetum purpureum* (L.) Schumach). *Grass and Forage Science, 65*(4), 421–430.

Wadi, A., Ishii, Y., & Idota, S. (2004). Effects of cutting interval and cutting height on dry matter yield and overwintering ability at the established year in Pennisetum species. *Plant Production Science, 7*(1), 88–96.

Wally, O., Daayf, F., Iranpour, M., Adam, L., Elliott, B., Shinners-Carnelley, T., … Khadhair, A. (2004). Incidence and molecular detection of yellows-type disease in carrots, associated with leafhoppers in southern Manitoba, Canada. *Canadian Journal of Plant Pathology, 26*(4), 498–505.

Waloff, N., & Jervis, M. (1987). Communities of parasitoids associated with leafhoppers and planthoppers in Europe. In *Advances in ecological research* (Vol. 17, pp. 281–376). San Diego, CA: Elsevier.

Wamalwa, N., Midega, C. A., Ajanga, S., Omukunda, N., Muyekho, F., Asudi, G., … Khan, Z. R. (2017). Screening Napier grass accessions for resistance to Napier grass stunt disease using the loop-mediated isothermal amplification of DNA (LAMP). *Crop Protection, 98*, 61–69.

Wambua, L., Schneider, B., Okwaro, A., Wanga, J. O., Imali, O., Wambua, P. N., … Masiga, D. (2017). Development of field-applicable tests for rapid and sensitive detection of Candidatus Phytoplasma oryzae. *Molecular and Cellular Probes, 35*, 44–56.

Wangchuk, K., Rai, K., Nirola, H., Dendup, C., & Mongar, D. (2015). Forage growth, yield and quality responses of Napier hybrid grass cultivars to three cutting intervals in the Himalayan foothills. *Tropical Grasslands-Forrajes Tropicales, 3*(3), 142–150.

Weintraub, P. G., & Beanland, L. (2006). Insect vectors of phytoplasmas. *Annual Review of Entomology, 51*, 91–111.

Chapter 6
Mineral and Plant Oils as Management Tools to Control Insect Vectors of Phytoplasmas

Philippe Giordanengo, Sébastien Boquel, Julien Saguez, and Charles Vincent

Abstract Phytoplasmas are small unculturable and wall-less prokaryotes that are transmitted by phloem-feeder insects. Depending on the economic value of the plants they infect and/or the lifespan of the infected plants, phytoplasma infection can have devastating economic impacts in term of fruit/seed production and quality. To date, there is no commercially available methods to control phytoplasma in plants or insects. Phytoplasma disease management is frequently achieved by spraying insecticides to control the insect vector population. However, this method is not environmentally friendly and have several shortcomings such as favoring insect resistance. Plant and mineral oils have been used to control insect pests for many decades, whether as solvent, adhesives or because of their direct insecticidal properties. This chapter reviews elements relevant to direct or indirect management of insect vectors of phytoplasmas using mineral or plant oils, focusing on oil structure, composition and properties as well as mode of actions. Factors influencing their efficacy as insectides is also discussed.

Keywords Mineral and plant oil · Insecticide · Insect vector · Phytoplasma disease management

P. Giordanengo (✉)
Université Côte d'Azur, INRA, CNRS, ISA, Sophia Antipolis Cedex, France

Université de Picardie Jules Verne, Amiens Cedex, France
e-mail: philippe.giordanengo@inra.fr

S. Boquel · J. Saguez
CÉROM, Québec, Canada
e-mail: sebastien.boquel@cerom.qc.ca; julien.saguez@cerom.qc.ca

C. Vincent
Saint-Jean-sur-Richelieu Research and Development Centre, Agriculture Agri-Food Canada, Saint-Jean-sur-Richelieu, Quebec, Canada
e-mail: charles.vincent2@canada.ca

C. Y. Olivier et al. (eds.), *Sustainable Management of Phytoplasma Diseases in Crops Grown in the Tropical Belt*, Sustainability in Plant and Crop Protection 12,
https://doi.org/10.1007/978-3-030-29650-6_6

6.1 Introduction

Phytoplasmas are small unculturable and wall-less prokaryotes (related to bacteria) belonging to the Phylum Tenericutes, the class Mollicutes, and the Family Acholeplasmataceae. They are obligate parasites that need host plants and insect vectors. While feeding on infected plants, phloem-feeding pests acquire phytoplasmas and vector them after a latent period. Phytoplasmas are transmitted in a persistent manner. Depending on the economic value of the infected plants and/or lifespan of plants, phytoplasma infection have short or long-term impacts.

Phytoplasmas, such as the Flavescence Dorée (*Candidatus Phytoplasma vitis*) and Bois Noir (*Candidatus Phytoplasma solani*) of grapevines, are quarantine organisms that are strictly regulated to prevent further dissemination (Olivier et al., 2012). Management of diseases and their vector pests are therefore important concerns and challenges for growers and regulatory agencies. For example, in northwest Italy, a control program for the Flavescence Dorée costed €1.5 million per year between 1999 and 2003, while in 2005 grower compensation for yield losses and replanting was €34 million (Belli et al., 2010). Other phytoplasmas, such as aster yellows, are not regulated although they may have some economic impact. Regulation of phytoplasmas is usually implemented at the national level and depends on a number of factors, notably potential economic losses.

Several species of Auchenorrhyncha (planthoppers, leafhoppers, treehoppers, cicadas and spittlebugs) and Sternorrhyncha (mealybugs, aphids, whiteflies and psyllids) are important insect vectors of plant pathogens. Their piercing-sucking mouthparts allow them to collect organic fluids while causing physical damage when they are inserted in plant tissues. Moreover, several piercing-sucking insects are tissue-specific feeders that acquire and transmit pathogens directly within plant tissues. The epidemiology of such pathogens depends upon the population dynamics, dispersal and feeding behavior of the vectors, and relies on a complex set of interactions involving the pathogen, vector, plant and environment.

To date, no curative treatment for managing plant pathogen epidemics caused by phytoplasmas or viruses has been successfully implemented once the pathogen is in the plant (Perring et al., 1999). Management is frequently achieved indirectly and preventively by targeting vectors with synthetic insecticides that provide an economic, convenient and timely management method. However, synthetic insecticides have several shortcomings. Resistance to insecticides may develop, leading to ineffective results (Bass & Field, 2011; Field et al., 1999; Wang et al., 2018). Insecticides may induce hyperactivity of insect vectors which in turn enhance pathogen transmission (Simons & Zitter, 1980). Finally, insecticides may not kill the insect vector fast enough as some pathogens (especially non-persistent viruses) may be transmitted following a single insertion of insect vector mouthparts within plant tissues.

Concerns for human health and environment have fueled increasing consumer demand for pesticide-free agricultural products as well as non-chemical alternatives. In addition, stringent regulations preclude the use of synthetic insecticides against insect vectors in some agricultural and urban settings (see Franco-Lara and Perilla-Henso, this volume). Several cities that have completely banned the use of synthetic insecticides still must manage pests of ornamentals with technologies that comply with regulations aiming for minimal impact to human beings and urban environments.

Among non-chemical alternatives, physical control methods offer an array of science-based technologies that can be successfully implemented, including mineral and plant oils (Vincent et al., 2003). While mineral oils have been successfully used since the end of the nineteenth century to manage arthropod pests in several crops, plant oils have been recently considered as natural alternatives to petroleum-based pesticides (Agnello, 2002; Singh et al., 2015). Weekly application of mineral or plant oil was an effective tool for managing a wide range of insect-transmitted pathogens (Bradley et al., 1966; Butler & Henneberry, 1990; Orozco-Santos et al., 2016; Ouyang et al., 2013; Powell, 1992; Powell et al., 1998; Samara et al., 2016; Simons et al., 1977). Currently, oils are recommended to manage several pests such as mites, scale insects, aphids and leafhoppers in various crops, notably ornamentals (University of California, 2018) and citrus (Beattie & Hardy, 2005).

In plant protection, oils have been historically used as adhesives in the formulation of sprays and dusts, solvents acting as carriers for toxicants, and active ingredients having insecticidal properties (Smith, 1952). An important property of oils in favor of their use is their non-toxicity towards mammals, including humans (Beattie & Hardy, 2005) even though they can be phytotoxic depending on dose or plant species.

This chapter reviews elements relevant to direct or indirect management of insect vectors of phytoplasmas using mineral or plant oils. The scientific literature with respect to mineral oils is much more abundant than that of plant oils. Most of the published literature deals with the use of oils to manage insects that vector plant pathogens. We excluded essential oils from this chapter because they were reviewed by Regnault-Roger et al. (2012), and Nollet & Rathore (2017), and their insecticidal activities are essentially due to the presence of terpenoids and phenylpropanoids, compounds that are absent or in very low concentrations in plant oils extracted from seeds. Neem oil (Schmutterer, 1990), in which the azadirachtin terpenoid is considered the most important active insecticidal ingredient, was excluded for the same reasons. Finally, as the literature on oils and insects vectors of viruses is more abundant, we draw on it to discuss some points that are relevant to vectoring phytoplasmas.

6.2 Oil Composition and Properties

6.2.1 *Mineral Oil Structure and Activities*

Mineral oils are petroleum-derived products composed of carbon atoms. Median equivalent *n*-paraffin carbon number, *n*C*y*, where *n* denotes a normal paraffin (alkane) molecule, and *y* denotes the number of carbon atoms, is used to describe the properties of mineral oils. Mineral oils are essentially composed of a mixture of paraffinic, asphaltic and aromatic hydrocarbons in various proportions typically ranging from *n*C16 to *n*C32 (Ouyang et al., 2013). Because of the number and complexity of components present in mineral oils, their chemical classification is tentative and based on the main components such as paraffinic chains, naphtenic rings, aromatic rings, unsaturated hydrocarbons, or asphaltic materials (Agnello, 2002) (Fig. 6.1).

Molecular structures and weight of mineral oils affect their physico-chemical properties such as saturation, viscosity, distillation temperature (Al-Mrabeh et al., 2010) and bioactivities. Their size and number of carbon composition are important properties conferring insecticidal activities. Paraffinic oils with >60% of carbon atoms in paraffin chains (PC) are more effective than naphthenic oils (< 55% PC) (Agnello, 2002). Pesticidal efficacy of mineral oils increases with increasing distillation temperature, which is related to molecular structure and weight (Liu et al., 2001).

Mineral oils having a lower range of paraffinic hydrocarbons (up to decane) produced fumigant action in insects (Taverner et al., 2001), while mineral oils having heavy fractions remained on arthropod cuticles without insecticidal effect (Agnello, 2002). Small molecular size fractions have reduced insecticidal efficacy, while

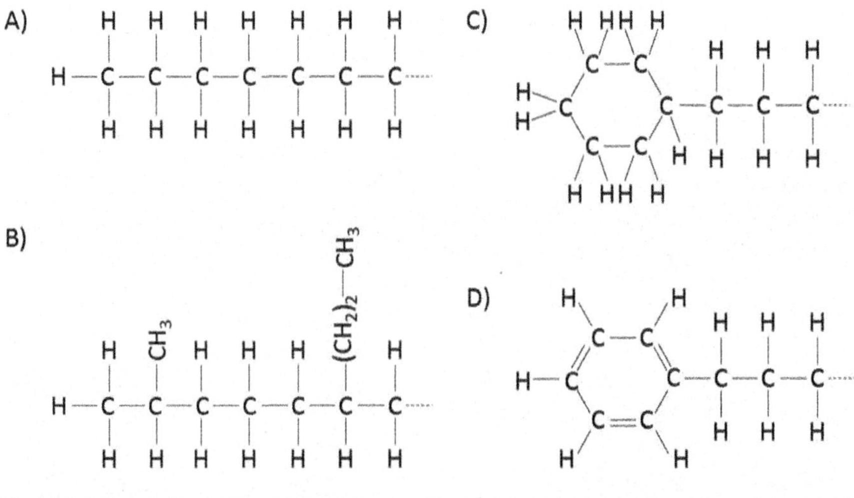

Fig. 6.1 Mineral base oils: (**a**) straight paraffin, (**b**) branched paraffin, (**c**) naphthene, (**d**) aromatic

larger-sized fractions could be phytotoxic (Adhvaryu et al., 2005; Davidson et al., 1991; Furness et al., 1987).

A *nC21* mineral oil caused complete defoliation while soybean oil primarily composed of *nC18* and *nC16* (Ivanov et al., 2010) caused only slight leaf scorching (Lancaster et al., 2002). Oxidation of unsaturated molecules (sulfonated residue), especially aromatic hydrocarbons, forms acidic compounds that cause membrane disruption inducing cell death. Acute effects like burns and necrotic lesions shortly appear after mineral oil spraying on plants (Tan et al., 2005).

Although applications of mineral oils were once limited to early-season or dormant sprays, notably applied on apple trees to suffocate overwintered mite eggs (Davidson et al., 1991) to avoid oil injury to green plant tissues caused by their significant content of phytotoxic sulfonated residue, modern mineral oils used in horticulture are highly refined (92 to 99% pure), generally with linear molecules composed of hydrocarbons with a range between *nC21* and *nC24* that combine good insecticidal activity with low phytotoxicity (Buteler & Stadler, 2011).

6.2.2 *Plant Oils*

Primarily derived from seeds, plant oils are acylglycerides that mainly consist of glycerol and fatty acids (Adhvaryu et al., 2005). Due to the higher molecular weight of the triacylglycerol molecule, plant oils have much larger molecules. For example, rapeseed oil composition has median *nC55* while mineral oil has median *nC24* (Nicetic et al., 2011), conferring a higher boiling point, greater viscosity and lower volatility (Adhvaryu et al., 2005).

Differences in plant oils result from the composition on the fatty acids attached to the three sites on the glycerol, that can differ in the number of carbons, in the hydrocarbon chain and in the number of double bonds between carbons (saturated or unsaturated). Large variations in oil composition are found across plant species, the most commonly fatty acids being palmitic (16:0), stearic (18:0), oleic (18:1) and linoleic (18:2) acids (Sams & Deyton, 2002) (Fig. 6.2). Despite strong phytotoxic reaction symptoms in some cases (Wojdyła, 2002), plant oils are preferred as horticultural treatments because they are biodegradable, non-toxic, perceived as eco-responsible and, unlike mineral oils, renewable resources (Battersby et al., 1998). In addition, plant oils are good solvents of polar contaminants and additive molecules (Adhvaryu et al., 2005). However, plant oils are sensitive to hydrolysis, thermal degradation and oxidation because of the numerous unsaturated bonds of their fatty acids (Schneider, 2006). Lower release of volatile fractions explained the lower deterrent effect of plant oils on insects (Nicetic et al., 2011), and repellent effects were correlated with the carbon mass percentages of C*22*, C*23* and C*24* (Ouyang et al., 2013).

Even though plant oils have different molecular structures than mineral oils, it appears that the different physical characteristics resulting from different molecular structures, namely, a high boiling point and viscosity, may explain lower efficacy

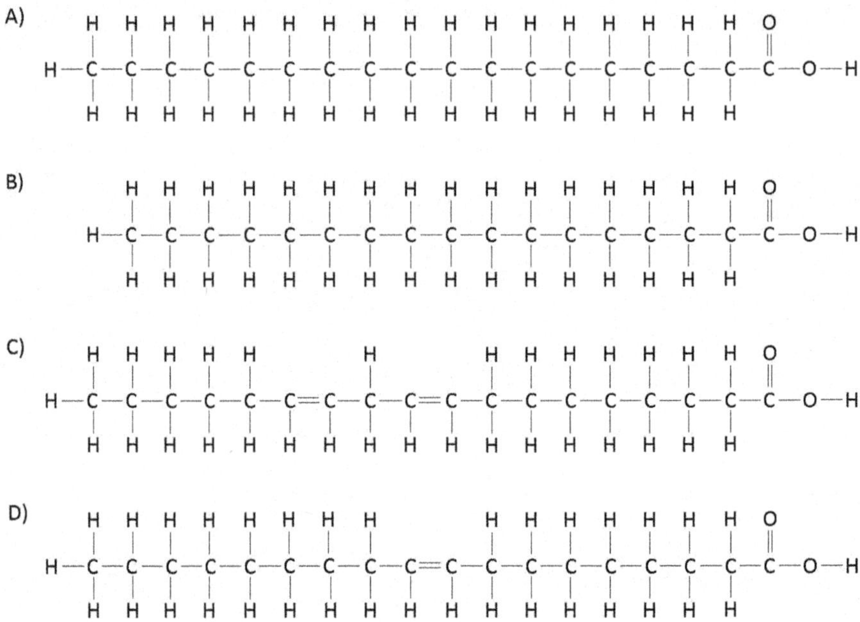

Fig. 6.2 Common plant oil fatty acids: (**a**) stearic acid, (**b**) palmitic acid, (**c**) linoleic acid, (**d**) oleic acid

against insects compared to mineral oils (Nicetic et al., 2011). Plant oils' toxicity to insects depends on their fatty-acid content, corresponding to the unsaturated hydrocarbons of mineral oils (Agnello, 2002). Their pesticidal properties can be enhanced when combined with detergents (Agnello, 2002).

Variations in their efficiency are also related to the species of the protected plant, the nature of oil used, the applied concentration, and the frequency and conditions of the application affecting droplet size and coverage (Wróbel, 2012).

6.3 Modes of Action of Oils

In contrast to synthetic insecticides that generally have one target site, oils may exert their effects through various mechanisms. When sprayed in fields, oils may directly affect the insect physiology by interacting with their cuticle and their respiratory organs. Oils may also have indirect effects by disturbing insect oviposition and feeding behaviors. Oils have been shown to modulate the pathogen transmission during the insect-plant interaction as well as plant defense mechanisms. Other indirect effects could be due to the interaction with plant cuticle and waxes.

6.3.1 Physiological Effects on Insect Vectors

The insect cuticle is a complex mixture of long-chain compounds, including hydrocarbons, wax esters, free fatty acids, alcohols, ketones, aldehydes, and cyclic compounds, in particular involved in preventing water losses (Dekker et al., 2000). As shown in cotton boll weevils (*Anthonomus grandis* Boh.,Coleoptera: Curculionidae), mineral and plant oils can compete with some components of the wax layer and induce structural changes leading to softening the insect cuticle. Teratogenic effects on the epidermis and aberrant molts were observed in *A. grandis* when treated with sub-lethal doses of mineral or plant oils (Stadler et al., 2002). Furthermore, a correlation was also found between cuticle softening and oil toxicity in laboratory bioassays, where a greater softening was associated with increased mortality (Stadler et al. 2002).

Because the insect respiratory system is connected to the cuticle, oils may enter the insect body through the spiracles. Mineral and plant oils were traditionally used as suffocants against small insects, such as aphids, mites and scale insects, inducing mortality on contact (Davidson et al., 1991; Riehl, 1969; Rongai et al., 2008). Observations showed a *n*C*15* mineral oil formulation rapidly penetrate the tracheal system of the light brown apple moth (*Epiphyas postvittana* Walker, Lepidoptera: Tortricidae) (Taverner et al., 2001). Taverner et al. (2001) suggested that saturated hydrocarbons have a direct effect on the neural lipid membranes, increasing membrane thickness and tension, thus altering its permeability to ions exchange. The possibility that oil blocks gas exchange inducing carbon dioxide accumulation leading to narcosis cannot be ruled out. However, death by suffocation through oil by blocking the spiracles and accumulating in the tracheae may take hours in insects (de Ong et al., 1927) and does not explain rapid death, within minutes (Martoub et al., 2011; Najar-Rodríguez et al., 2007a).

Stadler and Buteler (2009) concluded that suffocation occurs when insects are over-sprayed or dipped in oil. Upon contact with the cuticle, capillary forces and an array of physical interactions exert changes in the melting point and permeability of cuticle waxes in such that waterproofing is affected and oil would reach various lipophilic tissues. Narcosis is generally associated with decreased activity through anoxia, while treated *E. postvittana* showed increased nervous activity (Taverner et al., 2001). *In vitro*, oils have been shown to penetrate cell membranes and accumulate into the cytoplasm and cause cell dehydration and DNA condensation (Najar-Rodríguez et al., 2007a). Additional symptoms, such as darkening of the hemolymph and flaccidity of treated insects, suggest that oils disrupt other systems through membrane alterations (Taverner et al., 2001). The rapid knockdown of the cotton aphid (*Aphis gossypii* Glover, Hemiptera: Aphididae) and the oriental leafworm (*Spodoptera litura* F., Lepidoptera: Noctuidae) treated with a commercial *n*C*24* mineral oil and the absence of any sign of oil penetration into tracheae along with a range of cellular disruptions support such hypothesis (Najar-Rodriguez et al., 2008). However, the extent of tracheal penetration varies with oils and it was suggested that lower surface tension between a light hydrocarbon mixture (<*n*C15) and the tracheal lining confer a better penetration than heavier oils (>*n*C23) (Taverner, 2002; Taverner et al., 2001).

6.3.2 Behavioral Effects on Insect Vectors

Mineral oil sprayed on plants was reported to induce a lack of attractiveness of treated plants for the potato aphid (*Macrosiphum euphorbiae* Thomas, Hemiptera: Aphididae). As the repellent effect was observed only shortly after plant treatment and oil itself did not induce repellency, Ameline et al. (2009) suggested a masking effect of oil on foliage caused by the more volatile components (<*n*C21, 13.7% of the hydrocarbon mixture). Similar interferences with host plant location was reported on greenhouse whitefly (*Trialeurodes vaporarium* Westwood, Hemiptera: Aleyrodidae) (Larew & Locke, 1990) and cotton bollworm (*Helicoverpa armigera* Hübner, Lepidoptera: Noctuidae) (Mensah et al., 2005). Conversely, *A. gossypii* was not affected by spraying plants with *n*C24 or *n*C27 mineral oil (Najar-Rodríguez et al., 2007b).

Following mineral oil spraying of their host plant, oviposition deterrence was reported on citrus leafminer (*Phyllocnistis citrella* Stainton, Lepidoptera: Gracillariidae) (Beattie et al., 1995), codling moth (*Cydia pomonella* L., Lepidoptera: Tortricidae) (Riedl et al., 1995), European corn borer (*Ostrinia nubilalis* Hübner, Lepidoptera: Crambidae), *H. armigera* (Mensah et al., 2005), citrus psylla (*Diaphorina citri* Kuwayama, Hemiptera: Liviidae) (Rae et al., 1997), greenhouse whitefly *T. vaporarium* (Pavela & Herda, 2007), queensland fruit fly (*Bactrocera tryoni* Froggatt, Diptera: Tephritidae) (Liu et al., 2002), and European cherry fruit fly (*Rhagoletis cerasi* L., Diptera: Tephritidae) (Daniel, 2014). For instance, Fernandez et al. (2001) have shown that mineral oil emulsion as low as 1% prevents the oviposition of white apple leafhopper (*Typhlocyba pomaria* McAtee, Hemiptera: Cicadellidae) for at least 3 days. However, oviposition deterrence was not observed in *A. gossypii* after a mineral oil spray since the treatment did not affect nymphs production (Najar-Rodríguez et al., 2007b).

Oil treatments also impact feeding behavior of insects. Delays in the initiation of the first probing (mouthparts penetration of leaf tissues) were reported in the green peach aphid (*Myzus persicae* Sulzer, Hemiptera: Aphididae) (Powell, 1992; Powell et al., 1998; Simons et al., 1977) and in the bird cherry oat aphid (*Rhopalosiphum padi* L., Hemiptera: Aphididae) (Boquel et al., 2016) on mineral oil-sprayed plants. Cen et al. (2002) showed that mineral oil sprayed on sweet oranges repelled the citrus red mite (*Panonychus citri* Mc Gregor, Arachnida: Tetranychidae). Conversely, the initial probing was shortened in *M. euphorbiae*, suggesting facilitated mouthparts insertion (Ameline et al., 2009). On mineral oil-treated plants, *M. persicae* (Simons et al., 1977; Wróbel, 2014) and *M. euphorbiae* (Ameline et al., 2009) exhibited a delayed ingestion phase, suggesting an antixenotic effect of the sprayed plant. Conversely, Bradley (1963), Vanderveken (1977), and Qiu and Pirone (1989) reported no effect of sprayed plants on different aphid species including *M. persicae*. Interestingly, recent results obtained on *M. euphorbiae* showed probiotic effects regarding the dose of *n*C22-C27 oil used: the highest concentrations increased aphid survival, while the lowest concentrations increased fecundity (Martoub et al., 2011). Such probiotic effects were also showed on *M. persicae* and buckthorn aphid (*Aphis nasturtii* Kaltenbach, Hemiptera: Aphididae) whose density on plants were increased after spraying different rapeseed oils (Wróbel, 2012).

Host plant location, feeding, oviposition and avoidance of plants unsuitable as hosts are dependent upon perception by insects of a complex blend of acoustic, visual, mechanic and olfactive cues emitted by the plant (Bruce & Pickett, 2011). Epicuticular waxes of most plants form three-dimensional thin structures responsible for the maintenance of wettability and self-cleaning properties, sliding of insects, reflection of visible light, adsorption of UV-radiation, and reduction of particle adhesion. These waxes are composed of a mixture of long hydrocarbon chains and, in some waxes, cyclic hydrocarbons. The cuticle itself is composed of a cutin network built up of hydroxyl and hydroxyepoxy fatty acids and sometimes also by cutan (composed of polymethylene chains) (Koch et al., 2008). As specified above for insect cuticles, oils could also interfere with the structure and composition of plant cuticle and/or epicuticular waxes, thus altering signals emitted by the treated plant. It is well known that the morphological and chemical characteristics of plant or leaf surface affect the performance and host preference of phytophagous insects (Nikolakakis et al., 2003). Large differences in chemical composition and microstructure of the waxes and the cutin network have been found across different plant species (Koch et al., 2008). It is not surprising to observe such variability in insect behavioral responses to oil-sprayed plants, whether they result from the oil layer itself forming a barrier impeding plant signals perception by insect or from modified plant cues as a result of oil interaction with components of plant surface. Besides these alterations, oil applications have been reported to induce physiological changes in the plant, such as alteration of gas exchanges, photosynthesis (Wróbel, 2012) and plant defense responses (Khelifa, 2017). Sprayed oils have been reported to penetrate plant cuticle (Boquel, 2014; Sporleder & Lacey, 2013) (Fig. 6.3) and epidermic cells then accumulate in leaves, stems and fruits in citrus (Rohrbaugh, 1934), apples and vegetables (Young, 1936, 1941). Following mineral oil sprays on citrus, confocal observations showed that oil droplets can move through apoplastic compartment as well as between cells through plasmodesmata and accumulated in xylem and cell protoplasm including phloem (Tan et al., 2005). These observations suggest accumulation and persistence of oil within plant tissue, blocking translocation and distribution of metabolites and nutrients. This could in turn represent the major contributing factor to chronic phytotoxicity. However, Fageria et al. (2014) found that mineral oil stays within the sprayed potato leaves

Fig. 6.3 Potato plant sprayed with an emulsion of Nile Red dye and mineral oil at 10L/ha and visualized under UV light to study the penetration of oil in plant. Cover of *American Journal of Potato Research*, vol. 91 no. 4. Original photo provided by Sébastien Boquel. (Source: https://static-content.springer.com/cover/journal/12230/91/4.jpg)

without moving to untreated tissue. Similarly, Rohrbaugh (1934) did not observe any translocation of oil droplets between sprayed and unsprayed parts of citrus trees.

Whether translocation of oil droplets between plant tissues occurs or not, their presence within plant tissues can partly explain behavioral alterations of insects on treated plants (physico-chemical interaction with mouthparts and/or deterrent effect), or through plant physiological modifications induced by oil accumulation within plant tissues. Although oils could provide effective control against pathogen spreads in fields (Boiteau et al., 2009; Boiteau & Singh, 1982; Bradley et al., 1962; Kirchner et al., 2014), behavioral alterations appear not strong enough to account for plant protection (Ameline et al., 2009; Boquel et al., 2017).

6.3.3 Effects on Pathogen/Vector Interaction

The main preventive effect of mineral oil sprays in controlling the spread of non-persistent viruses is thought to result from the interference with the virus-aphid mouthparts interaction. When an aphid penetrates the oil layer deposited on the plant surface, a hydrophobic film of oil can be formed in the food canal of the aphid mouthparts. Such film of oil prevents non-persistent virus retention on the aphid mouthparts thus reducing acquisition by non viruliferous aphids (Wang & Pirone, 1996). Since the pathway of mouthparts to vascular bundles is apoplastic, i.e. between cells (Tjallingii & Hogen Esch, 1993), oil droplets could be encountered when aphids probe plant tissue and a film of oil could then be formed over the tip of the mouthparts where non-persistent viruses bind (Boquel et al., 2013). Most of the literature indicates that mineral oils reduce transmission of non-persistent viruses, but have no effect on persistent viruses.

Fageria et al. (2014) found that translocation between sprayed and unsprayed parts of the potato plant does not occur. In addition, the inhibition of the acquisition of *Potato virus Y* (Potyviridae: Potyvirus) by aphids was reported to be a function of the concentration of oil in leaves (Boquel et al., 2013) suggesting that mineral oil is limiting virus colonization within the plant by constraining the virus to a few leaves (Boquel et al., 2017). Moreover, the persistence of oil on leaf surface is limited (Boiteau & Wood, 1982) and the inhibition effect of *Potato virus Y* acquisition by aphids induced by oil decreases with time from application (Boquel et al., 2013). Together, these findings suggest that the efficacy of oil sprays was not due to the presence of an oily layer on the plant surface but rather to the presence of oil within plant tissue, especially in the apoplastic compartment.

6.3.4 Effects on Host Plant

Resistant plants inhibit long distance movement of viruses through various ways such as blocking viral particles translocation to the phloem, reducing viral content in phloem cells, and inhibiting viral particles from unloading into systemic tissues (Vuorinen et al., 2011). Mineral oils are also thought to inhibit virus replication and accumulation in inoculated plants. For instance, oil-treated tobacco plants exhibit fewer symptoms of *Tobacco Mosaic Virus* (Virgaviridae: Tobamovirus) after mechanical inoculation (Loebenstein et al., 1964; Peters & Lebbink, 1975). Another study showed that mineral oil seems to inhibit *Potato virus Y* movement thus reducing virus amount in systemic tissue of older potato plants but not of younger ones (Al-Daoud et al., 2014). The different effects on plant age may result from varying temperature and daylight or physiological changes induced by tuberization (Al-Daoud et al., 2014). Together, these studies suggest that mineral oils activate plant defense responses in mineral oil-treated plants as illustrated by increased defense genes expression (Ameline et al., 2010; Lin et al., 1996).

Recent works on potato plants sprayed with *n*C22-C27 mineral oil demonstrated over-expression of various marker genes involved in plant-virus interaction (Khelifa, 2017) such as *PR1,* a defense-related gene involved in plant response to *Potato virus Y* (Vuorinen et al., 2010), *PAL1* inducing the phenylpropanoid pathway leading to production of many secondary metabolites (Huang et al., 2010), *2-ODD* involved in flavone and antimicrobial compounds production (Hausinger, 2004), and *IVR-1* known to act in tobacco as a cysteine proteinase inhibitor blocking virus replication (Gholizadeh et al., 2005). Interestingly, *PAL1* and *2-ODD* are highly expressed in a resistant potato cultivar when challenged with *Potato virus Y* (Baebler et al., 2009). The over-expression of these genes after preventive oil-treatment could elicit plant defenses which could in turn make the plant less susceptible to pathogens at the time of inoculation (Khelifa, 2017).

6.4 Physical Factors Affecting Oil Efficacy

Because oils are insoluble in water, they must be emulsified to enhance the adhesion of oil droplets to plant surface. Most horticultural oils contain a nonionic emulsifier (surfactant) dissolved in the oil at a concentration of up to 15% (Ebbon, 2002). Such concentrations can have drastic effects on plant waxes and cuticle and affect insects as well. When exposed to emulsifiers like glyceryl monooleate, glycol oleate and glyceryl monolaurate, various arthropods rapidly desiccate (Taverner, 2002). Wigglesworth (1941) suggested that saturated fatty acids dissolved the lipidic surface layer covering the cuticle, allowing polar compounds to draw out moisture.

Emulsifiers used to decrease surface tension allow oils to penetrate deeper in the tracheal system inducing insect narcosis (Taverner, 2002). However, the concentration of nonionic emulsifiers had no effect on the efficacy of various combinations of paraffin fractions *nC17*, *n*C22 and *n*C25 in preventing oviposition of the citrus leafminer *P. citrella* (Liu et al., 2001). Liu et al. (2001) concluded that efficacy of mineral oils is related to the timing of spray application, the dose of oil deposited and the persistence of oil on plant surface.

The use of biologically active emulsifiers questions the need to pay special attention to the toxicology and ecotoxicology of these emulsifiers. Although the relatively high concentration of emulsifier in some of the formulations (17% vs 3%) did not increase their efficacy, the emulsifier alone decreased the spread of *Potato virus Y* (Boiteau & Singh, 1982). Conversely, Vanderveken (1968) reported that emulsifiers used in commercial preparations had no inhibitory effect.

To prevent *Potato virus Y* spread in pepper crops, Simons & Zitter (1980) recommended applications of 1–2% oil concentration at 200–400 psi spray pressure, adding 0.75–1.25% emulsifier and using spray nozzles to achieve small droplet size. Concentrations <1% are reported to be less effective, whereas concentrations >3% are toxic to the plants and significantly decrease crop yields (Groves et al., 2009).

Spray conditions may also contribute to oil efficacy. *Potato virus Y* transmission in potato fields is affected by oil concentration and, to a lesser extent, by its delivery rate, while spray pressure had no effect (Boiteau & Singh, 1982). In contrast, Simons & Zitter (1980) reported that oil efficiency was improved by increasing delivery rates and spray pressure. These contrasting results could be explained by the variation of plant architecture between species. Oil sprays at low concentrations (<5% (v/v)) on maize leaves (at an angle of 45–50° to the main stem) run off quickly onto the ground leading to few oil residues left on the leaves to deter *O. nubilalis* egg laying. Conversely, a 2% (v/v) oil emulsion inhibited *H. armigera* oviposition on cotton where leaves stood almost horizontal thus allowing oil sprays to remain on the leaves after application long enough for residue to be built up even at lower concentrations (Mensah et al., 2005).

As a good coverage of plant surface is needed to control pathogens spread, weekly application or twice-weekly application for rapidly growing plants is required to protect newly grown leaves and stems. However, frequencies of oil sprays remain controversial. Shands (1977) reported that six weekly applications gave best result in controlling *Potato virus Y* than three, whereas Bradley et al. (1966) found little variation between their efficacy.

As used nozzles can significantly affect spraying efficacy, nozzles that allow to produce small oil droplets are recommended to ensure a better coverage (Al-Mrabeh et al., 2010).

Weather conditions affect the efficacy and persistence of oils sprays. After application, oils photodegrade (especially under ultra-violet light) to undefined carboxylic acids which can in turn induce phytotoxicity (Hodgkinson et al., 2002).

Some studies have suggested that the loss of efficacy of mineral oils could be due to rainfall that washes off the oil from the leaves. Under light rain, oils persist for 10–14 days after spraying (Simons et al., 1977). However, other studies have shown that natural rainfall up to 3 cm and artificial rinses up to 10 cm did not remove any significant amount of mineral oil from the leaves (Boiteau & Wood, 1982; Szatmari-Goodman & Nault, 1983).

Temperature also influences oil persistence on plant surface, lasting longer at 16 °C compared with 24 °C or 32 °C (Simons et al., 1977). However, these latter results are to consider with precaution as varying temperatures highly impact plant growing and leaves expansion leading to decreased oil coverage (Al-Mrabeh et al., 2010).

The effect of mineral oils in reducing pathogen transmission can be enhanced by mixing insecticides to the oil (Boiteau et al., 2009). Raccah et al. (1983) were the first to report the enhanced effect between mineral oils and pyrethroids in controlling non-persistent viruses. Because of their low surface tension, oils can be used as adjuvants to improve the coverage on and persistence of contact pesticides on targeted surface. As their water solubility is low, they also protect pesticides from wash off. Oils also facilitate penetration of pesticides through insect cuticle or increase the rate of absorption of systemic pesticides into plants by modifying epicuticular waxes (MacKenzie et al., 2016, 2017). Martín-López et al. (2006) reported that adding a low dose of imidacloprid (1/5 the recommended dose) to 1% mineral or rapeseed oil emulsion significantly decreased *Potato virus Y* transmission in potato fields. However, incompatibility can occur between mineral oil and some pesticides. Simons & Zitter (1980) reported such incompatibility between mineral oil and various fungicides. Similarly, Boiteau & Singh (1982) suggested that fungicide application should be delayed at least 24 h after oil application, as application of fungicides directly after oil may result in greater phytotoxicity.

6.5 Conclusion

In this chapter, we reviewed the modes of action and advantages and disadvantages of mineral and plant oils. This chapter present several beneficial effects that may alter the physiology and the behavior of insect vectors and that can be extended to leafhoppers, psyllids and other piercing-sucking pests that transmit pathogens to plants including phytoplasmas. Oil applications can also induce plant defense responses that could reduce pathogen movements in plants. As phytoplasma vectors lay their eggs on the cuticle of their host plant (Fig. 6.4; (Saguez et al., 2015)), oil applications could also reduce their populations, due to their ovicidal activities or their deterrent effects on oviposition (Fernandez et al., 2001). In such cases, where eggs are exposed to sprays, such as some cicadellid species associated with vines (Fig. 6.4), they should be impacted by sprays.

Until now, researches were essentially conducted on aphids and virus transmission and there are several gaps of information with respect to the management of phytoplasma diseases with oils. More work should be done to fully understand the biological and physical parameters that modulate the efficacy of oils to prevent phytoplasma dissemination.

Fig. 6.4 Leafhopper eggs (**a**) eggs of *Erythroneura ziczac* laid in clusters under the epidermis of a grapevine leaf; (**b**) close-up of *Erythroneura ziczac* eggs; (**c**) egg of *Macrosteles quadrilineatus* laid on the epidermis of wheat leaf; (**d**) eggs of *Erythroneura vitis* laid singly along the veins of a grapevine leaf. (Photos from Saguez et al., 2015)

References

Adhvaryu, A., Liu, Z., & Erhan, S. (2005). Synthesis of novel alkoxylated triacylglycerols and their lubricant base oil properties. *Industrial Crops and Products, 21*(1), 113–119.

Agnello, M. A. (2002). Petroleum derived spray oils: Chemistry, history, refining and formulation. In G. A. C. Beattie, D. M. Watson, M. L. Stevens, D. J. Rae, & R. N. Spooner-Hart (Eds.), *Spray oils beyond 2000* (pp. 2–18). Sydney, Australia: University of Western Sydney.

Al-Daoud, F., Fageria, M. S., Zhang, J., Boquel, S., & Pelletier, Y. (2014). Mineral oil inhibits movement of potato virus Y in potato plants in an age-dependent manner. *American Journal of Potato Research, 91*(4), 337–345.

Al-Mrabeh, A., Anderson, E., Torrance, L., Evans, A., & Fenton, B. (2010). *A literature review of insecticide and mineral oil use in preventing the spread of non-persistent viruses in potato crops*. Retrieved from Agriculture and Horticulture Development Board, Kenilworth, UK. https://ahdb.org.uk/

Ameline, A., Couty, A., Martoub, M., & Giordanengo, P. (2009). Effects of mineral oil application on the orientation and feeding behaviour of *Macrosiphum euphorbiae* (Homoptera: Aphidae). *Acta Entomologica Sinica, 52*(6), 617–623.

Ameline, A., Couty, A., Martoub, M., Sourice, S., & Giordanengo, P. (2010). Modification of *Macrosiphum euphorbiae* colonisation behaviour and reproduction on potato plants treated by mineral oil. *Entomologia Experimentalis et Applicata, 135*(1), 77–84.

Baebler, S., Krecic-Stres, H., Rotter, A., Kogovsek, P., Cankar, K., Kok, E. J., Gruden, K., Kovac, M., Zel, J., Pompe-Novak, & Ravnikar, M. (2009). PVYNTN elicits a diverse gene expression response in different potato genotypes in the first 12 h after inoculation. *Molecular Plant Pathology, 10*(2), 263–275. https://doi.org/10.1111/j.1364-3703.2008.00530.x.

Bass, C., & Field, L. M. (2011). Gene amplification and insecticide resistance. *Pest Management Science, 67*(8), 886–890. https://doi.org/10.1002/ps.2189

Battersby, N. S., Pack, S. E., & Watkinson, R. J. (1998). A correlation between the biodegradability of oil products in the CEC L-33-T-82 and modified strum tests. *Chemosphere, 24*, 1998–2000.

Beattie, G. A. C., & Hardy, S. (2005). Using petroleum-based spary oils in citrus. *Agfacts*, p. 7.

Beattie, G. A. C., Liu, Z. M., Watson, D. M., Clift, A. D., & Jiang, L. (1995). Evaluation of petroleum spray oils and polysaccharides for control of Phyllocnistis citrella Stainton (Lepidoptera: Gracillariidae). *Journal of the Australian Entomological Society, 34*, 349–353.

Belli, G., Bianco, P. A., & Conti, M. (2010). Grapevine yellows in Italy: Past, present and future. *Journal of Plant Pathology, 92*, 303–326.

Boiteau, G., Singh, M., & Lavoie, J. (2009). Crop border and mineral oil sprays used in combination as physical control methods of the aphid-transmitted potato virus Y in potato. *Pest Management Science, 65*, 255–259.

Boiteau, G., & Singh, R. P. (1982). Evaluation of mineral oil sprays for reduction of virus Y spread in potatoes. *American Potato Journal, 59*(6), 253–262. https://doi.org/10.1007/bf02856561

Boiteau, G., & Wood, F. A. (1982). Persistence of mineral-oil spray deposits on potato leaves. *American Potato Journal, 59*, 55–63.

Boquel, S. (2014). *Penetration kinetics of Superior 70 mineral oil in a potato plant*. Retrieved from https://open.canada.ca/data/en/dataset/7344deb1-c372-4dec-999a-efd710489d62

Boquel, S., Giguère, M. A., Clark, C., Nanayakkara, U., Zhang, J., & Pelletier, Y. (2013). Effect of mineral oil on Potato virus Y acquisition by *Rhopalosiphum padi*. *Entomologia Experimentalis et Applicata, 148*(1), 48–55.

Boquel, S., Giguère M. A., & Pelletier, Y. (2016) Effect of mineral oils on host plant selection and probing behavior of *Rhopalosiphum padi*. *Entomologia Experimentalis et Applicata, 160*, 241–250.

Boquel, S., Hepat, R., & Nie, X. (2017). Advances in elucidating the mode of action of mineral oil on reduction of aphid-mediated PVY transmission. *Canadian Journal of Plant Pathology, 39*(1), 90–90.

Bradley, R. (1963). Some ways in which a paraffin oil impedes aphid transmission of potato virus Y. *Canadian Journal of Microbiology, 9*(3), 369–380.

Bradley, R., Moore, C., & Pond, D. (1966). Spread of potato virus Y curtailed by oil. *Nature, 209*(5030), 1370–1371.

Bradley, R., Wade, C., & Wood, F. (1962). Aphid transmission of potato virus Y inhibited by oils. *Virology, 18*(2), 327–329.

Bruce, T. J. A., & Pickett, J. A. (2011). Perception of plant volatile blends by herbivorous insects – Finding the right mix. *Phytochemistry, 72*(13), 1605–1611. https://doi.org/10.1016/j.phytochem.2011.04.011

Buteler, M., & Stadler, T. (2011). A review on the mode of action and current use of petroleum distilled spray oils. In M. Stoytcheva (Ed.), *Pesticides in the modern world – Pesticides use and management* (pp. 119–136). Croatia: InTech.

Butler, G. D., & Henneberry, T. J. (1990). Pest control in vegetables and cotton with household cooking oils and liquid detergents. *Southwestern Entomologist, 15*, 123–131.

Cen, Y. J., Tian, M. Y., Pang, X. F., & Rae, D. J. (2002). Repellency, anti feeding effect and toxicity of an horticultural mineral oil against citrus red mite. In G. A. C. Beattie, D. M. Watson, M. L. Stevens, D. J. Rae, & R. N. Spooner-Hart (Eds.), *Spray oils beyond 2000* (pp. 134–141). Sydney, Australia: University Of Western Sydney.

Daniel, C. (2014). *Rhagoletis cerasi*: Oviposition reduction effects of oil products. *Insects, 5*(2), 319–331.

Davidson, N., Dibble, J., Flint, M., Marer, P., & Guye, A. (1991). *Managing insects and mites with spray oils*. Oakland, CA: University of California.

de Ong, E. R., Knight, H., & Chamberlin, J. C. (1927). A preliminary study of petroleum oil as an insecticide for citrus trees. *Hilgardia, 2*, 351–384.

Dekker, M. H., Piersma, T., & Sinninghe Damsté, J. S. (2000). Molecular analysis of intact preen waxes of *Calidris canutus* (Aves: Scolopacidae) by gas chromatography/mass spectrometry. *Lipids, 35*(5), 533–541.

Ebbon, G. P. (2002). Environmental and health aspects of agricultural spray oils. In G. A. C. Beattie, D. M. Watson, M. L. Stevens, D. J. Rae, & R. N. Spooner-Hart (Eds.), *Spray oils beyond 2000* (pp. 232–246). Sydney, Australia: University of Western Sydney.

Fageria, M., Boquel, S., Leclair, G., & Pelletier, Y. (2014). Quantification of mineral oil accumulation and movement in potato plants and its significance in potato virus Y management. *Pest Management Science, 70*(8), 1243–1248. https://doi.org/10.1002/ps.3682

Fernandez, D. E., Beers, E. H., Brunner, J. F., Doerr, M., & Dunley, J. E. (2001). Mineral oil inhibition of white apple leafhopper (Homoptera: Cicadellidae) oviposition. *Journal of Economic Entomology, 36*(3), 237–243.

Field, L., Blackman, R. L., Tyler-Smith, C., & Devonshire, A. (1999). Relationship between amount of esterase and gene copy number in insecticide-resistant *Myzus persicae* (Sulzer). *Biochemical Journal, 339*(3), 737–742.

Furness, G. O., Walker, D. A., Johnson, P. G., & Riehl, L. A. (1987). High resolution g.l.c. specifications for plant spray oils. *Pesticide Science, 18*(2), 113–128. https://doi.org/10.1002/ps.2780180205

Gholizadeh, A., Santha, I. M., Kohnehrouz, B. B., Lodha, M. L., & Kapoor, H. C. (2005). Cystatins may confer viral resistance in plants by inhibition of a virus-induced cell death phenomenon in which cysteine proteinases are active: Cloning and molecular characterization of a cDNA encoding cysteine-proteinase inhibitor (celostatin) from *Celosia cristata* (crested cock's comb). *Biotechnology and Applied Biochemistry, 42*(3), 197–204.

Groves, R., Charkowski, A., Crockford, A., Coltman, R., Hafner, R., & Bula, K. (2009). Integrated pest and disease management: Reducing current season spread of potato virus Y in potato. *Phytopathology, 99*, S47.

Hausinger, R. P. (2004). Fe(II)/α-Ketoglutarate-dependent hydroxylases and related enzymes. *Critical Reviews in Biochemistry and Molecular Biology, 39*(1), 21–68. https://doi.org/10.1080/10409230490440541

Hodgkinson, M., Johnson, D., & Smith, G. (2002). Using FTIR to predict the potential for petroleum-derived spray oils to photodegrade. In G. A. C. Beattie, D. M. Watson, M. L. Stevens, D. J. Rae, & R. N. Spooner-Hart (Eds.), *Spay oils beyond 2000* (pp. 72–76). Sydney, Australia: University of Western Sydney.

Huang, J., Gu, M., Lai, Z., Fan, B., Shi, K., Zhou, Y.-H., Yu, J.-Q., & Chen, Z. (2010). Functional analysis of the Arabidopsis PAL gene family in plant growth, development, and response to environmental stress. *Plant Physiology, 153*(4), 1526–1538. https://doi.org/10.1104/pp.110.157370

Ivanov, D. S., Lević, J. D., & Sredanović, S. A. (2010). Fatty acid composition of various soybean products. *Food and Feed Research, 2*, 65–70.

Khelifa, M. (2017). Possible induction of potato plant defences against potato virus Y by mineral oil application. *European Journal of Plant Pathology, 147*(2), 339–348. https://doi.org/10.1007/s10658-016-1006-7

Kirchner, S. M., Hiltunen, L. H., Santala, J., Döring, T. F., Ketola, J., Kankaala, A., Virtanen, E., & Valkonen, J. P. T. (2014). Comparison of straw mulch, insecticides, mineral oil, and birch extract for control of transmission of potato virus Y in seed potato crops. *Potato Research, 57*(1), 59–75. https://doi.org/10.1007/s11540-014-9254-4

Koch, K., Bhushan, B., & Barthlott, W. (2008). Diversity of structure, morphology and wetting of plant surfaces. *Soft Matter, 4*(10), 1943–1963. https://doi.org/10.1039/B804854A

Lancaster, A. l., Deyton, D., Sams, C., Cummins, J., Pless, C., & Fare, D. (2002). Soybean oil controls two-spotted spider mites on burning bush. *Journal of Environmental Horticulture, 20*(2), 86–92.

Larew, H. G., & Locke, J. C. (1990). Repellency and toxicity of a horticultural oil against whiteflies on chrysanthemum. *HortScience, 25*, 1406–1407.

Lin, K. C., Bushnell, W. R., Szabo, L. J., & Smith, A. G. (1996). Isolation and expression of a host response gene family encoding thaumatinlike proteins in incompatible oat-stem rust fungus interactions. *Molecular Plant-Microbe Interactions, 9*, 511–522.

Liu, Z., Beattie, G., Hodgkinson, M., Rose, H., & Jiang, L. (2001). Influence of petroleum-derived spray oil aromaticity, equivalent n-paraffin carbon number and emulsifier concentration on oviposition by citrus leafminer, *Phyllocnistis citrella* Stainton (Lepidoptera: Gracillariidae). *Australian Journal of Entomology, 40*(2), 193–197.

Liu, Z., Beattie, G., Johnson, D., & Spooner-Hart, R. (2002). Influence of deposits of a horticultural mineral oil and selected fractions of paraffinic and naphthenic petroleum-derived oils on oviposition by Queensland fruit fly on tomato fruit. *Spray Oils Beyond, 2000*, 142–146.

Loebenstein, G., Alper, M., & Deutsch, M. (1964). Preventing aphid-spread Cucumber mosaic virus with oils. *Phytopathology, 54*, 960–962.

MacKenzie, T. D., Lavoie, J., Nie, X., & Singh, M. (2017). Effectiveness of combined use of mineral oil and insecticide spray in reducing Potato Virus Y (PVY) spread under field conditions in New Brunswick, Canada. *American Journal of Potato Research, 94*(1), 70–80.

MacKenzie, T. D., Nie, X., & Singh, M. (2016). Crop management practices and reduction of on-farm spread of potato virus Y: A 5-year study in commercial potato fields in New Brunswick, Canada. *American Journal of Potato Research, 93*(6), 552–563.

Martín-López, B., Varela, I., Marnotes, S., & Cabaleiro, C. (2006). Use of oils combined with low doses of insecticide for the control of *Myzus persicae* and PVY epidemics. *Pest Management Science: Formerly Pesticide Science, 62*(4), 372–378.

Martoub, M., Couty, A., Giordanengo, P., & Ameline, A. (2011). Opposite effects of different mineral oil treatments on *Macrosiphum euphorbiae* survival and fecundity. *Journal of Pest Science, 84*(2), 229–233.

Mensah, R. K., Frerot, B., & Al Dabel, F. (2005). Effects of petroleum spray oils on oviposition behaviour and larval survival of *Helicoverpa armigera* Hubner (Lepidoptera: Noctuidae) and *Ostrinia nubilalis* Hubner (Lepidoptera: Pyralidae). *International Journal of Pest Management, 51*(2), 111–119.

Najar-Rodriguez, A. J., Lavidis, N. A., Mensah, R. K., Choy, P. T., & Walter, G. H. (2008). The toxicological effects of petroleum spray oils on insects - evidence for an alternative mode of action and possible new control options. *Food and Chemical Toxicology, 46*(9), 3003–3014. https://doi.org/10.1016/j.fct.2008.05.042

Najar-Rodríguez, A. J., Walter, G. H., & Mensah, R. K. (2007a). The efficacy of a petroleum spray oil against *Aphis gossypii* Glover on cotton. Part 1: Mortality rates and sources of variation. *Pest Management Science: Formerly Pesticide Science, 63*(6), 586–595.

Najar-Rodríguez, A. J., Walter, G. H., & Mensah, R. K. (2007b). The efficacy of a petroleum spray oil against *Aphis gossypii* glover on cotton. Part 2: Indirect effects of oil deposits. *Pest Management Science: Formerly Pesticide Science, 63*(6), 596–607.

Nicetic, O., Cho, Y. R., & Rae, D. J. (2011). Impact of physical characteristics of some mineral and plant oils on efficacy against selected pests. *Journal of Applied Entomology, 135*(3), 204–213. https://doi.org/10.1111/j.1439-0418.2010.01553.x

Nikolakakis, N., Margaritopoulos, J., & Tsitsipis, J. (2003). Performance of *Myzus persicae* (Hemiptera: Aphididae) clones on different host-plants and their host preference. *Bulletin of Entomological Research, 93*(3), 235–242.

Nollet, L. M. L., & Rathore, H. S. (2017). *Green pesticides handbook: Essential oils for pest control.* Boca Raton, FL: CRC Press.

Olivier, C., Vincent, C., Saguez, J., Galka, B., Weintraub, P. G., & Maixner, M. (2012). Leafhoppers and planthoppers: Their bionomics, pathogen transmission and management in vineyards. In *Arthropod management in Vineyards* (pp. 253–270). Dordrecht, The Netherlands: Springer.

Orozco-Santos, M., Robles-González, M., Hernández-Fuentes, L. M., Velázquez-Monreal, J. J., de Jesús Bermudez-Guzmán, M., Manzanilla-Ramírez, M., Manzo-Sanchez, G., & Nieto-Ángel, D. (2016). Uso de Aceites y Extractos Vegetales para el Control de *Diaphorina citri* Kuwayama1 en *Lima Mexicana* en el Trópico Seco de México. *Southwestern Entomologist, 41*(4), 1051–1067.

Ouyang, G., Fang, X., Lu, H., Zhou, X., Meng, X., Yu, S., Guo, M., & Xia, Y. (2013). Repellency of five mineral oils against *Diaphorina citri* (Hemiptera: Liviidae). *Florida Entomologist, 96*(3), 974–983.

Pavela, R., & Herda, G. (2007). Repellent effects of pongam oil on settlement and oviposition of the common greenhouse whitefly *Trialeurodes vaporariorum* on chrysanthemum. *Insect Sci., 14*(3), 219–224.

Perring, T. M., Gruenhagen, N. M., & Farrar, C. A. (1999). Management of plant viral diseases through chemical control of insect vectors. *Annual Review of Entomology, 44*(1), 457–481. https://doi.org/10.1146/annurev.ento.44.1.457

Peters, D., & Lebbink, G. (1975). The inhibitory action of mineral oil on the number of local lesions on *Nicotiana glutinosa* L. leaves inoculated with tobacco mosaic virus. *Virology, 65*(2), 574–578.

Powell, G. (1992). The effect of mineral oil on stylet activities and potato virus Y transmission by aphids. *Entomologia Experimentalis et Applicata, 63*(3), 237–242.

Powell, G., Hardie, J., & Pickett, J. (1998). The effects of antifeedant compounds and mineral oil on stylet penetration and transmission of potato virus Y by *Myzus persicae* (Sulz.)(Hom., Aphididae). *Journal of Applied Entomology, 122*(1–5), 331–333.

Qiu, J. Y., & Pirone, T. P. (1989). Assessment of the effect of oil on the Potyvirus aphid transmission process. *Journal of Phytopathology, 127*(3), 221–226. https://doi.org/10.1111/j.1439-0434.1989.tb01132.x

Raccah, B., Antignus, Y., & Cohen-Braun, M. (1983). Effect of a combination of a mineral oil and a pyrethroid on the transmission of CMV in laboratory and on the natural infection of MDMV in a corn field. In: *Proceedings of the 4th International Congress of Phytopathology, Melbourne,* p. 231.

Rae, D., Liang, W., Watson, D., Beattie, G., & Huang, M. (1997). Evaluation of petroleum spray oils for control of the Asian citrus psylla, *Diaphorina citri* (Kuwayama)(Hemiptera: Psyllidae), in China. *International Journal of Pest Management, 43*(1), 71–75.

Regnault-Roger, C., Vincent, C., & Arnason, J. T. (2012). Essential oils in insect control: Low-risk products in a high-stakes world. *Annual Review of Entomology, 57*(1), 405–424. https://doi.org/10.1146/annurev-ento-120710-100554

Riedl, H., Halaj, J., Kreowski, W. B., Hilton, R. J., & Westigard, P. H. (1995). Laboratory evaluation of mineral oils for control of codling moth (Lepidoptera: Tortricidae). *Journal of Economic Entomology, 88*(1), 140–147. https://doi.org/10.1093/jee/88.1.140

Riehl, L. (1969). Advances relevant to narrow-range spray oils for citrus pest control. *Proceedings of the First International Citrus Symposium, 2*, 897–907.

Rohrbaugh, P. W. (1934). Penetration and accumulation of petroleum spray oils in the leaves, twigs, and fruit of Citrus trees. *Plant Physiology, 9*(4), 699–730. https://doi.org/10.1104/pp.9.4.699

Rongai, D., Cerato, C., Lazzeri, L., Palmieri, S., & Patalano, G. (2008). Vegetable oil formulation as biopesticide to control California red scale (*Aonidiella aurantii* Maskell). *Journal of Pest Science, 81*(4), 179–185.

Saguez, J., Olivier, C., Lasnier, J., Hamilton, A., Stobbs, L., & Vincent, C. (2015). Biology and integrated management of leafhoppers and phytoplasma diseases in vineyards of eastern Canada. *Technical Bulletin*, A59-32/2015E-PDF, Agriculture and Agri-Food Canada, 67p.

Samara, R., Lowery, D., Stobbs, L., Vickers, P., & Bittner, L. (2016). Horticultural mineral oil influences Plum pox virus transmission by *Myzus persicae*. *Journal of Applied Entomology, 140*(9), 688–696.

Sams, C. E., & Deyton, D. E. (2002). Botanical and fish oils: Histiory, chemistry, refining, formulation and current uses. In G. A. C. Beattie, D. M. Watson, M. L. Stevens, D. J. Rae, & R. N. Spooner-Hart (Eds.), *Spray oils beyond 2000* (pp. 19–28). Sydney, Australia: University of Western Sydney.

Schmutterer, H. (1990). Properties and potential of natural pesticides from the neem tree, *Azadirachta indica*. *Annual Review of Entomology, 35*(1), 271–297.

Schneider, M. P. (2006). Plant-oil-based lubricants and hydraulic fluids. *Journal of the Science of Food and Agriculture, 86*(12), 1769–1780. https://doi.org/10.1002/jsfa.2559

Shands, W. A. (1977). Control of aphid-borne potato virus Y in potatoes with oil emulsions. *American Potato Journal, 54*, 179–187.

Simons, J. N., McLean, D., & Kinsey, M. (1977). Effects of mineral oil on probing behavior and transmission of stylet-borne viruses by *Myzus persicae*. *Journal of Economic Entomology, 70*(3), 309–315.

Simons, J. N., & Zitter, T. A. (1980). Use of oils to control aphid-borne. *Plant Disease, 64*(6), 543.

Singh, O., Rathore, H. S., & Nollet, L. M. L. (2015). Biochemical pesticides: Oil pesticides. In L. M. L. Nollet & H. S. Rathore (Eds.), *Biopesticides handbook* (pp. 183–224). Boca Raton, FL: CRC Press.

Smith, E. H. (1952). Tree spray oils. *Advances in Chemistry, Series, 7*, 3–11.

Sporleder, M., & Lacey, L. A. (2013). Biopesticides. In P. Giordanengo, C. Vincent, & A. Alyokhin (Eds.), *Insect pests of potato: Global perspectives on biology and management* (pp. 463–497). Waltham, MA: Elsevier.

Stadler, T., & Buteler, M. (2009). Modes of entry of petroleum distilled spray-oils into insects: A review. *Bulletin of Insectology, 62*(2), 169–177.

Stadler, T., Zerba, M. A., & Buteler, M. (2002). Toxicity and cuticle softening effect of mineral and vegetable oils on the cotton boll weevil. In G. A. C. Beattie, D. M. Watson, M. L. Stevens, D. J. Rae, & R. N. Spooner-Hart (Eds.), *Spray oils beyond 2000* (pp. 152–155). Sydney, Australia: University of Western Sydney.

Szatmari-Goodman, G., & Nault, L. R. (1983). Tests of oil sprays for suppression of aphid-borne maize dwarf mosaic virus in Ohio sweet corn. *Journal of Economic Entomology, 76*(1), 144–149. https://doi.org/10.1093/jee/76.1.144

Tan, B. L., Sarafis, V., Beattie, G. A. C., White, R., Darley, E. M., & Spooner-Hart, R. (2005). Localization and movement of mineral oil in plants by fluorescence and confocal microscopy. *Journal of Experimental Botany, 56*(420), 2755–2763. https://doi.org/10.1093/jxb/eri269

Taverner, P. (2002). Drowning or just waving? A perspective on the ways petroleum-derived oils kill arthropod pests of plants. In G. A. C. Beattie, D. M. Watson, M. L. Stevens, D. J. Rae, & R. N. Spooner-Hart (Eds.), *Spray oils beyond 2000* (pp. 78–86). Sydney, Australia: University of Western Sydney.

Taverner, P., Gunning, R., Kolesik, P., Bailey, P., Inceoglu, B., Hammock, B., & Roush, R. (2001). Evidence for direct neural toxicity of a "light" oil on the peripheral nerves of Lightbrown apple moth. *Pesticide Biochemistry and Physiology, 69*, 153–165. https://doi.org/10.1006/pest.2000.2527

Tjallingii, W. F., & Hogen Esch, T. (1993). Fine structure of aphid stylet routes in plant tissues in correlation with EPG signals. *Physiological Entomology, 18*, 317–328.

University of California. (2018). *Leafhoppers and sharpshooters*. How to Manage Pests. UC Pest Management Guidelines. Retrieved from http://ipm.ucanr.edu/PMG/r280301711.html

Vanderveken, J. (1968). Effects of mineral oils and lipids on aphid transmission of beet mosaic and beet yellows viruses. *Virology, 34*, 807–809.

Vanderveken, J. J. (1977). Oils and other inhibitors of nonpersistent virus transmission. In K. Maramorosch (Ed.), *Aphids as virus vectors* (pp. 435–454). New York, NY: Academic.

Vincent, C., Hallman, G., Panneton, B., & Fleurat-Lessard, F. (2003). Management of agricultural insects with physical control methods. *Annual Review of Entomology, 48*(1), 261–281. https://doi.org/10.1146/annurev.ento.48.091801.112639

Vuorinen, A., Gammelgård, E., Auvinen, P., Somervuo, P., Dere, S., & Valkonen, J. (2010). Factors underpinning the responsiveness and higher levels of virus resistance realised in potato genotypes carrying virus-specific R genes. *Annals of Applied Biology, 157*(2), 229–241.

Vuorinen, A., Kelloniemi, J., & Valkonen, J. P. (2011). Why do viruses need phloem for systemic invasion of plants? *Plant Science, 181*(4), 355–363.

Wang, K., Zhang, M., Huang, Y., Yang, Z., Su, S., & Chen, M. (2018). Characterisation of imidacloprid resistance in the bird cherry-oat aphid, *Rhopalosiphum padi*, a serious pest on wheat crops. *Pest Management Science, 74*(6), 1457–1465.

Wang, R., & Pirone, T. (1996). Mineral oil interferes with retention of tobacco etch potyvirus in the stylets of *Myzus persicae*. *Phytopathology (USA), 86*, 820.

Wigglesworth, V. B. (1941). Oils aiding loss of water from the cuticle. *Nature, 147*, 116.

Wojdyła, A. T. (2002). Oils activity in the control of rose powdery mildew. *Mededelingen (Rijksuniversiteit te Gent. Fakulteit van de Landbouwkundige en Toegepaste Biologische Wetenschappen), 67*(2), 369–376.

Wróbel, S. (2012). Comparison of mineral oil and rapeseed oil used for the protection of seed potatoes against PVY and PVM infections. *Potato Research, 55*(1), 83–96. https://doi.org/10.1007/s11540-012-9210-0

Wróbel, S. (2014). Efficacy of mineral oil-insecticide mixtures for protection of potato tubers against PVY and PVM. *American Journal of Potato Research, 91*(6), 706–713. https://doi.org/10.1007/s12230-014-9403-y

Young, P. A. (1936). Distribution and effect of petroleum oils and kerosenes in potato, cucumber, turnip, barley, and onion. *Journal of Agricultural Research, 51*, 925–934.

Young, P. A. (1941). Physiological and physical effects of spray oils on deciduous trees. *Journal of Economic Entomology, 34*, 838–844.

Chapter 7
Phytoplasma Diseases Affecting Cassava

Elizabeth Álvarez

Abstract Cassava (*Manihot esculenta* Crantz) is an important energy source in tropical regions of the world and a staple in the diet of 17.8 million people worldwide. Phytoplasma diseases of cassava are a threat to its production as an export crop, and as food and income for smallholder farmers. As such, phytoplasma diseases are a potential threat to food security for millions of people. In Asia, phytoplasma from the groups 16SrI 16SrII, 16SrIII, 16SrVI, 16SrX, 16SrXII and 16SrXV were detected in cassava and in Africa, phytoplasma from the groups 16SrII and 16SrXXII have been reported in cassava plants. In Latin America and the Caribbean, phytoplasma belonging to groups, 16SrI, 16SrII and 16SrIII have been associated with cassava diseases. In this chapter, we present information on the most important cassava phytoplasma diseases in Asia, Latin America and Africa such as cassava witches'-broom (CWB) and cassava frogskin disease (CFSD), including their biology, genetic diversity, economic importance, symptomatology, epidemiology and control.

Keywords Cassava frogskin disease · Cassava witches'-broom disease · Phytoplasma · Africa · Southeast Asia · South and Central America · Oceania

7.1 Introduction

Cassava (*Manihot esculenta* Crantz) (Euphorbiaceae), maize, sugarcane, and rice are the most important energy sources in tropical regions of the world. Native to South America, cassava was domesticated 5000 years ago and has been cultivated extensively since then (Ceballos, 2002). Today, it is a staple in the diet of 17.8 million people worldwide (Food and Agriculture Organisation of the United Nations, 2016) as well as being an industrial crop with high potential to generate permanent income for small and medium farmers, and a crop that contributes to the social development of the regions where it is produced. Cassava is primarily a vegetatively

E. Álvarez (✉)
Plant Pathology Program, Centro Internacional de Agricultura Tropical, Cali, Colombia

C. Y. Olivier et al. (eds.), *Sustainable Management of Phytoplasma Diseases in Crops Grown in the Tropical Belt*, Sustainability in Plant and Crop Protection 12,
https://doi.org/10.1007/978-3-030-29650-6_7

propagated crop, mainly through stem cuttings. Consequently, it is vulnerable to infection by pathogens that pass from one cropping cycle to the next through planting material. Cassava phytoplasma diseases are a potential threat to food security for millions of people (Jackson, 2014).

Phytoplasmas are small wall-less pleomorphic pathogenic prokaryotes that are surrounded by a single membrane. They possess a single circular or linear chromosome and may additionally contain up to four different plasmids. The chromosome has a reduced size as compared to the ancestral gram-positive bacteria. Phytoplasmas are restricted to, and multiply in the phloem of their host plants and the organs of their insect vectors (Rao, Álvarez, & Yadav, 2018). Recently, their cultivation in axenic culture has become possible (Álvarez, Betancourth, & Muñoz, 2017). The most common symptoms seen in diseases caused by phytoplasmas are phyllodes (expanded and/or flattened petioles resembling and having the function of a leaf), witches'-broom, virescence, dwarfism, sterility of floral organs, wrinkled leaves, and yellowing (Bertaccini & Duduk, 2009). The distribution and accumulation of phytoplasmas are irregular in host plants (Pardo, 2013).

Phytoplasmas are becoming increasingly important in cassava crops grown in Asia, Latin America, and the Caribbean. Losses caused by cassava phytoplasma diseases vary greatly depending on the regions, local environmental conditions, and cassava varieties. Existing reports have documented yield losses in commercial crops of up to 90% (Álvarez et al., 2009). The most important cassava diseases associated with phytoplasma infection are cassava frogskin disease (CFSD) reported in Colombia as early as 1971 (Pineda, Jayasinghe, & Lozano, 1983) and cassava

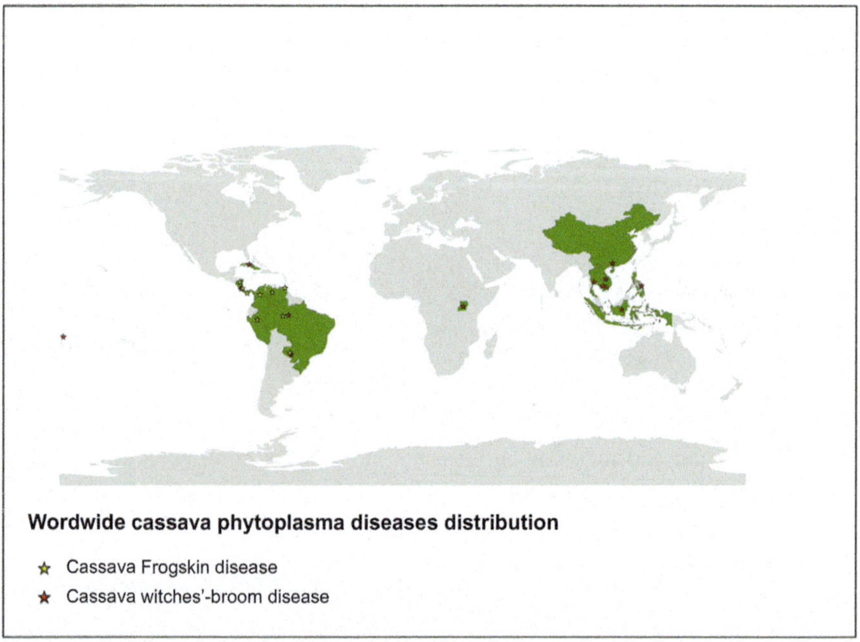

Fig. 7.1 Cassava Frogskin and witches'-broom disease distribution world-wide

witches'-broom (CWB) discovered in 2010 in Southeast Asia (Fig. 7.1) (Álvarez et al., 2013). Phytoplasmas are spread through vegetative propagules such as stem cuttings, but also through interactions with insect vectors. Changes in abundance levels and/or patterns of distribution of the vectors directly impact the transmission and geographic spread of the pathogens. Since insects can be particularly sensitive to abiotic factors, such as temperature, this also means that anticipated climate change will have a significant impact on transmitted pathogens (Legg & Álvarez, 2017).

7.2 Cassava Phytoplasma Disease in Latin America and the Caribbean

7.2.1 Cassava Frogskin Disease (CFSD)

7.2.1.1 Distribution, Symptoms and Epidemiology

The origin of CFSD is probably the Amazon region of Colombia, Peru, or Brazil. CFSD was described for the first time in Colombia in 1971, in the cassava-growing region of Quilcacé, Cauca Department, and is considered to be one of the most problematic diseases for cassava cultivation. CFSD affects the production of roots, causing losses in crop yield of more than 90% in Colombia (Pineda et al., 1983). CFSD has been reported in Colombia (Álvarez et al., 2009), Costa Rica (Pardo, Truke, Cardozo Téllez, Varela, & Álvarez, 2014), Paraguay (Cardozo Téllez, Pardo, Zacher, Torres, & Álvarez, 2016), Brazil (Oliveira et al., 2014), Panama, Peru, and Venezuela (Calvert, 1994; Chaparro-Martínez & Trujillo-Pinto, 2001; Pineda et al., 1983). In Suriname, cassava farmer field visits were conducted in 2018. Diseased cassava plants exhibiting CFSD-like symptoms were sampled in fields located in two farmer communities in Maho and Kampong Baroe in the Saramaca district. Farmers mentioned an increasing incidence of CFSD in their area and also in other cassava growing areas in Suriname (Alvarez, 2018 unpublished). A phytoplasma was detected in CFSD-infected cassava plants and identified as belonging to the group 16SrIII subgroup L (Álvarez et al., 2009).

The disease causes deep lesions in roots, decreasing their diameter, and increasing their woodiness. More specifically, symptoms consist of small longitudinal fissures distributed throughout the root. As roots increase in diameter, the fissures tend to heal, giving the injuries a lip form. Symptomatic roots have a woody aspect and a thickened peel that is cork-like, fragile, and opaque. The peel also presents lip-like slits that create a net-like or honeycomb pattern when joined (Álvarez et al., 2009; Cardozo Téllez et al., 2016; Pardo et al., 2014). The expression of CFSD symptoms is influenced by temperature and host genotype. Depending on the severity of symptoms, the depth and number of lesions increase until the root becomes deformed (Álvarez, Mejía, Loke, Hernández, & Llano, 2003; Pineda et al., 1983). The diseased cassava roots contained very low level of starch.

The disease is exponentially propagated through asexual propagation of infected host plants. Indeed, since stems of diseased plants are thicker than those of healthy ones, diseased plants tend to be selected by growers to provide seeds, hence increasing the disease spread. Moreover, most cassava varieties infected with CFSD express no leaf or stem symptoms and farmers usually remove the stems before harvesting the roots. This means that symptoms are observed only when the plants are harvested (Betancourth, 2016; Rao et al., 2018). Therefore, the production of disease-free planting materials to prevent the dissemination of CFSD is greatly needed as well as the introduction of more rigorous screening practices (Pardo et al., 2014). The disease can also be transmitted via dodder (*Cuscuta* sp. L., Convolvulaceae) (Fig. 7.2a–d) and grafting (CIAT, 2005).

Research confirmed that CFSD is not transmitted by soil, but by aerial vectors of the Hemiptera families, such as Cicadellidae and Delphacidae (Mejía et al., 2011). The *Scaphytopius* genus includes species such as *S. fuliginosus* Osborn (Hemiptera: Cicadellidae), which transmits the phytoplasma causing shoot proliferation in soybean (Granada, 1979) and *S. marginelineatus* Stål, which carries phytoplasmas associated with CFSD (CIAT, 2004, 2005). Recently, it has been demonstrated that the leafhopper *S. marginelineatus* is able to acquire the CFSD phytoplasma from infected plants and successfully transmit it to healthy plants (Álvarez & Betancourth, 2016). Further studies are required to determine how significantly this vector transmission affects the field and regional-level epidemiology of CFSD.

7.2.1.2 Molecular Characterisation of Phytoplasma

The CFSD-associated phytoplasma strains were identified by PCRs, restriction fragment length polymorphism (RFLP) and sequence analyses of amplified rDNA products. Results were corroborated by PCRs employing group 16SrIII-specific rRNA gene or ribosomal protein (rp) gene primers (Álvarez et al., 2009). A real-time PCR and a TaqMan probe were also developed, based on the *rp* gene (16SrIII-L phytoplasma) and increased detection sensitivity from 100- to 1000-fold higher than that obtained from conventional PCR tests (Álvarez, Mejía, & Pardo, 2010;

Fig. 7.2 **(a–d)** (**a** and **b**) Process of transmission in cassava plants infected by phytoplasma 16SrIII-L to healthy plants of *Cuscuta* sp. (**c**) Plant of healthy *Catharanthus roseous* before transmission. (**d**) Symptoms in plants of *C. roseous* after the transmission process from the phytoplasma

Rao et al., 2018). In Brazil, a phytoplasma 16SrIII subgroup A was reported in cassava (De Souza, Da Silva, Bedendo, & Carvalho, 2014). Also in Brazil, a phytoplasma causing the same type of symptoms reported by Álvarez et al. (2009) and belonging to the group 16SrIII subgroup L was also described (Oliveira et al., 2014). In Paraguay, phytoplasma belonging to group 16SrI and 16SrIII were detected in Caaguazú, Isla Sola, and Barrientos Cue (Cardozo Téllez et al., 2016). In Costa Rica, phytoplasma from group 16SrIII-L and associated with CFSD were detected in Aguas Zarcas, Pital, Los Lagos, La Merced, Los Chiles, Alajuela, Fortuna, and Guápiles (Pardo et al., 2014). In addition, a phytoplasma 16SrI group was detected in San Carlos, Costa Rica (Álvarez, Pardo, Contaldo, et al., 2014). Using phylogenetic analysis on 16SrDNA and tuf genes, Mejia (2014) also identified phytoplasma associated to CFSD and belonging to group 16SrI and 16SrX in Paraguay (San Pedro province) and group 16SrI, 16SrIII, 16SrX-A, 16SrX-B and 16SrXII in Costa Rica (Guanacaste, Alajuela and Limòn). Mixed infection were also observed with 16SrI phytoplasma found with 16SrX-A, 16SrX-B, 16SrIII or 16SrXII, and 16SrIII found with 16SrXII in Costa Rica. No mixed infection was observed in Paraguay (Mejia, 2014). The results from Mejia (2014) allowed the association of symptoms of slightly opaque roots with reduced diameter/volume and inflexible brittle peel with many fissures or lip-like splits, with phytoplasmas belonging to groups 16SrIII and -XII in Costa Rica and 16SrI and -X in Costa Rica and Paraguay. Interestingly, disease symptoms observed in the field in Paraguay were less severe than those observed in Costa Rica fields, where phytoplasma group 16SrIII-L showed greater disease severity (Pardo et al., 2014). PCR and real- time tests using specific CFSD primers as well as a tagman probe previously developed at CIAT showed that the Suriname CFSD strain was associated with phytoplasma belonging to the 16Sr III-L subgroup (Álvarez, 2018 unpublished). Disease symptoms observed in the field in Suriname were severe. A summary of the subgroups and strains found in Cassava is indicated in Tables 7.1 and 7.2.

7.2.1.3 Phytoplasma Isolate and Pathogenicity Tests

In Colombia, *in vitro* isolation of phytoplasmas associated with CFSD was successfully accomplished using an artificial culture medium described by Contaldo, Bertaccini, Paltrinieri, Windsor, and Windsor (2012). Fragments of roots, petioles, stems, leaves, and embryos from diseased cassava plants were placed in liquid medium (P*iv*L) that changes colors when pH is below 6.8, indicating potential

Table 7.1 Phytoplasma subgroups detected in Cassava grown in Latin America and Caribbean countries

Country	Group
Brazil	III-A, III-B, III-L
Colombia	III-L
Costa Rica	I, III-L, X-A, X-B, XII
Cuba	II, I-C
Paraguay	I, III, X

Table 7.2 Phytoplasma strains detected in cassava grown in Latin America and the Caribbean

Disease	Country	16S ribosomal RNA gene		Institution	References
		Clasification	GenBank Accesion		
Frogskin	Colombia	16SrIII-L	KC999115	CIAT	(Álvarez et al., 2013)
Frogskin	Colombia	16SrIII-L	KC816736.1	CIAT	(Álvarez et al., 2013)
Frogskin	Colombia	16SrIII-L	EU346761.1	CIAT	(Álvarez, Llano, et al., 2012)
Frogskin	Colombia	16SrIII-L	AY737646.1	CIAT	(Álvarez, Llano, et al., 2012)
Frogskin	Colombia	16SrIII-L	AY737647.1	CIAT	(Álvarez, Llano, et al., 2012)
Frogskin	Colombia	16SrIII-L	KC999116	CIAT	(Álvarez et al., 2013)
Frogskin	Costa Rica	16SrIII	KC816736.1	TEC-INTA-CIAT	(Álvarez et al., 2013)
Frogskin	Paraguay	16SrIII	KF701485	IPTA-CIAT	(Cardozo Téllez et al., 2016)
Frogskin	Paraguay	16SrIII	KF701497	IPTA-CIAT	(Cardozo Téllez et al., 2016)
Frogskin	Paraguay	16SrIII	KF701498	IPTA-CIAT	(Cardozo Téllez et al., 2016)
Frogskin	Brazil	16SrIII-L	KF019185	EMBRAPA	(Oliveira et al., 2014)
Frogskin	Brazil	16SrIII-L	KF019184	EMBRAPA	(Oliveira et al., 2014)

growth of phytoplasma. When the P*iv*L changed color from red to orange, an aliquot of 50 μL was transferred to solid medium (P*iv*S) (Betancourth, 2016; Rao et al., 2018). The P*iv*L medium changed color 5 to 10 days post-inoculation, when root and embryo tissues were used, and colonies on P*iv*S were observed 12 to 15 days after transfer to solid medium. Optical and transmission electron microscopy were used to examine the solid medium (Figs. 7.3a–e and 7.4a–d). Colonies of 1-mm diameter (20X) constituted of pleomorphic cells without walls and measuring between 0.5 and 2 microns were observed in the solid medium. The presence of phytoplasmas was verified using molecular techniques such as qPCR technique and sequencing. DNA polymorphic patterns obtained from the microorganism grown in both media, were similar to those of group III, and sequences showed 99% homology with CFSD phytoplasma. This is the first report of the isolation of a phytoplasma associated with CFSD using an artificial culture medium (Betancourth, Pardo, Truke, Muñoz Flórez, & Álvarez, 2014; Rao et al., 2018).

Phytoplasma isolated from diseased cassava tissues and grown in P*iv*L media were evaluated for pathogenicity. *In vitro* plants of cassava clones CM 2952, Col 1, Col 896, Bra 184, SM 909–25, Per 334, Per 326, Per 534, and Per 1999 (Roca, Beltrán, Jorge, & Fernández, 1984) were grown in soil in a greenhouse for 60 days and inoculated by stem and root injection with phytoplasma grown in P*iv*L media. Inoculation consisted of 1 mL of phytoplasma cells inoculum grown in P*iv*L medium, injected either into the cassava stem, at the point of insertion of the petiole in the stem, or directly into the first storage root. The negative control consisted in

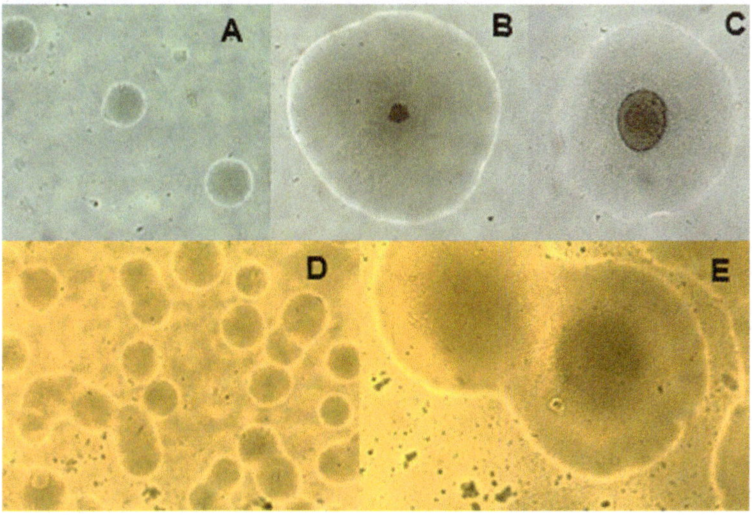

Fig. 7.3 (**a–e**) Colonies of phytoplasmas grown in P*iv*S medium from P*iv*L, in an atmosphere of 95% N_2 and 5% CO_2, photographs 40 and 100 X. Colonies obtained from cassava tissue (**a–c**) or from Vinca tissue (**d, e**). (**a**) Colonies at 2 days of incubation. (**b**) Single colony at 5 days of incubation. (**c**) Individual colony at 12 days of incubation. (**d**) Colonies at 3 days of incubation. (**e**) Colonies at 8 days of incubation

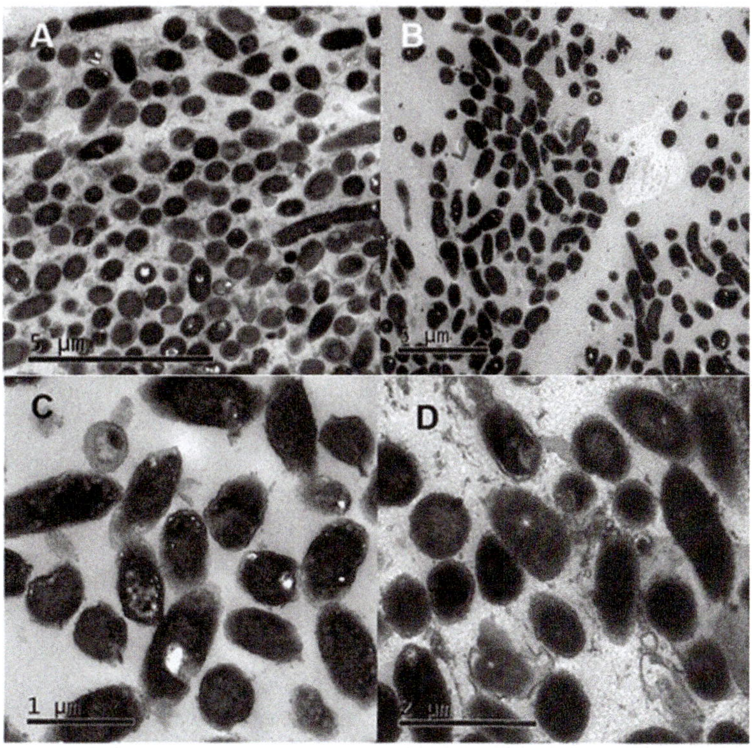

Fig. 7.4 (**a–d**) (**a, b**) Proliferation of phytoplasma cells with different shapes and sizes. (**c, d**) Cells pleomorphic without cell wall and irregular cell membrane

the inoculation of phytoplasma-free P*iv*L medium. Once inoculated with phytoplasma-free P*iv*l or phytoplasma-infected P*iv*l, the plants were maintained in an insect-proof greenhouse until root symptoms developed (Betancourth, 2016). The inoculated plants were evaluated every 2 months. Pathogenicity was confirmed in five genotypes (CM 2952, Col 1, Col 896, Bra 184, SM 909–25) that exhibited severe CFSD symptoms on roots, 6 months after inoculation. This study reports for the first time the pathogenicity of phytoplasma isolated on artificial culture medium originated from cassava roots (Álvarez et al., 2017).

7.2.2 Cassava witches'-broom disease

Phytoplasmas have been associated with an important disease known as cassava witches'-broom (CWB) in diverse tropical regions. In Latin America, the disease has been reported in Brazil, Cuba, Argentina, Venezuela, Mexico and Peru (Arocha, Piñol, et al., 2009; Fernández, Uset, Baumgratz, & Conci, 2018; Lozano, 1992). Here, we are presenting data about the biology of CWB affecting cassava in Cuba (Arocha, Piñol, et al., 2009), Brazil (Lozano, 1992) and Argentina (Fernández et al., 2018).

7.2.2.1 Biology

CWB was first described in the 1940s in the southeast region of Brazil (Silberschmidt & Campos, 1944) and was later observed in plants grown in the central (Kitajima & Costa, 1971), northeast (Mariano, Laranjeiras, Santos, Padovan, & Peixoto, 1991) and northern regions of Brazil (Lozano, 1992). Molecular analysis using specific primers evidenced that the phytoplasma found was affiliated with group 16SrIII (X-disease group). Phylogenetic analysis showed that this phytoplasma is closely related to the subgroup 16SrIII-B (Flôres, Haas, Canale, & Bedendo, 2013). In Cuba, subgroup 16SrI-C was reported in organoponic culture of cassava (Arocha, Piñol, et al., 2009). Cassava plants displaying symptoms of dwarfing, witches'-broom, and chlorosis, were also detected in production fields located in Misiones province in Argentina, where Fernández et al. (2018) identified the phytoplasma as belonging to the 16SrIII-J subgroup.

7.2.2.2 Symptoms and Effects on Yield

In Brazil, the disease has been associated with several types of symptoms such as stunting and excessive proliferation of branches, small rougher leaves and shortened internodes, proliferation of shoots from the cutting with generally weak growth, stunted shoots germinating from the cuttings and black necrotic spots (Jayasinghe, Pineda, & Lozano, 1984; Lozano, 1992). In the northeast regions of Brazil, CWB was present in 85% of the fields and caused production losses up to 70% (Fukuda et al., 1990). The disease also caused yield losses of cuttings for replanting, as affected plants displayed reduced size and excess budburst (Da Silva Souza & Fialho, 2012). Recently, the disease was observed in fields located in the state of São Paulo (SP), southeastern Brazil, with symptoms of generalized stunting, shoot proliferation, foliar chlorosis, and leaf malformation (Flôres et al., 2013). Cassava plants exhibiting symptoms of leaf yellowing, small roots, and stunting were also observed in Cuba (Acosta et al., 2014).

7.2.3 Cassava Phytoplasma Antholysis

7.2.3.1 Biology

Jayasinghe, Pineda, and Lozano (1983) observed antholysis in cassava grown in southwestern Colombia in 1981, with severe symptoms being observed in the field. Cassava antholysis occurs in Brazil, the Caribbean, Central America, Colombia, and Venezuela (Frison & Feliu, 1991), but does not seem to have high economic importance. The causal agent is a phytoplasma and its dissemination is mainly by vegetative propagation (Jayasinghe et al., 1984).

7.2.3.2 Symptoms

The symptoms of the disease appear in the inflorescence, with a characteristic petal virescence in which pigment turns from pink to green. Hypertrophy of the petals is later observed and they become structures similar to leaves (phyllody). The floral racemes lose their normal appearance and resemble sprouts, giving this syndrome its name, "antholysis" (*antho* = flower; *lysis* = dissolve, loosen). Infected flowers commonly exhibit a very swollen gynophore and develop internodes in the floral receptacle, a phenomenon known as apostasis (Fig. 7.5a, b). Furthermore, elongation of the receptacle occurs above the insertion of the pistil, with development of sprouts. Flower fertility is lost, resulting in nonfunctional flowers that abort prematurely. Affected plants do not present symptoms in other organs; moreover, germination does not differ between infected and healthy stakes (Jayasinghe et al., 1983).

7.2.3.3 Epidemiology

Transmission is 100% by stakes. Under greenhouse conditions, symptoms of antholysis appear within one month of planting (flowering is induced more rapidly), contrasting with healthy plants which take five months to flower. The disease is reduced by selecting stakes from healthy plants. Varietal resistance also exists. Treatment with penicillin (500 to 1000 ppm) did not reduce symptoms, whereas

Fig. 7.5 **(a, b)** (**a**) Cassava inflorescence severely affected by antholysis, where flowers transformed into green leaf-like structures can be observed. (**b**) Normal inflorescence of a healthy plant. (Source: Jayasinghe, 1983)

tetracycline reduced antholysis by 90%. Tetracycline sensitivity and detection by Dienes' stain confirmed that the causal agent is a phytoplasma (Jayasinghe et al., 1983).

7.2.4 CFSD, CWB and Antholysis Control

Today, phytoplasma diseases are managed principally by taking cuttings only from symptomless plants, by avoiding planting new plots of cassava next to those that have the disease and removing and burning any plants with symptoms as soon as they are seen, as well as avoiding moving asexual seeds from infested plots to new areas. However, more could be done locally to prevent disease propagation. For instance, cassava roots should be inspected for symptoms at harvest and only cuttings from apparently healthy plants that bear normal roots should be selected. Also, plant health surveillance could be strengthened to prevent entry of planting materials from areas with the disease. Quarantine procedures are in place for CFSD and CWB- infected cassava tubers in shipment (Frison & Feliu, 1991), but procedures need improvement to prevent long-range disease spread within Latin America.

Using clean planting material, resistant genotypes and resistance inducers, as well as early disease diagnosis are reliable strategies that need to be developed and implemented in order to sustainably manage phytoplasma diseases.

A heat treatment has been developed and implemented in Colombia for the mass-propagation of clean cassava planting material and the same approach was subsequently validated in Brazil, Costa Rica, and Paraguay (Álvarez et al., 2017). Thermal chamber systems reportedly raised efficiency of cassava production to 90% as compared to traditional farming systems and also increased the availability of seeds of commercial genotypes for farmers, with an estimated monthly production of about 150 clean cassava plants per square meter. Thermotherapy also reduced the risk of disease dissemination and stabilized or increased food security, crop productivity, sustainability and competitiveness for farmers. Thermotherapy coupled or not with other *in vitro* technologies or antibiotic treatments can also be used to produce seeds free of phytoplasma, insects and other pathogens such as bacteria or viruses (Cassells, 2012; Laimer & Bertaccini, 2019). However, thermotherapy can be complicated for farmers to manage and can lead to high initial losses if protocols are not accurately followed (Coyne, Claudius-Cole, Kenyon, & Baimey, 2010).

Phytoplasma-free plant materials can also be obtained with tissue culture technologies, such as meristem cultures, embryo rescues and the use of somatic embryos and synthetic seeds (Chavarriaga-Aguirre et al., 2016). The drawback of tissue culture of cassava is the somaclonal variation occurring during meristem (or microcutting) culture that leads to alteration of the methylation patterns and therefore to phenotypic, genetic and epigenetic changes (Kitimu et al., 2015). These changes can be unwanted but can also be considered as a source of exploitable genetic variations. Other drawbacks are the costs associated with the use of biotechnology, the lack of disease-free certification for the initial material and the need for the support of official agricultural agencies to ensure low cost of clean materials (Chavarriaga-Aguirre et al., 2016). Nursery entrepreneurs and phytosanitary National Plant Protection Institutions in Latin America have improved their production and certification processes, but more needs to be done. Programs for producing certified planting materials should be increased in the principal production regions of Colombia, Brazil, and Costa Rica, particularly for those materials with desirable traits such as higher yield, industrial usefulness, and resistance to diseases. In addition, an optimal system of efficient mass production of planting materials should be set up, consonant with farmers' needs for establishing cash crops.

During germplasm evaluation studies in Colombia, cassava genotypes HMC-1 and CM 1495-7 showed low incidence of CFSD (5%) and expressed mild symptoms, whereas genotypes MCOL 1505 and NIG 11 were severely affected and showed disease incidence of 60%. In Costa Rica, genotype Per 183 is promising, as disease symptoms were not observed in this genotype over the two years of evaluation (Álvarez, Gómez, Zacher, & Campo, 2014). In Colombia, genotype Per 183 showed resistance to CFSD whereas genotypes HMC-1, CM 4574-7, and local genotype Cumbre 3 showed tolerance to the disease. These genotypes demonstrate strong potential to replace commonly used susceptible genotype MBRA 383.

Resistance inducers (RIs) work indirectly on the pathogen by activating plant defense responses, therefore reducing the use of toxic pesticides. RIs have the potential to be more environmentally sustainable, with less impact on human health (Burketova, Trda, Ott, & Valentova, 2015). In Colombia, the use of resistance inducer Kendal® produced a 20% increase in yield and reduced disease incidence by 10%, whereas, in Paraguay, this inducer increased the weight of commercial roots. In Costa Rica, symptoms were not observed in plants treated with Kendal® (Álvarez, Gómez, et al., 2014).

Genetically modified cassava is being studied and GM varieties already exist, but are still at the proof-of-concept stage with several field trails reported in the last few years, and therefore, will not be available commercially before a few years (Zhang et al., 2017).

7.3 Cassava Phytoplasma Disease in Asia

7.3.1 Cassava witches'-broom disease

Asia's cassava sector has been affected by a number of insect-vectored diseases, including cassava witches'-broom (CWB) disease. CWB disease was first observed in Southeast Asia, in 2010 (Graziosi et al., 2016). At present, CWB occurs in over 60% of Asian cassava fields at average incidence levels of 30–35% (Graziosi et al., 2016). CWB has been reported in Cambodia, China, Indonesia, Laos, the Philippines, Thailand and Vietnam and is still spreading (Jackson, 2014). RFLP analyses of nested PCR amplified fragments from Vietnamese and Cambodian CWB phytoplasmas indicated the presence of different strains all related to the 16SrI group (Álvarez et al., 2013). Other phytoplasma strains were also identified in Asia, such as '*Ca.* P. aurantifolia' (16SrII) in the Philippines (Aliaga, Álvarez, & Pardo, 2017) and Thailand (Graziosi et al., 2016) and '*Ca.* P. pruni' (16SrIII) in cassava in Japan in 2018 (Koinuma et al., 2018).

In Cambodia, in 2012, cassava farms were affected by a disease associated with CWB, causing significant crop losses of up to 50% (Álvarez, Pardo, & Truke, 2014). The disease was observed in the Kampong Cham, Kratie and Prey Veng provinces of Cambodia. Symptoms included yellowing, leaf proliferation, and stunting of cassava plants, in addition to reductions in starch content (Álvarez, Llano, & Mejía, 2012; Rao et al., 2018). Total DNA from samples collected from symptomatic and asymptomatic plants were extracted using a modified methodology from Doyle and Doyle (1990). Direct PCR assays were performed using phytoplasma universal primer P1A/P7A (Lee et al., 2004). PCR products were then used in nested PCR with primers R16F2n/R16R2 (Gundersen-Rindal & Lee, 1996). Samples were also evaluated, using real-time PCR and a TaqMan® probe specific for detecting the 16SrI gene (Tables 7.3 and 7.4 and Fig. 7.6). Sequence analyses of partial 16S rRNA fragments showed that Cambodian CWB phytoplasmas have 99% to 100%

Table 7.3 Sequences of the CWB-F and CWB-R primers as well as CWB-P Taqman probe for PCR tests

CWB-F	CATAGGGGGCAAGCGTTATCC
CWB-R	TGAGCATTGCACTTAGACCATAAAC
CWB-P	/5'-FAM/ TGGGCGTAAAGGGTGCGTAGGCTGT/ BHQ2-'3

Table 7.4 Table of Ct values obtained through real-time PCR for samples positive for 16SrI gene (witches'-broom phytoplasma)

0 fluorescence threshold		0.5 fluorescence threshold			
Sample code no.	Ct (dRn)	Sample code no.	Ct (dRn)	Sample code no.	Ct (dRn)
+ Control (16SrI–B)	*26.37*	*+ control (16SrI–B)*	*25.89*		
3–7	29.39	8–3	29.85	6–6	33.07
5–2	31.36	6–5	30.19	7–5	33.07
6–3	31.85	9–3	30.47	3–3	33.14
6–4	31.97	3–1	30.90	6–7	33.76
3–2	32.65	3–5	31.35	3–4	33.99
6–2	32.77	5–4	31.97	3–9	34.20
9–1	33.42	8–4	32.04	9–4	34.57
8–3	33.89	5–4	32.06	6–1	34.89
4–4	34.06	5–2	32.13		
7–1	34.20	9–2	32.44		
8–1	34.57	6–3	32.50	*– Control*	*No Ct*
4–3	34.66	4–1	32.54		
8–5	34.67	4–2	32.78		
7–2	34.70	6–5	32.84		
7–4	35.00				
– Control	*No Ct*				

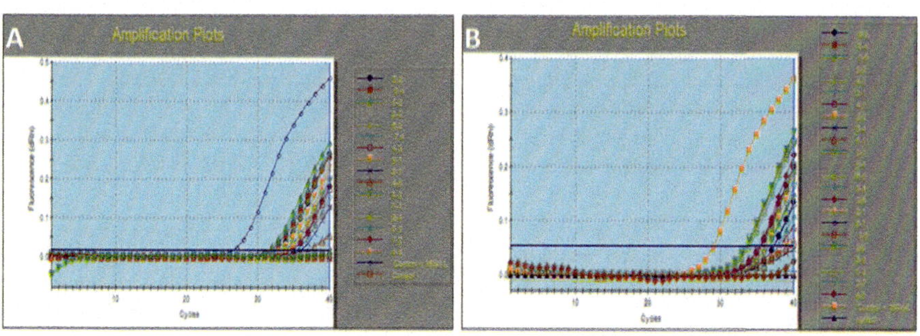

Fig. 7.6 Amplification curves for the 16S rRNA gene, using real-time PCR, for 39 cassava samples (16 samples in A and 23 samples in B) that had tested positive for the presence of witches'-broom phytoplasma (R2 = 0.98, slope – 3.35). Positive control phytoplasma 16SrI

similarity with '*Ca.* P asteris'(16SrI phytoplasma group) (Álvarez et al., 2013). *In silico* RFLP analyses, cloning, sequencing and phylogenetic analyses using Maximum Likelihood method based on the Kimura 2-parameter model (Kimura, 1980) and MEGA6 (Tamura, Stecher, Peterson, Filipski, & Kumar, 2013) demonstrated that phytoplasma affecting cassava in Cambodia belonged to ribosomal subgroups 16SrI-C (Figs. 7.7 and 7.8).

In Thailand, CWB was first observed in 2008 (Álvarez, unpublished). In 2012, cassava farms were affected by a phytoplasma disease that caused significant crop yield losses and reductions in starch content. Symptoms were similar to those found in Cambodia and were associated with CWB. Affected plants show bunches of shoots with short internodes, small yellowish leaves at the top of the plants, brown vascular tissues and poor storage root development (Jackson, 2014; Rao et al., 2018). Since 2008, the disease spread rapidly and is now widely distributed across the region.

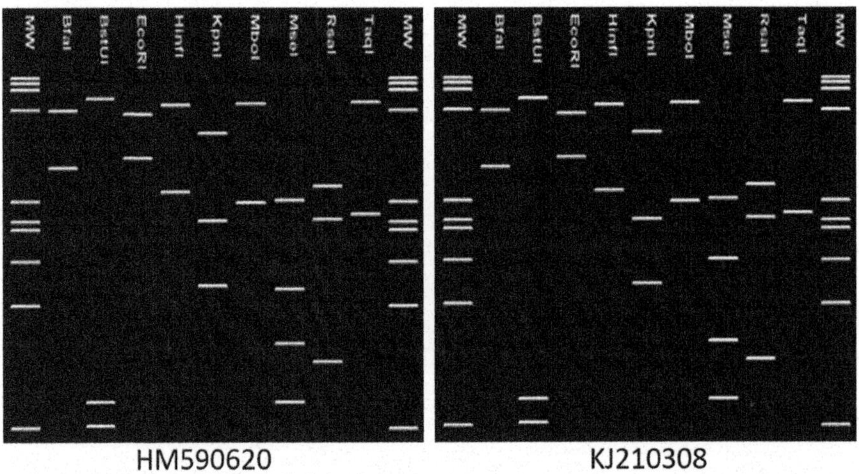

Fig. 7.7 Virtual RFLP patterns from *in silico* digestions of 16S rDNA R16(I)F1/R1 fragments. Patterns of Cassava witches'-broom phytoplasma strain (accession number KJ210308) and representative strain belong to 16SrI-C subgroup (accession number HM590620). Recognition sites for the following 9 restriction enzymes were used in the simulated digestions: *BfaI, BstUI, EcoRI, HinfI, KpnI, HhaI, MboI, HpaI, HpaII, MseI, RsaI, TaqI*. MW: ΦX174 DNA-*HaeIII* digestion

Fig. 7.8 (continued) model (Kimura, 1980). The tree with the highest log likelihood (−1548.5687) is shown. The percentage of trees in which the associated taxa clustered together is shown next to the branches. Initial tree(s) for the heuristic search were obtained automatically by applying Neighbor-Join and BioNJ algorithms to a matrix of pairwise distances estimated using the Maximum Composite Likelihood (MCL) approach, and then selecting the topology with superior log likelihood value. The tree is drawn to scale, with branch lengths measured in the number of substitutions per site. The analysis involved 45 nucleotide sequences. There were a total of 744 positions in the final dataset. Evolutionary analyses were conducted in MEGA6 (Tamura, Stecher, Peterson, Filipski, & Kumar, 2013)

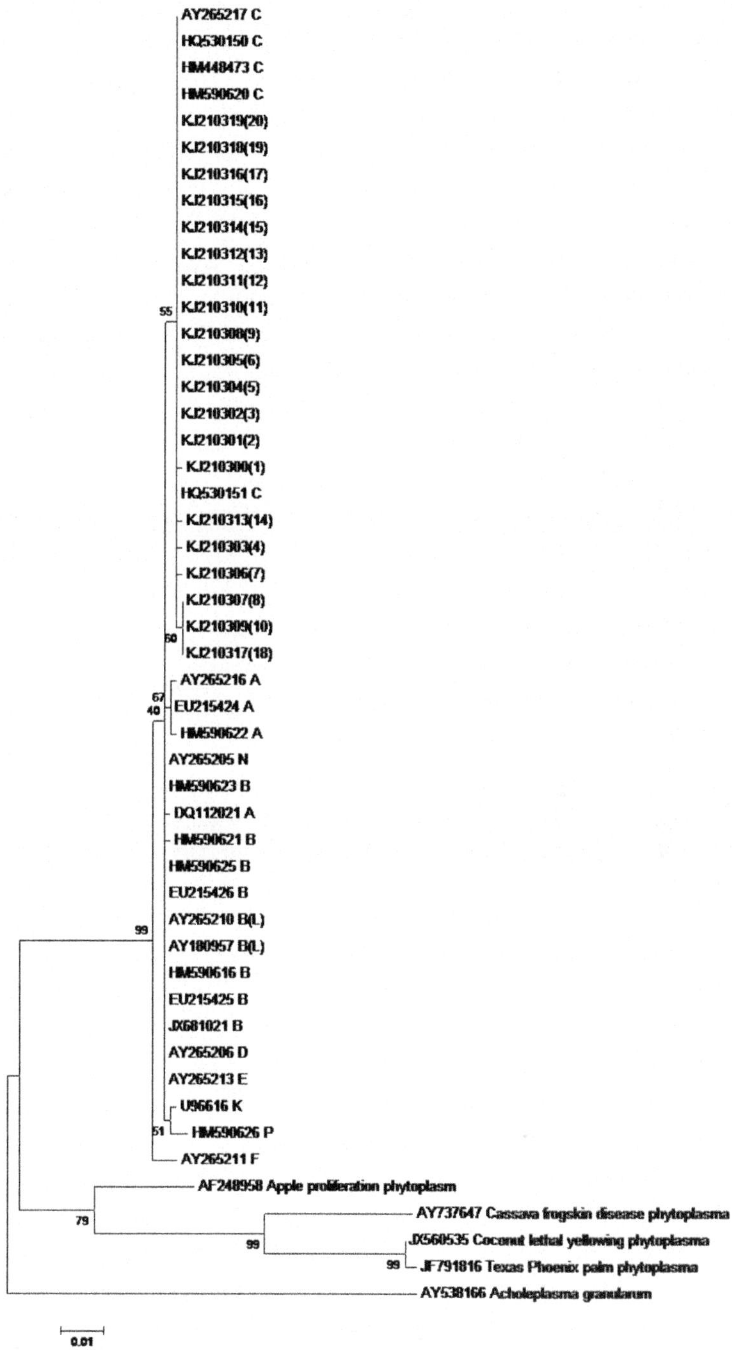

Fig. 7.8 Molecular Phylogenetic analysis by Maximum Likelihood method . The evolutionary history was inferred by using the Maximum Likelihood method based on the Kimura 2-parameter

Table 7.5 Selected samples from Thailand evaluated by nested PCR and RFLP methods

Sample no.	Genotype	Source	Nested PCR R16(I)F1/R1 R16F2n	RFLP/ sequencing	Accession no.
11–1	Rayong 5	Leaves	+	16SrI-C	KJ880973
4–1	Rayong 86–13	Leaves	+	16SrI-C	KJ910010
9–2	Huay bong 60	Leaves	+	16SrI-C	KJ910009
2–1	Rayong 86–13	Leaves	+	16SrI-C	KJ880974
5–3 (1)	Rayong 11	Roots	+	16SrI-C	KJ910016
13–1	Rayong 9	Leaves	+	16SrI-C	KJ910017
4	Rayong 86–13	Leaves	+	16SrI-C	KJ880975
7	Rayong 7	Leaves	+	16SrI-C	KJ880976
2	Rayong 86–13	Leaves	+	16SrI-C	KJ910014
6–1	Rayong 11	Leaves	+	16SrI-C	KJ910011
11	Rayong 5	Leaves	+	16SrI-C	KJ880977
8	Huay bong 60	Leaves	+	16SrI-C	KJ910015
6–3	Rayong 11	Roots	+	16SrI-C	KJ910018
5–1	Rayong 11	Leaves	+	16SrI-C	KJ910012
7–1	Rayong 7	Leaves	+	16SrI-C	KJ910013

Samples of diseased cassava tissues were collected in the Rayong province in Thailand and total DNA was extracted, using the Dellaporta, Wood, and Hicks (1983) protocol with modifications. The 16S gene from rDNA was amplified, using the nested PCR technique with pairs of universal primers for direct PCR, P1A/P7A (Schneider, Seemüller, Smart, & Kirkpatrick, 1995) and R16F2n/R2 for nested PCR (Lee, Gundersen-Rindal, Davis, & Bartoszyk, 1998). The nested PCR products were amplified in a second nested PCR with primers pair R16(I)F1/R1 (Anfoka & Fattash, 2004) specifically targeting 16Sr I phytoplasma group. PCR tests revealed that 21 out of 31 samples, were positive (Table 7.5). Sequences obtained from the products amplified by primers R16(I)F1/R1 were analysed using the BLASTn tool, and showed high levels of homology (99%–100%) with sequences of the 16SrI group ('*Ca.* Phytoplasma asteris') (Table 7.6). *In silico* RFLP analyses of the thai samples and representatives of other phytoplasma subgroups, indicated that the cassava CWB phytoplasma isolates from Thailand are related to the 16SrI-C subgroup (Fig. 7.9). The phylogenetic tree illustrated in Fig. 7.10 shows that the sequences obtained from phytoplasma isolates clustered in the same clade as Group 16SrI-C. This result coincides with those from the sequence analyses obtained with the BLASTn tool and RFLP analysis. This is the first report of cassava infection with a phytoplasma related to the 16SrI-C group in Thailand. Sampling in the Chacheoengsao and Rayong regions indicated that phytoplasma belonging to

Table 7.6 Homology found in order of appearance for DNA sequences of phytoplasmas detected in cassava samples with those of phytoplasmas reported in GenBank

Pairing with GenBank[a]	GenBank code[b]	Homologous bases[c]	Probability of higher homology[d]	Homology (%)
Cassava witches'-broom phytoplasma clone 9–3, 16S ribosomal RNA gene partial sequence	KC009838	956/959	0	100
Clover phyllody phytoplasma strain CPh, clone S1, 16S rRNA gene	KC283218	955/956	0	100
'Euscelis lineolatus' phytoplasma isolate EUC4, 16S rRNA gene (1025)	KC243397	947/947	0	100
Cassava witches'-broom phytoplasma clone 3–3, 16S ribosomal RNA gene partial sequence	KJ210317	965/966	0	99
Cassava witches'-broom phytoplasma clone 5–3, 16S ribosomal RNA gene partial sequence	KJ210313	965/966	0	99
Cassava witches'-broom phytoplasma clone 8–4, 16S ribosomal RNA gene partial sequence	KJ210307	909/909	0	99

[a]Values in parentheses indicate number of total bases reported in GenBank
[b]Code assigned by GenBank
[c]Number of homologous bases in different regions of the sequence reported in GenBank
[d]The value 0 is expected for high homology percentages

16SrIII, VI and XV were also present in CWB-infected cassava plants, with groups I, VI and XV found in mixed infections in a few plants (Mejia, 2014).

In 2015 and 2016, plants with typical CWB symptoms (yellowing, extensive leaf proliferation, shortened internodes, stem necrosis and stunting) were collected from Rayong, Prachinburi, Chachoengsao, Srakaew, Kanchanaburi, and Kampeangphet cassava-growing provinces in Thailand (Graziosi et al., 2016). Sequencing results showed a 99% similarity between Thai samples and 17 CWB disease accession KM280680. RFLP profiles were identical to accession L33765 (Peanut witches'-broom 16SrII) (Lee et al., 1998) (Fig. 7.11), and phylogenetic tree showed sequences in the same clade of phytoplamas belonging to 16SrII group (Fig. 7.12). This is the first report of a 16SrII phytoplasma in cassava in Thailand.

In the **Philippines,** CWB disease-like symptoms (yellowing or reddening of leaves, shrinking of the internodes resulting in stunted growth, excessive multiplication of shoots, smaller leaves, and premature death) were observed in cassava fields (Argana, 2016). Direct and nested PCR products indicated that 16 out of the 29 samples collected from Visayas and Mindanao regions were positive for phyto-

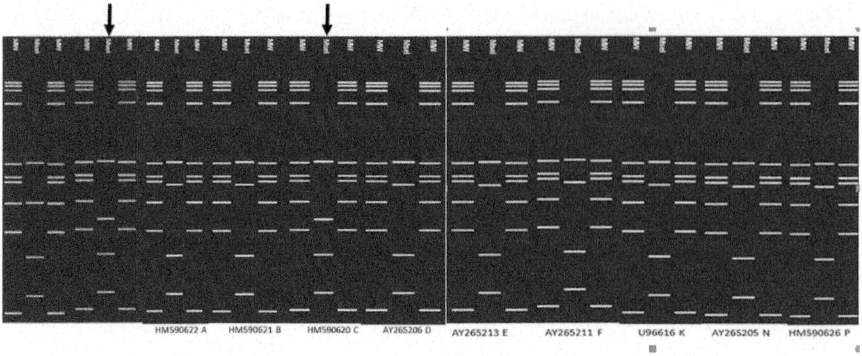

Fig. 7.9 *In silico* restriction analysis of the 16S rRNA gene of two positive samples from Thailand selected as patterns for this restriction (Lanes 1 and 2, arrows) and selected representatives of the 16SrI subgroups (Lanes lanes 3-11), using enzyme *MseI*. The molecular weight markers used are represented by the first and last lanes (ϕ174 and HyperLadder II, respectively and HiperLader II)

plasma infection. RFLP profiles of the cassava symptomatic leaf samples and phylogenetic analyses demonstrated that the phytoplasma belonged to the 16SrII phytoplasma group (Figs. 7.13 and 7.14) (Aliaga, Álvarez, et al., 2017). This is the first report of a 16SrII phytoplasma in cassava in the Philippines.

Field observations and molecular tests showed that cassava varieties KU-50, Rayong 5 and 72 are highly susceptible to CWB. CWB resulted in 100% yield loss if symptoms are observed during plant establishment, and 40–60% if symptoms appear 5 months after planting until harvest (ProMED-mail, 2015).

In Vietnam, phytoplasma-associated symptoms including small yellow leaves, short internodes, sprout proliferation, stem vascular necrosis, and/or stunting, were observed in cassava (Álvarez et al., 2013; Mejia, 2014). Other symptoms included stakes producing only a few stunted, spindly sprouts and thinner, smaller roots with rough-textured skins and drastically reduced starch content (Álvarez, Llano, et al., 2012). In 2010, the disease was observed in Quang Ngai, Dong Nai, and Yen Bai provinces of Vietnam. More than 60,000 ha were affected, with crop losses as high as 80% and reductions in yield and starch content reaching 30% (Álvarez et al., 2013). Molecular tests indicated that phytoplasma belonging to groups 16SrI, III, V, VI, X-A, XII and XV were also identified in CWB-infected cassava plants, sometimes in mixed infections (Mejia, 2014).

In Japan, in 2018, symptomless cassava cuttings were tested for the presence of phytoplasma by LAMP and conventional PCR tests. Results indicated that 5 out of the 16 samples were positive for the presence of phytoplasma. A 16S rRNA gene sequence analysis showed that the phytoplasma was related to '*Ca.* P. pruni' (16SrIII). Analyses based on 16S rRNA, secY, and ribosomal protein gene sequences revealed that the phytoplasma was closely related to phytoplasmas found in cassava in Latin America (Koinuma et al., 2018).

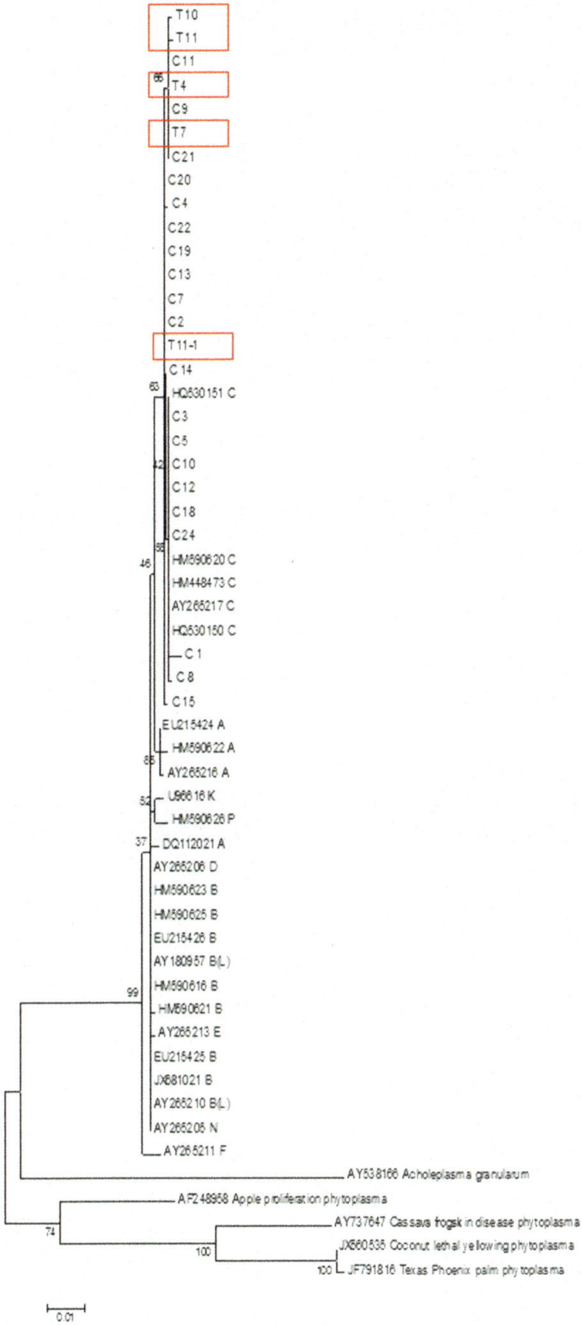

Fig. 7.10 Phylogenetic tree of the partial sequencing of the gene from subgroups A, B, C, D, E, F, K, N, P of 'Ca. Phytoplasma asteris' and phytoplasma isolated from cassava plant samples showing symptoms of witches'-broom disease in Thailand (plant samples with code C are from Cambodia and samples with code T are from Thailand)

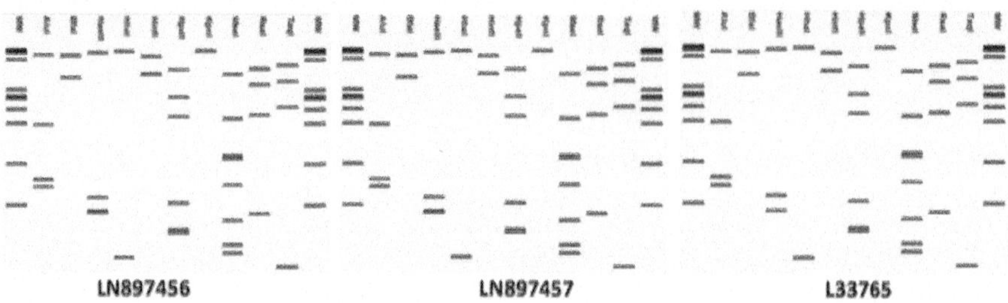

Fig. 7.11 Virtual RFLP patterns from *in silico* digestions of 16S rDNA fragments. Patterns of Cassava witches'-broom phytoplasma strains (accession numbers LN897456-LN897457) and representative strain belonging to 16SrII group Peanut witches'-broom (accession number L33765). Recognition sites for the following 10 restriction enzymes were used in the simulated digestions: *AluI, BfaI, HaeIII, HhaI, HinfI, HpaII, KpnI, MseI, RsaI, TaqI*. MW: ΦX174 DNA-*HaeIII* digestion

7.3.1.1 Epidemiology

CWB has no known vector, but spread is mediated by the use of planting material obtained from infected stems. Key aspects of disease etiology, epidemiology, and control require further investigation (Graziosi et al., 2016).

7.3.1.2 Control

During crop growth, immediate removal and disposal by burning, of any plant with symptoms of phytoplasma diseases is critical to limit phytoplasma propagation (Jackson, 2014). Disease spread can also be reduced by selecting stakes from healthy plants and by restricting the movement of cassava planting stakes, as well as related species such as jatropha, from infected areas. Phytoplasma are susceptible to antibiotics, and antibiotic treatments have been developed to eliminate phytoplasma. For example, in the Philippines, dipping cassava cuttings for six hours in streptomycin solution before planting was the most effective method in lowering plant mortality and increasing germination rate. The treatment also increased root yield, dry matter, and starch content as compared to untreated cassava cuttings (Argana, 2016). Varietal resistance to CWB exists (Álvarez, Llano, et al., 2012). In Thailand, studies

Fig. 7.12 (continued) model (Kimura, 1980). The tree with the highest log likelihood (−8101.1070) is shown. Initial tree(s) for the heuristic search were obtained automatically by applying Neighbor-Join and BioNJ algorithms to a matrix of pairwise distances estimated using the Maximum Composite Likelihood (MCL) approach, and then selecting the topology with superior log likelihood value. A discrete Gamma distribution was used to model evolutionary rate differences among sites (5 categories (+G, parameter = 0.3350)). The tree is drawn to scale, with branch lengths measured in the number of substitutions per site. The analysis involved 94 nucleotide sequences. Codon positions included were 1st + 2nd + 3rd + Noncoding. There were a total of 1179 positions in the final dataset. Evolutionary analyses were conducted in MEGA6 (Tamura et al., 2013)

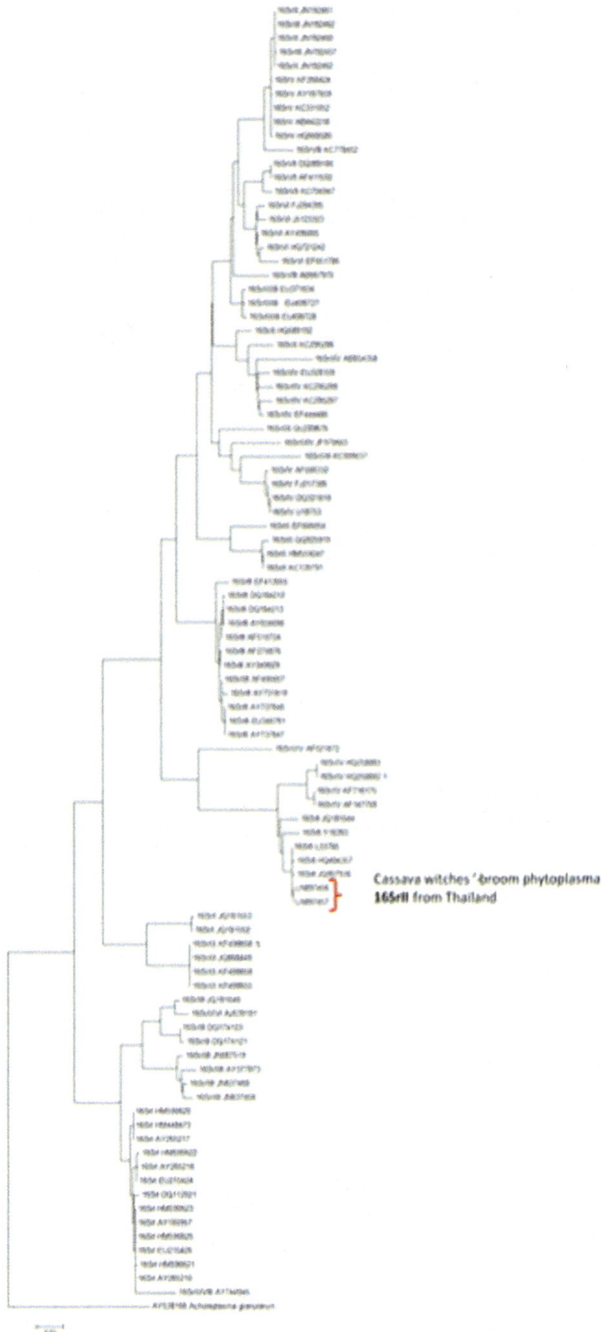

Fig. 7.12 Molecular Phylogenetic analysis by Maximum Likelihood method . The evolutionary history was inferred by using the Maximum Likelihood method based on the Tamura 3-parameter

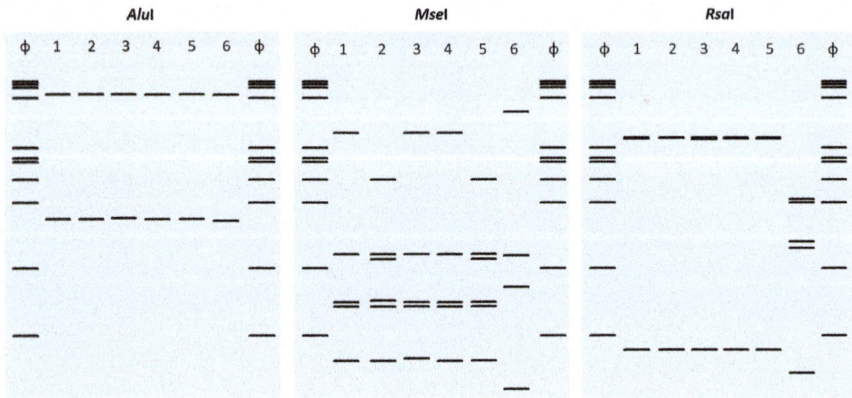

Fig. 7.13 Virtual RFLP patterns from *in silico* digestion of partial 16S rRNA sequences of '*Ca.* Phytoplasma asteris' with *AluI*, *MseI* and *RsaI*. ϕ: phiX174DNA-HaeIII digest DNA molecular weight marker. Lane 1: Cassava witches'-broom Phytoplasma Philippines (MG008911 and MG008912). Lane 2: 16SrII–A (L33765). Lane 3: 16SrII–B (U15442). Lane 4: 16SrII–C (KR059875). Lane 5: 16SrII–D (KY449416). Lane 6: *Acholeplasma palmae* (L33734)

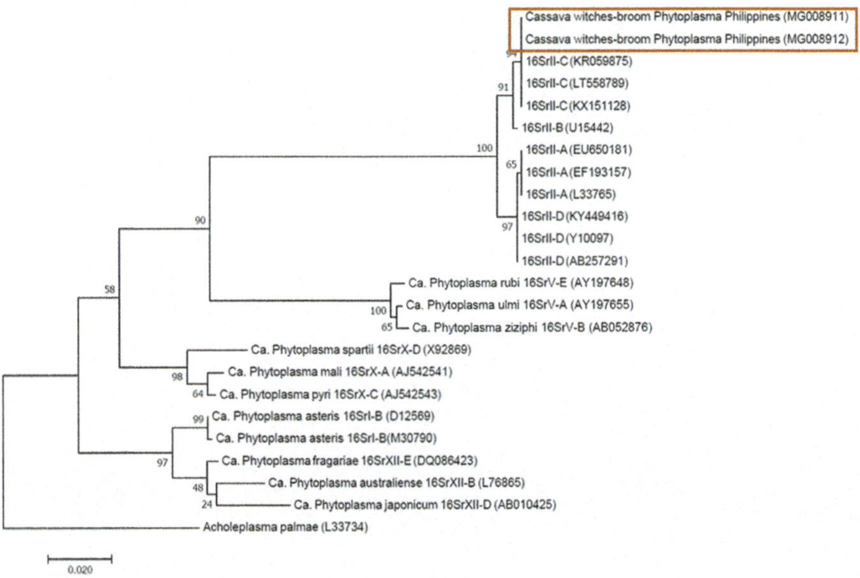

Fig. 7.14 Phylogenetic tree of partial 16S rRNA sequences showing the relationships between Cassava witches'-broom phytoplasma Philippines (MG008911 and MG008912), 16SrII and other reference groups. Tree constructed using MEGA 7.0 by the neighbour-joining method. GenBank accession numbers are in parentheses with ribosomal groups. Bootstrap values obtained for 1000 replicates are shown on the branches. Bar indicates number of substitutions per nucleotide position. *Acholeplasma palmae* (L33734) is used as the outgroup

on disease resistance described variation in the reaction of cassava genotypes to phytoplasma infection and identified MPer183 as a source of resistance to CWB diseased (Moonjuntha, 2016, personal communication). Research at Rayong Field Crops Research Center (Thailand) has generated new cassava genotypes with tolerance to CWB phytoplasma. In addition to disease control an accurate and rapid field diagnosis is a key component of disease management strategies. For example, a loop-mediated isothermal amplification (LAMP) assay has been developed and allowed specific detection of phytoplasma causing CWB from field-collected samples (Vu et al., 2016).

Other control methods developed for annual or perennial crops can be adapted for cassava crops and developed in the future. Heat therapy has been successfully used to control phytoplasma diseases in woody perennials such as grapevines and horticultural crops (Caudwell, Larrue, Boudon-Padieu, & McLean, 1997; Laimer & Bertaccini, 2019) and could become a simple, low-cost, practical approach to cassava stake disinfection and CWB management. Abiotic and biotic resistance inducers could carry ample potential for phytoplasma control in cassava. As a systemic pathogen, phytoplasma can modulate plant hormones and down-regulate plant defenses, thus opening an avenue to external application of phytohormones to prime defense systems and induce resistance (Graziosi et al., 2016). Moreover, once the identity and ecology of CWB vectors have been described, vector management could constitute an effective way of managing this disease and its spread.

7.4 Genetic Diversity among CFSD and CWB Strains

Genetic variability was observed in strains of CFSD and CWB phytoplasmas detected in heavily symptomatic samples in Latin America and Asia (Álvarez, Pardo, Contaldo, et al., 2014; Cardozo Téllez et al., 2016; Mejia, 2014). Based on SNP positions and the presence of unique restriction sites, strain polymorphism was found in CFSD phytoplasma belonging to groups 16SrI, -III, X, and –XII and sampled in Costa Rica and Paraguay as well as in CWB phytoplasma belonging to groups 16SrI, -VI, –X, -XII, and –XV and sampled in Thailand and Vietnam (Mejia, 2014). For example, in Costa Rica two different profiles related to 16SrIII group phytoplasmas in mixed infection with 16SrI and 16SrXII group phytoplasmas were distinguished, after *in silico* and phylogenetic analyses on the elongation factor gene (tuf) with both strains clearly differentiated in five SNPs positions and three unique restrictions sites (Mejia, 2014). The detected genetic variability was not always related to geographic strain distribution but was probably related to the ability of these phytoplasma strains to rapidly modify their genomes (Bertaccini, 2007). More research is needed to verify epidemiological relevance of the unique restriction profiles detected in cassava strains belonging to ribosomal groups 16SrI (Álvarez et al., 2017), 16SrII (Moonjuntha et al. 2018), 16SrIII, 16SrVI, 16SrX, 16SrXII and 16SrXV and to use them as a tool for subgroups and strain differentiation in plant and insect vectors (Tables 7.7 and 7.8).

Table 7.7 Cassava Phytoplasma groups detected in Asian countries

Country	Group
Cambodia	I, II
Phillipines	II
Thailand	I, II, VI, XV
Vietnam	I, V, VI, XII, XV

7.5 Cassava Phytoplasma Disease in Africa

In Uganda, symptoms of leaf yellowing, chlorosis, shortening of internodes, and slight stunting were observed in cassava fields in Kawanda. Restriction profiles obtained after RFLP of PCR amplicons with *Sau3* AI, *Hpa*II, and *Hae*III enzymes were all similar to those of '*Ca.* P. aurantifolia' (16SrII group). The 16SrRNA sequences of phytoplasmas detected in cassava (EU315317) and four other plant species growing nearby (*Malvaviscus arborus* Cav (Malvaceae), *Codiaeum variegatum* (L.) A. Juss (Euphorbiaceae), *Hibiscus rosa-sinensis* L. (Malvaceae) and *Passiflora edulis* Sims, Passifloraceae), showed 98% identity with that of the cactus witches'-broom phytoplasma (AJ293216) (Arocha, Echodu, et al., 2009).

In Côte d'Ivoire, cassava production is second to yams, and cassava is widely used as a typical food side dish called "attieké." Moreover, cassava-based flour or starch is widely used by private companies to food products, and has made cassava one of the most important Ivorian industrial crops. Cassava has also become an alternative income source for women in the south littoral of the coconut-growing area of Grand-Lahou, since the lethal yellowing phytoplasma (CILY, 16SrXXII-B, '*Ca.* P. palmicola'-related strains) devastated more than 400 hectares of coconut plantations (Arocha et al., 2017). In the coconut-growing villages of Badadon and Braffedon in Grand-Lahou, plants of cassava exhibiting leaf curling, and yellowing were observed. DNA was extracted from symptomatic and asymptomatic plant samples (Kra, Toualy, Kouamé, Diallo, & Arocha, 2017). Universal primers targeting phytoplasma 16S rRNA gene, R16mF2/R1 for the first PCR reaction and R16F2n/R2 for the nested reaction, were used. A second PCR assay was performed with primers P1 and P7 followed by a nested PCR with primers G813 and AwkaSR that specifically target the 16S rRNA gene and part of the intergenic region 16/23S rRNA gene of the Cape St. Paul wilt disease phytoplasma (16SrXXII-B) from Ghana. All samples shared 100% sequence identity and were 99% similar to those of the 16SrXXII-B phytoplasmas. A consensus sequence was deposited in GenBank (Accession No. KY563222), (Kra et al., 2017). Phylogenetic analysis based on the 16S rDNA sequence supported the grouping of the Côte d'Ivoire cassava phytoplasma within the 16SrXXII-B cluster (Kra et al., 2017).

Table 7.8 Cassava phytoplasmas detected in Asia

Disease	Strain or isolate	Country	16S ribosomal RNA gene		Institute	References
			Classification	GenBank Accesion		
Cassava witches'-broom	BRVT-02	Vietnam	Undetermined	KC295285	PPRI	(Nguyen, Mai, Vu, Ha, & Trinh, 2013)
Cassava witches'-broom	YB-xaemroi.2012	Vietnam	Undetermined	KC295283	PPRI	(Nguyen, Mai, Vu, Ngo, et al., 2013)
Cassava witches'-broom	KT-TT1.7	Vietnam	Undetermined	KC295284	PPRI	(Nguyen, Ngoc, et al., 2013)
Cassava witches'-broom	Hung Loc	Vietnam	16SrVI	JX857696	CIAT	(Álvarez, Pardo, & Ganan, 2012b)
Cassava witches'-broom	Chachoengsao	Thailand	16SrXV	JX857697	CIAT	(Álvarez, Pardo, & Ganan, 2012a)
Cassava witches'-broom	Rayong	Thailand	16SrI	JX857695	CIAT	(Álvarez, Pardo, & Ganan, 2012c)
Cassava witches'-broom	Rayong	Thailand	16SrI	KJ910018	CIAT	(Álvarez, Pardo, Truke, et al., 2014)
Cassava witches'-broom	Chatuchak Bangkok	Thailand	16SrII	LN897459	Department of Agriculture Thailand	(Moonjuntha et al., 2015b)
Cassava witches'-broom	Chatuchak Bangkok	Thailand	16SrII	LN897458	Department of Agriculture Thailand	(Moonjuntha et al., 2015a)
Cassava witches'-broom	Yen Bai	Vietnam	16SrI	KJ173677	CIAT	(Álvarez, Truke, & Pardo, 2014b)
Cassava witches'-broom	Yen Bai	Vietnam	16SrI	KJ173679	CIAT	(Álvarez, Truke, & Pardo, 2014c)
Cassava witches'-broom	Tay Ninh	Vietnam	16SrI	KJ210320	CIAT	(Álvarez, Truke, & Pardo, 2014a)
Cassava witches'-broom	alp4	China	Undetermined	JQ957931	CATAS	(Zheng, 2013)

(continued)

Table 7.8 (continued)

Disease	Strain or isolate	Country	16S ribosomal RNA gene		Institute	References
			Classification	GenBank Accesion		
Vietnamese cassava phytoplasma	Kon Tum-17	Vietnam	Undetermined	JQ973105	PPRI	(Álvarez et al., 2013)
Vietnamese cassava phytoplasma	Dong Nai-34	Vietnam	Undetermined	JQ973106	PPRI	(Álvarez et al., 2013)
Cassava witches'-broom	H2TB	Vietnam	Undetermined	JN251019	PPRI	(Nguyen, Ngo, & Trinh, 2011)
Cassava witches'-broom	Clone AHTB12	Vietnam	16SrI	JN381550	PPRI	(Nguyen, Ngo, et al., 2011a)
Cassava witches'-broom	Clone AHTB20-1	Vietnam	16SrI	JN381548	PPRI	(Nguyen, Ngo, et al., 2011b)
Cassava witches'-broom	Clone AHTB20-2	Vietnam	Undetermined	JN381549	PPRI	(Nguyen, Ngo, et al., 2011c)
Cassava witches'-broom	Clone AHTB31	Vietnam	16SrI	JN381547	PPRI	(Nguyen, Ngo, et al., 2011d)
Cassava witches'-broom	Kratie	Cambodia	16SrI	KJ210300	CIAT	(Álvarez, Pardo, & Truke, 2014)
Cassava witches'-broom	Kampong Cham	Cambodia	16SrI	KJ210301	CIAT	(Álvarez, Pardo, & Truke, 2014)
Cassava witches'-broom	Kratie	Cambodia	16SrI	KJ210302	CIAT	(Álvarez, Pardo, & Truke, 2014)
Cassava witches'-broom	Prey Veng	Cambodia	16SrI	KJ210303	CIAT	(Álvarez, Pardo, & Truke, 2014)
Cassava witches'-broom	Prey Veng	Cambodia	16SrI	KJ210304	CIAT	(Álvarez, Pardo, & Truke, 2014)

Cassava witches'-broom	Prey Veng	Cambodia	16SrI	KJ210305	CIAT	(Álvarez, Pardo, & Truke, 2014)
Cassava witches'-broom	Prey Veng	Cambodia	16SrI	KJ210306	CIAT	(Álvarez, Pardo, & Truke, 2014)
Cassava witches'-broom	Prey Veng	Cambodia	16SrI	KJ210307	CIAT	(Álvarez, Pardo, & Truke, 2014)
Cassava witches'-broom	Prey Veng	Cambodia	16SrI	KJ210308	CIAT	(Álvarez, Pardo, & Truke, 2014)
Cassava witches'-broom	Kampong Cham	Cambodia	16SrI	KJ210309	CIAT	(Álvarez, Pardo, & Truke, 2014)
Cassava witches'-broom	Kampong Cham	Cambodia	16SrI	KJ210310	CIAT	(Álvarez, Pardo, & Truke, 2014)
Cassava witches'-broom	Kratie	Cambodia	16SrI	KJ210311	CIAT	(Álvarez, Pardo, & Truke, 2014)
Cassava witches'-broom	Kratie	Cambodia	16SrI	KJ210312	CIAT	(Álvarez, Pardo, & Truke, 2014)
Cassava witches'-broom	Kratie	Cambodia	16SrI	KJ210313	CIAT	(Álvarez, Pardo, & Truke, 2014)
Cassava witches'-broom	Prey Veng	Cambodia	16SrI	KJ210314	CIAT	(Álvarez, Pardo, & Truke, 2014)
Cassava witches'-broom	Kratie	Cambodia	16SrI	KJ210315	CIAT	(Álvarez, Pardo, & Truke, 2014)
Cassava witches'-broom	Kratie	Cambodia	16SrI	KJ210316	CIAT	(Álvarez, Pardo, & Truke, 2014)
Cassava witches'-broom	Kratie	Cambodia	16SrI	KJ210317	CIAT	(Álvarez, Pardo, & Truke, 2014)

(continued)

Table 7.8 (continued)

Disease	Strain or isolate	Country	16S ribosomal RNA gene		Institute	References
			Classification	GenBank Accesion		
Cassava witches'-broom	Kratie	Cambodia	16SrI	KJ210318	CIAT	(Álvarez, Pardo, & Truke, 2014)
Cassava witches'-broom	Kratie	Cambodia	16SrI	KJ210319	CIAT	(Álvarez, Pardo, & Truke, 2014)
Cuban cassava phytoplasma	Las Tunas	Cuba	16SrI	EU328256	CENSA	(Arocha, Piñol, et al., 2009)
Cassava witches'-broom	CaWB-Br01	Brazil	16SrIIIB	GU193976	U. of Sao Paulo	(Flôres et al., 2013)
Cassava witches'-broom	CaWB-Br02	Brazil	16SrIIIB	GU193977	U. of Sao Paulo	(Flôres et al., 2013)
Cassava witches'-broom		Wallis and Futuna	16SrI	AY787139	Rothamsted research	(Davis et al., 2005)
Cassava witches'-broom Phytoplasma	Bukidnon, Northern Mindanao	Philippines	16SrII	MG008911	CIAT	(Aliaga, Álvarez, & Pardo, 2017)
Cassava witches'-broom Phytoplasma	Bukidnon, Northern Mindanao	Philippines	16SrII	MG008912	CIAT	(Aliaga, Álvarez, et al., 2017)

7.5.1 Control

Management of these diseases is difficult: methods of spread are unknown and commercial growers and smallholders lack access to clean planting material. Careful selection of cuttings is recommended in addition to early removal of symptomatic plants.

7.6 Cassava Phytoplasma Disease in Oceania

Plants with symptoms similar to those associated with CWB disease were sampled in Wallis and Futuna Islands located in the South Pacific. Phytoplasma were identified by nested polymerase chain reaction (PCR), restriction fragment length polymorphism analysis (RFLP) and sequence analysis. Phytoplasma DNA were shown to have 100% similarity with the '*Ca.* P. aurantifolia' (16SrII) (Davis, Arocha, Jones, & Malay, 2005).

7.7 Conclusion

Phytoplasma diseases have been recognized as one of the primary constraints to cassava production. Phytoplasma pathogens cause important yield losses wherever cassava is grown. Although some of these pathogens and the diseases they cause have attracted significant research attention, many are only partially characterized and understood, and many research gaps still need to be addressed. The limited progress in research and management of cassava diseases is partly due to the nature of cultivation of the crop and partly due to the fact that the majority of cassava growers do not have sufficient resources to invest in effective disease management.

In this chapter, we have presented information on the most important cassava phytoplasma diseases in the world and their causal agents. These are represented by cassava frogskin disease (CFSD) in Latin America and cassava witches'-broom (CWB) in Asia. We have also described the biology, economic importance, symptomatology, and epidemiology of cassava phytoplasma diseases in these regions. This and other related strategies from cassava stakeholder institutions are paving the way for an increased commitment to developing effective, affordable, and durable solutions to phytoplasma diseases. In Latin America, great progress has been made in recent years in untangling the complex set of pathogen interactions that give rise to the devastating CFSD. Another principal progress is the development of diagnostic tools that allow specific detection of disease-causing phytoplasmas.

Moreover, experience gained in investigating phytoplasmas is having a powerful spin-off effect on new research being conducted on a related pathogen that is causing a CWB epidemic in Southeast Asia. This highlights the importance of one of the most important themes in the management of cassava phytoplasma diseases: the prevention of phytoplasma spread between countries and continents through the vigorous application of phytosanitary protocols that prevent the movement of contaminated vegetative seed. Cassava offers great promise as a "climate change-proof" crop of the future, which could guarantee the food security of hundreds of millions of people living in the tropics for the foreseeable future.

References

Acosta, K. I., Zamora, L., Quiñones, M., Martínez, Z. Y., Miranda, I., Arocha, Y., … Martínez, R. M. L. (2014). *Avances en estudios de diversidad y diagnóstico de fitoplasmas asociados a cultivos de interés económico en Cuba*. Mayabeque, Cuba: Centro Nacional de Sanidad Agropecuaria (CENSA).

Aliaga, F., Álvarez, E., & Pardo, J. M. (2017). *Cassava witches'-broom disease phytoplasma clone F81 16S ribosomal RNA gene, partial sequence*. Retrieved March 15, 2017, from NCBI https://www.ncbi.nlm.nih.gov/nuccore/MG008911

Aliaga, F., Álvarez, E., & Pardo, J. M. (2017). *Cassava witches'-broom disease phytoplasma clone F82 16S ribosomal RNA gene, partial sequence*. Retrieved March 15, 2019, from NCBI https://www.ncbi.nlm.nih.gov/nuccore/MG008912.1

Álvarez, E., & Betancourth, C. (2016). Transmission of a 16SrIII-L phytoplasma by leafhoppers (*Scaphytopius marginelineatus*) to cassava (*Manihot esculenta* Crantz) in Colombia. *Phytopathology, 106*(12), 135–135.

Álvarez, E., Betancourth, C., & Muñoz, J. (2017). Pathogenicity of a 16SrIII-L phytoplasma associated with frogskin disease of cassava (*Manihot esculenta* Crantz) in Colombia. *Phytopathology, 107*(S2), S5.

Álvarez, E., Gómez, Y., Zacher, M., & Campo, R. (2014). *Innovaciones tecnológicas en el manejo integrado del cuero de sapo de la yuca (Manihot esculenta Crantz): Estrategias para reducir el impacto de la enfermedad por efectos del cambio climático en Colombia, Costa Rica y Paraguay*. Paper presented at the IX Taller de seguimiento técnico de proyectos FONTAGRO y 59 Reunión anual del PCCMCA (April 28–29, 2014), Fontagro, Managua, Nicaragua.

Álvarez, E., Llano, G., & Mejía, J. F. (2012). Cassava diseases in Latin America, Africa and Asia. In R. Howeler (Ed.), *The cassava handbook: A reference manual based on the Asian regional cassava training course, Bangkok, Thailand* (pp. 258–304). Cali, Colombia: CIAT.

Álvarez, E., Mejía, J. F., Llano, G. A., Loke, J. B., Calari, A., Duduk, B., & Bertaccini, A. (2009). Characterization of a phytoplasma associated with frogskin disease in cassava. *Plant Disease, 93*(11), 1139–1145.

Álvarez, E., Mejía, J. F., Loke, J. B., Hernández, L., & Llano, G. (2003). Detecting the phytoplasm-frogskin disease association in cassava (*Manihot esculenta* Crantz) in Colombia. *Phytopathology, 93*, 1139

Álvarez, E., Mejía, J. F., & Pardo, J. M. (2010). Development of a real-time PCR assay, to detect and quantify a 16SrIII-L phytoplasma associated with cassava frogskin disease (CFSD). *Phytopathology, 100*(6), S5.

Álvarez, E., Pardo, J. M., Contaldo, N., Paltrinieri, S., Bertaccini, A., & Mejia, J. F. (2014). *Phytoplasma infection in Cassava plant with frogskin disease from Costa Rica and Paraguay*. Paper presented at the 20th congress of the International Organization for Mycoplasmology (June 1–6, 2014), Blumenau, Brazil, p. 55–56.

Álvarez, E., Pardo, J. M., & Ganan, L. (2012a). *Cassava witches'-broom phytoplasma isolate Chacheoengsao 16S ribosomal RNA gene, partial sequence*. Retrieved March 19, 2019, from NCBI https://www.ncbi.nlm.nih.gov/nuccore/JX857697

Álvarez, E., Pardo, J. M., & Ganan, L. (2012b). *Cassava witches'-broom phytoplasma isolate Hung Loc 16S ribosomal RNA gene, partial sequence*. Retrieved March 29, 2019, from NCBI https://www.ncbi.nlm.nih.gov/nuccore/JX857696

Álvarez, E., Pardo, J. M., & Ganan, L. (2012c). *Cassava witches'-broom phytoplasma isolate Rayong 16S ribosomal RNA gene, partial sequence (partial gene sequence)*. Retrieved March 29, 2019, from NCBI https://www.ncbi.nlm.nih.gov/nuccore/JX857695

Álvarez, E., Pardo, J. M., Mejia, J. F., Bertaccini, A., Thanh, N. D., & Hoat, T. X. (2013). Detection and identification of '*Candidatus* Phytoplasma asteris'-related phytoplasmas associated with a witches' broom disease of cassava in Vietnam. *Phytopathogenic Mollicutes, 3*(2), 77–81.

Álvarez, E., Pardo, J. M., & Truke, M. (2014). Detection and identification of '*Candidatus* Phytoplasma asteris'-related phytoplasma associated with a witches' broom disease of cassava in Cambodia. *Phytopathology, 104*(S3), S3.7.

Álvarez, E., Pardo, J. M., Truke, M., Klingkong, S., Moonjuntha, P., & Wyckhuys, K. (2014). *Cassava witches'-broom phytoplasma isolate 6–3 16S ribosomal RNA gene, partial sequence*. Retrieved March 29, 2019, from NCBI https://www.ncbi.nlm.nih.gov/nuccore/KJ910018

Álvarez, E., Truke, M., & Pardo, J. M. (2014a). *Cassava witches'-broom phytoplasma clone 2–7 16S ribosomal RNA gene, partial sequence*. Retrieved March 18, 2019, from NCBI https://www.ncbi.nlm.nih.gov/nuccore/KJ210320

Álvarez, E., Truke, M., & Pardo, J. M. (2014b). *Cassava witches'-broom phytoplasma isolate 6–3 16S ribosomal RNA gene, partial sequence*. Retrieved March 29, 2019, from NCBI https://www.ncbi.nlm.nih.gov/nuccore/KJ173677

Álvarez, E., Truke, M., & Pardo, J. M. (2014c). *Cassava witches'-broom phytoplasma isolate 7–7 16S ribosomal RNA gene, partial sequence*. Retrieved March 29, 2019, from NCBI https://www.ncbi.nlm.nih.gov/nuccore/KJ173679

Anfoka, G., & Fattash, I. (2004). Detection and identification of Aster yellows (16SrI) Phytoplasma in peach trees in Jordan by RFLP analysis of PCR-amplified products (16S rDNAs). *Journal of Phytopathology, 152*(4), 210–214.

Argana, R. (2016). *Addressing the threats of Cassava phytoplasma disease*. Retrieved from Los Baños, Laguna, Philippines. http://www.pcaarrd.dost.gov.ph/home/portal/index.php/quick-information-dispatch/2761-addressing-the-threats-of-cassava-phytoplasma-disease

Arocha, Y., Diallo, H. A., Konan Konan, J., Yankey, N., Saleh, M., Pilet, F., … Scott, J. (2017). Detection and differentiation of the coconut lethal yellowing phytoplasma in coconut-growing villages of Grand-Lahou, Côte d'Ivoire. *Annals of Applied Biology, 170*(3), 333–347.

Arocha, Y., Echodu, R., Talengera, D., Muhangi, J., Rockefeller, E., Asher, O., … Tripathi, J. (2009). Occurrence of '*Candidatus* Phytoplasma aurantifolia' (16SrII group) in cassava and four other species in Uganda. *Plant Pathology, 58*(2), 390–390.

Arocha, Y., Piñol, B., Almeida, R., Acosta, K. I., Quiñones, M., Zayas, T., … Lucas, J. (2009). First report of phytoplasmas affecting organoponic crops in central and eastern Cuba. *Plant Pathology, 58*(4), 793–793.

Bertaccini, A. (2007). Phytoplasmas: Diversity, taxonomy, and epidemiology. *Frontiers in Bioscience, 12*(2), 673–689.

Bertaccini, A., & Duduk, B. (2009). Phytoplasma and phytoplasma diseases: A review of recent research. *Phytopathologia Mediterranea, 48*(3), 355–378.

Betancourth, C. (2016). *Aislamiento, patogenicidad y transmisión con Scaphytopius marginelineatus (Stal) de un fitoplasma asociado a cuero de sapo en yuca.* PhD. Universidade National de Colombia, Palmira, Colombia.

Betancourth, C., Pardo, J. M., Truke, M., Muñoz Flórez, J. E., & Álvarez, E. (2014). In vitro isolation of a phytoplasma associated with frogskin disease of cassava (*Manihot esculenta* Crantz). *Phytopathology, 105*(Suppl 1), S1.6.

Burketova, L., Trda, L., Ott, P. G., & Valentova, O. (2015). Bio-based resistance inducers for sustainable plant protection against pathogens. *Biotechnology Advances, 33*(6), 994–1004.

Calvert, L. (1994). *The safe movement of cassava germplasm.* Paper presented at the First Meeting of the International Network for Cassava Genetic Resources Report. International Crop Network Series 10 (August 18–23, 2014), CIAT, Cali, Colombia.

Cardozo Téllez, L., Pardo, J. M., Zacher, M., Torres, A., & Álvarez, E. (2016). First report of a 16SrIII phytoplasma associated with frogskin disease in cassava (*Manihot esculenta* Crantz) in Paraguay. *Plant Disease, 98*(1), 153.
Cassells, A. C. (2012). Pathogen and biological contamination management in plant tissue culture: Phytopathogens, vitro pathogens, and vitro pests. In V. Loyola-Vargas & N. Ochoa-Alejo (Eds.), *Plant cell culture protocols* (3rd ed., pp. 57–80). New York, NY: Springer.
Caudwell, A., Larrue, J., Boudon-Padieu, E., & McLean, G. (1997). Flavescence dorée elimination from dormant wood of grapevines by hot-water treatment. *Australian Journal of Grape and Wine Research, 3*(1), 21–25.
Ceballos, H. (2002). La yuca en Colombia y el mundo: Nuevas perspectivas para un cultivo milenario. In B. Ospina & H. Ceballos (Eds.), *La yuca en el tercer milenio. Sistemas modernos de producción, procesamiento, utilización y comercialización* (pp. 1–13). CIAT: Cali, Colombia.
Chaparro-Martínez, E., & Trujillo-Pinto, G. (2001). First report of frog skin disease in cassava (*Manihot esculenta*) in Venezuela. *Plant Disease, 85*(12), 1285–1285.
Chavarriaga-Aguirre, P., Brand, A., Medina, A., Prías, M., Escobar, R., Martinez, J., … Tohme, J. (2016). The potential of using biotechnology to improve cassava: A review. *Vitro Cellular & Developmental Biology-Plant, 52*(5), 461–478.
CIAT. (2004). *Integrated Pest and Disease Management in Major Agroecosystems: Project PE-1. Annual report* 2003 Retrieved from Cali, Colombia. https://cgspace.cgiar.org/handle/10568/68969
CIAT. (2005). *Project PE-1, executive summary – Annual report 2005.* Retrieved from CIAT, Cali, Colombia. https://cgspace.cgiar.org/handle/10568/68971?show=full
Contaldo, N., Bertaccini, A., Paltrinieri, S., Windsor, H. M., & Windsor, G. D. (2012). Axenic culture of plant pathogenic phytoplasmas. *Phytopathologia Mediterranea, 51*(3), 607–617.
Coyne, D. L., Claudius-Cole, A. O., Kenyon, L., & Baimey, H. (2010). Differential effect of hot water treatment on whole tubers versus cut setts of yam (*Dioscorea* spp.). *Pest Management Science: Formerly Pesticide Science, 66*(4), 385–389.
Da Silva Souza, L., & Fialho, J. (2012). *Cultivoda Mandioca para a Região do Cerrado.* Retrieved from Cruz das Almas, Brazil. http://sistemasdeproducao.cnptia.embrapa.br/FontesHTML/Mandioca/mandioca_cerrados/doencas.htm
Davis, R., Arocha, Y., Jones, P., & Malay, A. (2005). First report of the association of phytoplasmas with plant diseases in the territory of Wallis and Futuna. *Australasian Plant Pathology, 34*(3), 417–418.
De Souza, A. N., Da Silva, F. N., Bedendo, I. P., & Carvalho, C. M. (2014). A phytoplasma belonging to a 16SrIII-A subgroup and dsRNA virus associated with cassava frogskin disease in Brazil. *Plant Disease, 98*(6), 771–779.
Dellaporta, S. L., Wood, J., & Hicks, J. B. (1983). A plant DNA minipreparation: Version II. *Plant Molecular Biology Reporter, 1*(4), 19–21.
Doyle, J. J., & Doyle, J. L. (1990). Isolation of plant DNA from fresh tissue. *Focus, 12*(13), 39–40.
Fernández, F., Uset, A., Baumgratz, G., & Conci, L. (2018). Detection and identification of a 16SrIII-J phytoplasma affecting cassava (*Manihot esculenta* Crantz) in Argentina. *Australasian Plant Disease Notes, 13*(1), 24.
Flôres, D., Haas, I. C., Canale, M. C., & Bedendo, I. P. (2013). Molecular identification of a 16SrIII-B phytoplasma associated with cassava witches' broom disease. *European Journal of Plant Pathology, 137*(2), 237–242.
Food and Agriculture Organisation of the United Nations. (2016). FAOSTAT. Crops. Retrieved August 25, 2018, from FAOSTAT Statistics Division, http://www.fao.org/faostat/en/#data/QC
Frison, E., & Feliu, E. (1991). *FAO/IBPGR technical guidelines for the safe movement of cassava germplasm.* Retrieved from Rome, Italy. https://bioversityinternational.org/uploads/tx_news/Cassava_349.pdf
Fukuda, C., Filho, J., Sa, M., Carneiro, J., Paiva, A., Pereira, C., & Lima, H. (1990). *Levantamento da ocorrência do superbrotamento da mandioca na microrregião de Ibiapaba, CE.* Paper presented at the VI Congresso Brasileiro De Mandioca (July 16–21, 1990), Londrina, Brazil.

Granada, G. (1979). Machismo disease of soybeans. I. Symptomatology and transmission. *Plant Disease Reporter, 63*(1), 47–50.

Graziosi, I., Minato, N., Álvarez, E., Ngo, D. T., Hoat, T. X., Aye, T. M., ... Wyckhuys, K. A. (2016). Emerging pests and diseases of South-east Asian cassava: A comprehensive evaluation of geographic priorities, management options and research needs. *Pest Management Science, 72*(6), 1071–1089.

Gundersen-Rindal, D., & Lee, I.-M. (1996). Ultrasensitive detection of phytoplasmas by nested-PCR assays using two universal primer pairs. *Phytopathologia Mediterranea*, 144–151.

Jackson, G. (2014). Phytoplasma diseases of cassava: Various phytoplasmas. In CABI-Africa Soil Health consortium (Ed.). http://africasoilhealth.cabi.org/wpcms/wp-content/uploads/2015/02/6-tubers-phytoplasma-cassava.pdf

Jayasinghe, U., Pineda, B., & Lozano, J. (1983). Antólisis en yuca (*Manihot esculenta* Crantz), asociada con organismos similares a micoplasmas. *Fitopatologia Brasileira, 9*, 51–57.

Jayasinghe, U., Pineda, B., & Lozano, J. (1984). Antholysis in cassava (*Manihot esculenta* Crantz) possibly caused by mycoplasma-like organisms. *Phytopathologische Zeitschrift, 109*, 295–300.

Kimura, M. (1980). A simple method for estimating evolutionary rates of base substitutions through comparative studies of nucleotide sequences. *Journal of Molecular Evolution, 16*(2), 111–120.

Kitajima, E., & Costa, A. (1971). Corpúsculos do tipo micoplasma associados a moléstias de plantas, do grupo amarelo, no ESP. *Ciência e Cultura, 23*(3), 285–291.

Kitimu, S. R., Taylor, J., March, T. J., Tairo, F., Wilkinson, M. J., & Rodríguez López, C. M. (2015). Meristem micropropagation of cassava (*Manihot esculenta*) evokes genome-wide changes in DNA methylation. *Frontiers in Plant Science, 6*, 590.

Koinuma, H., Miyazaki, A., Wakaki, R., Fujimoto, Y., Iwabuchi, N., Nijo, T., ... Yamaji, Y. (2018). First report of '*Candidatus* Phytoplasma pruni' infecting cassava in Japan. *Journal of General Plant Pathology, 84*(4), 300–304.

Kra, K., Toualy, Y., Kouamé, A., Diallo, H., & Arocha, Y. (2017). First report of a phytoplasma affecting cassava orchards in cote d'Ivoire. *New Disease Reports, 35*, 21–21.

Laimer, M., & Bertaccini, A. (2019). Phytoplasma elimination from perennial horticultural crops. In A. Bertaccini, P. G. Weintraub, G. P. Rao, & N. Mori (Eds.), *Phytoplasmas: Plant pathogenic bacteria-II* (pp. 185–206). Singapore, Singapore: Springer.

Lee, I.-M., Gundersen-Rindal, D. E., Davis, R. E., & Bartoszyk, I. M. (1998). Revised classification scheme of phytoplasmas based on RFLP analyses of 16S rRNA and ribosomal protein gene sequences. *International Journal of Systematic and Evolutionary Microbiology, 48*(4), 1153–1169.

Lee, I.-M., Gundersen-Rindal, D. E., Davis, R. E., Bottner, K., Marcone, C., & Seemüller, E. (2004). 'Candidatus Phytoplasma asteris', a novel phytoplasma taxon associated with aster yellows and related diseases. *International Journal of Systematic and Evolutionary Microbiology, 54*(4), 1037–1048.

Legg, J., & Álvarez, E. (2017). Diseases affecting cassava. In C. Hershey (Ed.), *Achieving sustainable cultivation of cassava Volume 2: Genetics, breeding, pests and diseases*. Cambridge, UK: Burleigh Dodds Science Publishing Limited.

Lozano, J. (1992). Overview of integrated control of casava diseases. *Fitopatologia Brasileira, 17*, 18–18.

Mariano, R., Laranjeiras, D., Santos, E., Padovan, I., & Peixoto, A. (1991). Superbrotamento da mandioca em Pernambuco. *Fitopatologia Brasileira, 16*, 42.

Mejia, J. F. (2014). *Identification and genetic diversity in phytoplasmas associated with diseases of cassava and other agronomic relevant crops in south-east Asia and Latin America*. PhD, Università di Bologna, Bologna, Italy.

Mejía, J. F., Álvarez, E., Campo, A., Rodrigo Orlando, L. R., José Moises, L., Germán, A., . . . Restrepo-Pachecho, E. (2011). Determinación de vectores aéreos y/o del suelo, asociados con la transmisión de la enfermedad del cuero de sapo en yuca (*Manihot esculenta* Crantz) en los departamentos de Cauca y Sucre en Colombia. *Fitopatol. Colombiana, 35*(1), 15.

Moonjuntha, P., Athipunyakom, P., Kositcharoenkul, N., Thongkreng, R., Waravichanee, K., Kanhayat, T., … Wyckhuys, K. (2015a). *Cassava witches'-broom disease phytoplasma partial 16S rRNA gene, isolate Ry3*. Retrieved March 29, 2019, from NCBI https://www.ncbi.nlm.nih.gov/nuccore/LN897458

Moonjuntha, P., Athipunyakom, P., Kositcharoenkul, N., Thongkreng, R., Waravichanee, K., Kanhayat, T., … Wyckhuys, K. (2015b). *Cassava witches'-broom disease phytoplasma partial 16S rRNA gene, isolate Ry4*. Retrieved March 29, 2019. https://www.ncbi.nlm.nih.gov/nuccore/LN897459

Nguyen, T., Mai, Q., Vu, H., Ha, C., & Trinh, H. (2013). Cassava witches'-broom phytoplasma strain BRVT-02 16S ribosomal RNA gene, partial sequence. Retrieved March 29, 2019, from NCBI https://www.ncbi.nlm.nih.gov/nuccore/KC295285

Nguyen, T., Mai, Q., Vu, H., Ngo, B., Ha, C., & Trinh, H. (2013). *Cassava witches'-broom phytoplasma strain YB-xaemroi.2012 16S ribosomal RNA gene, partial sequence*. Retrieved March 29, 2019, from NCBI https://www.ncbi.nlm.nih.gov/nuccore/KC295283

Nguyen, T., Ngo, B., & Trinh, H. (2011). *Cassava witches'-broom phytoplasma strain H2TB 16S ribosomal RNA gene, partial sequence*. Retrieved April 02, 2019, from NCBI https://www.ncbi.nlm.nih.gov/nuccore/JN251019

Nguyen, T., Ngo, B., Trinh, H., Dang, T., Vu, H., Mai, Q., … Nguyen, H. (2011a). *Cassava witches'-broom phytoplasma clone AHTB12 16S ribosomal RNA gene, partial sequence*. Retrieved Aprril 2, 2019, from NCBI. https://www.ncbi.nlm.nih.gov/nuccore/JN381550

Nguyen, T., Ngo, B., Trinh, H., Dang, T., Vu, H., Mai, Q., … Nguyen, H. (2011b). *Cassava witches'-broom phytoplasma clone AHTB20-1 16S ribosomal RNA gene, partial sequence*. Retrieved April 02, 2019, from NCBI. https://www.ncbi.nlm.nih.gov/nuccore/JN381548

Nguyen, T., Ngo, B., Trinh, H., Dang, T., Vu, H., Mai, Q., … Nguyen, H. (2011c). *Cassava witches'-broom phytoplasma clone AHTB20-2 16S ribosomal RNA gene, partial sequence*. Retrieved April 2, 2019, from NCBI. https://www.ncbi.nlm.nih.gov/nuccore/JN381549

Nguyen, T., Ngo, B., Trinh, H., Dang, T., Vu, H., Mai, Q., … Nguyen, H. (2011d). *Cassava witches'-broom phytoplasma clone AHTB31 16S ribosomal RNA gene, partial sequence*. Retrieved April 2, 2019, from NCBI. https://www.ncbi.nlm.nih.gov/nuccore/JN381547

Nguyen, T., Ngoc, N., Mai, Q., Vu, H., Ha, C., & Trinh, H. (2013). *Cassava witches'-broom phytoplasma strain KT-TT1.7 16S ribosomal RNA gene, partial sequence*. Retrieved March 29, 2019, from NCBI. https://www.ncbi.nlm.nih.gov/nuccore/KC295284

Oliveira, S., Abreu, E., Araújo, T., Oliveira, E., Andrade, E., Garcia, J., & Álvarez, E. (2014). First report of a 16SrIII-L phytoplasma associated with frogskin disease in cassava (*Manihot esculenta* Crantz) in Brazil. *Plant Disease, 98*(1), 153–153.

Pardo, J. M. (2013). *Desarrollo de una metodología de PCR en tiempo real para detección y cuantificación de fitoplasma 16SrIII-L y reovirus CFSV asociados con la enfermedad del cuero de sapo en yuca.* (Maestría), Universidad Nacional de Colombia, Palmira, Colombia.

Pardo, J. M., Truke, M. J., Cardozo Téllez, L., Varela, I., & Álvarez, E. (2014). A real-time PCR assay to detect and quantify 16SRIII-L and 16SrI phytoplasmas associated with cassava frogskin disease in Costa Rica and Paraguay. *Phytopathology, 105*(Suppl 1), S1.3.

Pineda, B., Jayasinghe, U., & Lozano, J. (1983). La enfermedad cuero de sapo en yuca (*Manihot esculenta* Crantz). *ASIAVA (Colombia), 4*, 10–12.

ProMED-mail. (2015). *Witches' broom, cassava – Philippines (Mindanoa): Bukidnon under quarantine for cassava disease*. Retrieved from International Society for Infectious Diseases, Brookline, MA. http://www.promedmail.org/direct.php?id=20150712.3503394

Rao, G. P., Álvarez, E., & Yadav, A. (2018). Phytoplasma diseases of industrial crops. In G. P. Rao, A. Bertaccini, N. Fiore, & L. W. Liefting (Eds.), *Phytoplasmas: Plant pathogenic Bacteria-I, characterisation and epidemiology of Phytoplasma – Associated diseases* (pp. 91–121). Singapore, Singapore: Springer Nature Singapore Pte Ltd.

Roca, W. M., Beltrán, G., Jorge, A., & Fernández, O. (1984). *El cultivo de meristemas para la conservación de germoplasma de yuca in vitro* Retrieved from CIAT, Cali, Colombia. https://cgspace.cgiar.org/handle/10568/54663

Schneider, B., Seemüller, E., Smart, C., & Kirkpatrick, B. (1995). Phylogenetic classification of plant pathogenic mycoplasma-like organisms or phytoplasmas. In S. Razin & J. Tully (Eds.), *Molecular and diagnostic procedures in mycoplasmology* (Vol. 1, pp. 369–380). San Diego, CA: Academic.

Silberschmidt, K., & Campos, A. (1944). Estudos relativos à doença superbrotamento ou envassouramento da mandioca. *Arquivos do Instituto Biológico, 15*, 1–26.

Tamura, K., Stecher, G., Peterson, D., Filipski, A., & Kumar, S. (2013). MEGA6: Molecular evolutionary genetics analysis version 6.0. *Molecular Biology and Evolution, 30*(12), 2725–2729.

Vu, N. T., Pardo, J. M., Álvarez, E., Le, H. H., Wyckhuys, K., Nguyen, K.-L., & Le, D. T. (2016). Establishment of a loop-mediated isothermal amplification (LAMP) assay for the detection of phytoplasma-associated cassava witches' broom disease. *Applied Biological Chemistry, 59*(2), 151–156.

Zhang, P., Ma, Q., Naconsie, M., Wu, X., Zhou, W., & Yang, J. (2017). Advances in genetic modification of cassava. In C. Hershey (Ed.), *Achieving sustainable cultivation of cassava* (Vol. 2). Cambridge, UK: Burleigh Dodds Science Publishing.

Zheng, W. (2013). *Cassava witches'-broom disease phytoplasma isolate alp4 16S ribosomal RNA gene, partial sequence*. Retrieved March 29, 2019, from NCBI. https://www.ncbi.nlm.nih.gov/nuccore/JQ957931

Chapter 8
Management of Phytoplasmas in Urban Trees

Liliana Franco-Lara and Laura Perilla-Henao

Abstract Fifty four percent of the human population lives in cities. Urban trees offer important ecosystem services, such as reduction of the adverse effects of global climate change. In many parts of the world, trees are being affected by phytoplasmas, reducing their health status and their benefits to the communities. Since trees are perennial hosts, phytoplasmas remain in their hosts for long periods of time, becoming a perdurable source of inoculum for other plants. Management of phytoplasma diseases in trees represents a serious challenge. The control of phytoplasmas is difficult since there are no curative methods, and resistance or tolerance is rare. Classical approaches used to control phytoplasma diseases such as roguing, tetracycline treatment or insecticide are inapplicable for trees and often forbidden in cities. Therefore, the prevention of the acquisition of phytoplasma should be a priority for urban trees, since alternatives of management are limited once phytoplasmas are established. Planting phytoplasma-free material is part of the prevention. If the stock material is infected, recent advances in tissue culture should allow the production of clean planting materials. Once a phytoplasma disease is established in a region, the management strategies should consider an integrated approach. The classical phytopathological description of a plant disease consider three factors: a susceptible host, a virulent pathogen, and conducive environment. In the case of phytoplasma infecting trees, insect vectors (transmission agents) and host-pathogen-environment interactions should be included in the pathosystem. Induced defense mechanisms, host resistance, environmental conditions and vector management as part of integrated phytoplasma disease management for urban trees are discussed in this chapter.

L. Franco-Lara (✉)
Faculty of Basic and Applied Sciences, Universidad Militar Nueva Granada, Cajica, Colombia
e-mail: liliana.franco@unimilitar.edu.co

L. Perilla-Henao
Department of Plant Pathology, University of California, Davis, CA, USA

C. Y. Olivier et al. (eds.), *Sustainable Management of Phytoplasma Diseases in Crops Grown in the Tropical Belt*, Sustainability in Plant and Crop Protection 12,
https://doi.org/10.1007/978-3-030-29650-6_8

Keywords Phytoplasma · Urban tree · Ecosystemic services · Plant resistance · Vector management

8.1 Introduction

Trees are of vital importance for humans, whether as forests, tree crops or ornamentals, in cities or in countryside. It is estimated that forests cover about a third of the earth's land surface and trees provide important functions such as soil stabilization, prevention of erosion and management of water at the local and regional levels. Trees also help regulating climate and air quality (Luque & Iverson, 2016). More than 54% of the human population live in cities and the ecosystemic services of trees in urban areas are very important (United Nations DESA/Population Division, 2015). The so-called urban forests, provide ecosystemic services such as air purification, extreme climate palliation, urban temperature regulation, tree shade, food and wood supplies, noise reduction, habitat for wildlife and provision of aesthetic and cultural places (Escobedo, Kroeger, & Wagner, 2011; Reed et al., 2017). Urban vegetation may also play a role in the mitigation and adaptation to climate change, but correct planning and design is needed to maximize the beneficial effects of green coverages (Demuzere et al., 2014). Carbon sequestration, the process of fixing carbon during photosynthesis and storing it as biomass, is an important service for resilience against climate change since trees serve as sink for carbon dioxide (Nowak, Greenfield, Hoehn, & Lapoint, 2013).

Trees are associated with many pests and pathogens, such as phytoplasmas diseases (Marcone, 2015). The effect of pests and pathogens in trees can have negative environmental and economic impacts, especially if large areas are affected (Ayres & Lombardero, 2000; Pimentel, Zuniga, & Morrison, 2005). Under the present scenario of global climate change and rapid transcontinental movements of plants and animals, the risk of damage to trees due to emerging diseases and abiotic factors such as drought and wildfires, increases (Allen et al., 2010). Climate change potentially affects the development and survival of trees by producing physiological changes that alter their defenses and by changing the abundance and distribution of natural enemies, mutualists and competitors (Ayres & Lombardero, 2000). Additionally, new temperature and humidity conditions are very likely to allow the establishment of non-native pathogens and insect vectors, many of which could be accidentally introduced through the global trade of plants (Tubby & Webber, 2010).

8.2 Phytoplasma Diseases of Urban Trees

Phytoplasmas have been associated with many urban tree diseases and are affecting the tree ornamental attributes and ecological services. Symptoms vary depending on the species. Common symptoms include yellowing, premature autumn coloration, the presence of small leaves, leaf epinasty, leaf curl, chlorosis and premature casting of the leaves. Other symptoms encompass witches'-broom, stunting, sparse foliage, deliquescent branching due to loss of apical dominance, shortening or elongation of internodes, tufted foliage, epicormic growths and changes in the general architecture of branches and shoots (Fig. 8.1), in some cases leading to crown deformation. Reduced terminal growth, die-back and decline are also frequent in trees (Franco-Lara & Perilla-Henao, 2014; Marcone, 2015; Sinclair, Griffiths, & Davis, 1996).

In European urban areas, symptoms associated with phytoplasmas have been observed in trees species such as *Ulmus* spp. L. (Ulmaceae) and *Eucalyptus* spp. L'Hér (Myrtaceae) and attributed to Elm yellows phytoplasma (EY) (Marcone, Ragozzino, Cousin, Berges, & Seemüller, 1999; Seemüller & Lederer, 1988; Sfalanga, Martini, Surico, & Bertaccini, 2002), as well as in *Alnus* spp. Mill. (Betulaceae) and associated with EY or 16SrV-C phytoplasmas (Lee et al., 2004; Marcone et al., 1999). The Aster yellows group (AY) (group 16SrI) has been detected in *Populus nigra* 'Italica' L. (Salicaceae), *P. alba* L. and *P. tremula* L. in several European cities (Marcone et al., 1999; Seemüller & Lederer, 1988; Šeruga et al., 2003). Mixed infections of EY, 16SrI and 16SrXII are reported in elm hybrids in northern and central Italy (Lee et al., 1995). Ornamental fruit trees, such as *Malus* spp. Mill. (Rosaceae) and *Prunus serrulata* Lindl. (Rosaceae) growing in European cities are affected by the Apple proliferation (AP) '*Ca.* P. Mali' and European stone fruit yellows (ESFY) '*Ca.* P. prunorum' respectively. AP and ESFY phytoplasma belong to the 16SrX group and have also been detected in several species of symptomatic shrubs such as *Spartium junceum* L. (Fabaceae), *Sarothamnus scoparius* L. (Link) (Fabaceae) and *Rhamnus cathartica* L. (Rhamnaceae) in European cities (Marcone et al., 1999). In Asia, phytoplasma diseases known as Pawlonia witches'-brooms (PaWB) (16SrI-B) and Mulberry dwarf (MD) (16SrI-D), affecting *Pawlonia* spp. Ziebolt & Zucc (Paulowniaceae) and *Morus* spp. L. (Moraceae), respectively, were the first diseases to be found associated with phytoplasmas (Doi, Teranaka, Yona, & Asuyama, 1967; Lee et al., 2004).

In North America, the EY disease has been reported from the late 1800s, in many species of tree belonging to the genus *Ulmus* (Marcone & Rao, 2016). Phytoplasmas belonging to 16SrI and 16SrVI-C groups have been detected in the central United States associated with symptomatic *Ulmus* spp. (Jacobs, Lee, Griffiths, Miller Jr., & Bottner, 2003). Trees of several *Alnus* species are also affected with phytoplamas of subgroup 16SrV-C in the Washington State in the United States (Lee et al., 2004). In Europe, this disease is associated with '*Ca.* Phytoplasma ulmi' (subgroup 16SrV-A) (Lee et al., 2004).

Fig. 8.1 (**a–i**) Trees from Colombia showing symptoms associated with phytoplasmas. (**a**) Deformation of the crown, loss of apical dominance and elongation of branches in *Populus nigra*;

The epidemiology of phytoplasma tree diseases has been reviewed by Marcone (2015) and Marcone and Rao (2016). Since the 1930s, in North America, at least 12 *Fraxinus* spp. L. (Oleaceae) and 20 *Syringa* spp. L. (Oleaceae) have been reported affected with symptoms such as slow growth and decline by a disease called Ash yellows (Sinclair & Griffiths, 1994). The Ash yellows phytoplasma (AshY) associated with *Fraxinus* and *Syringa* in many areas of North America belongs to '*Ca.* Pytoplasmas fraxini', from the group 16SrVII (Bricker & Stutz, 2004; Griffiths, Sinclair, Smart, & Davis, 1999). The epidemiology of phytoplasma diseases in *Fraxinus* and *Syringa* has been reviewed by Sinclair and Griffiths (1994) and Sinclair et al. (1996).

In South America, witches' broom symptoms have been reported in *Ugni molinae* Turcz. (Myrtaceae) grown in Chile and infected with '*Ca.* P. Fraxini' and phytoplasmas belonging to the subgroups 16SrIII-B or 16SrVII-C have been identified in *Melia azedarach* L. (Meliaceae) in Argentina, Bolivia, Brazil and Paraguay (Arismendi, Andrade, Riegel, & Carrillo, 2010; Arneodo et al., 2005, 2007; Galdeano et al., 2004; Harrison, Boa, & Carpio, 2003). A remarkable case of urban trees affected by phytoplasmas occurred in Bogota, Colombia. In the last 15 years, the presence of phytoplasmas has been detected in symptomatic trees belonging to nine species growing in the city and its surroundings: *Acacia melanoxylon* R.Br. (Fabaceae), *Croton* spp. L. (Euphorbiaceae), *Eugenia neomyrtifolia* Gaertn (Myrtaceae), *Fraxinus uhdei* (Wenzig) Lingelsh, *Liquidambar styraciflua* L. (Altingiaceae), *Magnolia grandiflora* L. (Magnoliaceae), *Pittosporum* undulatum Vent. (Pittosporaceae), *Populus nigra* and *Quercus humboldtii* Bonpl. (Fagaceae). The predominant phytoplasmas found in these trees belong to the 16SrI and 16SrVII groups and were detected in all tested species (Franco-Lara et al., 2017; Franco-Lara & Perilla-Henao, 2014; Perilla-Henao, Dickinson, & Franco-Lara, 2012). Phytoplasmas from the 16SrV, 16SrIX and 16SrXII groups were also reported in *L. styraciflua* (Franco-Lara et al., 2017). Estimations of the incidence of the disease in the city were performed in 2013 and showed that the number of symptomatic trees per species varied from 93 to 100% for *F. uhdei, L. styraciflua* and *P. nigra* and from 85 to 66% for *E. neomyrtifolia, P. undulatum, M. grandiflora and Q. humboldtii*, whereas the least affected species were *A. melanoxylon* and *Croton* spp. with percentages of 47 and 36%, respectively (Franco-Lara & Perilla-Henao, 2014). So far, remission of symptoms has not been observed (L. Franco-Lara, personal observations). These facts suggest that a phytoplasma epidemy that affects several tree species is underway, and that the aesthetic value and ecosystemic services of the urban trees are being negatively impacted. In addition, phytoplasmas of the same groups

Fig. 8.1 (continued) (**b**) Loss of apical dominance and elongation of branches in *Magnolia grandiflora*; (**c**) Elongated branches in the top of the crown and tufted foliage in those branches in *Quercus humboldtii*; (**d**) Deformation of crown in *Pittosporum undulatum*; (**e**) "Leaflet" like structures emerging from trunk buds, densely covered with trichomes in *Croton* spp.; (**f**) Small leaves in an elongated branch (right) compared to a normal leaves in an asymptomatic branch in *Liquidambar styraciflua*; (**g**) Deliquescent branching in which apical dominance is lost in *Fraxinus uhdei*; (**h**) Witches'-broom in *Cordia alliodora* and (**i**) Epicormic shoots in *Salix humboldtiana* have been detected in symptomatic strawberry and potato crops in rural areas surrounding Bogota, suggesting that the pathogen is spreading beyond the urban area (Perilla-Henao & Franco-Lara, 2012; Varela, Quintero, Delgado, Arevalo, & Franco-Lara, 2017).

8.3 Management of Phytoplasma Diseases in Planted Urban Trees

8.3.1 Conventional Approaches

Control of phytoplasmas is difficult since there are no curative methods, and resistance or tolerance is rare. Classical approaches used to control phytoplasma diseases include roguing and insecticide treatments, are difficult to apply, and legally proscribed in several cities (Garnier et al., 2001; Weintraub & Beanland, 2006). The prevention of phytoplasma diseases spreading to new areas and hosts should be part of the managing strategy, since there are few management options once the disease is established in perennial plants.

The first step to avoid phytoplasma disease propagation is the production of phytoplasma-free planting material. Similar to other diseases, planting clean material should be considered a priority. Consequently, strategies to produce pathogen-free material and efficient methods to diagnose phytoplasmas before planting trees are needed. Diagnostic methods for the detection of phytoplasmas in plants have been reported and include nested PCR followed by RFLP or sequencing analysis (Bertaccini & Duduk, 2010; IRPCM Phytoplasma/Spiroplasma Working Team–Phytoplasma Taxonomy Group, 2004; Lee, Gundersen-Rindal, Davis, & Bartoszyk, 1998; Seemuller et al., 1994; Smart et al., 1996), qPCR (Aldaghi et al., 2009; Bianco, Casati, & Marziliano, 2004) and LAMP (De Jonghe, De Roo, & Maes, 2017; Dickinson, 2015). Molecular diagnostic of phytoplasmas should be a mandatory requirement before planting trees belonging to species known to be susceptible to phytoplasma.

Trees can be propagated by seeds or by vegetative propagation and are usually maintained in nurseries before being planted (Elevitch, 2004). It is generally admitted that phytoplasmas are not seed-borne (Lee, Davis, & Gundersen-Rindal, 2000). However, vertical transmission has been observed in several pathosystems and to our knowledge, the transmission of phytoplasmas through seeds in urban trees has not been studied (Botti & Bertaccini, 2006; Calari et al., 2011; Oropeza et al., 2017). Therefore, in addition to the use of clean planting material for seed growing or vegetative propagation, it is paramount to enforce stringent conditions to avoid transmission of phytoplasmas during the nursery stage.

Conventional vegetative propagation methods of cuttings usually employed for herbaceous plants have proven difficult for woody plants (Winkelmann, 2013). The *in vitro* tissue culture technology offers an opportunity to propagate woody plants and to eliminate phytoplasmas from valuable genotypes. The regeneration response of *in vitro* culture depends on the species, genotypes and physiological states of tissue and culture conditions, meaning that regeneration protocols must be adapted to each species and particular conditions. However new methods, many involving somatic embryogenesis, are now available for several tree species (Davey, 2017; Winkelmann, 2013).

In situations such as Bogota, where the infectious status is unknown for most dominant tree species, *in vitro* plant tissue culture methods might be used to remove phytoplasmas from infected plants and produce clean planting materials. Meristem tip culture is the most common method to eliminate viruses or phytoplasmas, as these pathogens do not infect meristematic tissues. Clean material can be obtained by dissecting the meristematic dome of one or two leaf primordia from the mother plant and proceed to its regeneration (Parmessur, Aljanabi, Saumtally, & Dookun-Saumtally, 2002). The elimination of phytoplasmas from mulberry plants (*Morus alba* L.) was achieved by tissue culture of stem segments propagated in basic Murashige and Skoog medium (1962) (Dai, He, & Liu, 1997). In Lebanon, almond trees (*Prunus dulci* Miller) explants were subjected to tissue culture treatments which resulted in the successful elimination of phytoplasmas; assays included the culture of stem cuttings treated by thermotherapy (35 ± 2 °C for 30 days), shoot tip cultures with or without thermotherapy, and shoot tip micrografting (Chalak et al., 2005).

Elimination of phytoplasmas from *Vitis vinifera* L. (Vitaceae) was obtained after bud meristem microcultivation of plants infected with the "Flavescence Dorée" (FD) phytoplasma, in 100% of explants and after six to nine months of culture. Conversely, the same tissue culture treatment eliminated the "Bois Noir phytoplasma" only in half of the treated explants (Gribaudo et al., 2007). However, Bois Noir phytoplasma was eliminated from *V. vinifera* using a combination of tissue culture and thermotherapy treatments. Stem cuttings treated with water at 50 °C for 15 or 30 min allowed the regeneration of 100% of phytoplasma-free explants (Chalak et al., 2013). Other biotechnological methods such as cryopreservation of plant materials have successfully eliminated viruses and phytoplasmas in important crops such as sweet potato (*Ipomea batata* (L.) Lam., Convolvulaceae). Briefly, plant material such as shoot tips containing leaf primordia, was encapsulated in alginate beads, subject to a pre-freezing treatment, immersed in liquid nitrogen and quickly thaw and cultured on a post-culture medium. The efficiency of the regeneration of the explants was over 80%, with all of them free of phytoplasma (Feng et al., 2011). The antibiotic oxytetracycline is known to induce symptom remission and its addition to the tissue culture media has been reported to enhance the production of phytoplasma free plantlets (Davies & Clark, 1994; Lee et al., 2000). Treatment of *Saccharum officinarum* L. (Poaceae) explants produced plantlets free of Sugarcane white leaf (SCWL) phytoplama when cultured *in vitro*. Tissue culture media without oxytretracycline induced the regeneration of phytoplasma free plantlets in 2% of the cases, whereas medium supplemented with 200–500 mg/ml oxytetracycline produced 70–100% of plantlets free of phytoplasmas. However, the use of oxytetracycline at concentrations of 500 mg/ml produced phytotoxicity for 50% of the explants cultured in the same conditions (Wongkaew & Fletcher, 2004). In contrast, Chalak et al. (2005) found that oxytetracycline at 50, 100 and 150 ug/ml completely inhibited the regeneration of *P. dulci* explants.

There is little published information on the application of tissue culture techniques for the production of phytoplasma-free plants belonging to species commonly used as urban trees. However, recent advances in tissue culture of woody plants and in somatic embryo regeneration described above, could be used for production of healthy urban trees. Tissue culture techniques such as meristem tip culture coupled with thermotherapy or chemotherapy are likely to be successful to generate clean material.

8.3.2 Integrated Management Approach

Once a phytoplasma disease is established in a region, the management strategy should consider an integrated approach. The classical phytopathological description of a plant disease considers three factors under the so-called 'disease triangle': a susceptible host, a virulent pathogen, and a conducive environment. In principle, disrupting any factor would lead to a management strategy (Agrios, 1997). However, phytoplasma disease involves more factors, as insect vectors are also included as transmission agents. In this tri-trophic interaction, the basic plant disease triangle representation is altered (Fig. 8.2), as the infection in both plants and insects is subject to host-pathogen-environment interactions. On a global scale, the incidence and spatial pattern distributions of documented pests and pathogens are increasing and holistic risk assessments are necessary for management strategies (Bebber, 2015; Boyd, Freer-Smith, Gilligan, & Godfray, 2013). Insect vectors determine the extent of the pathogen spread at the geographical, ecological and temporal scale (Eigenbrode, Bosque-Pérez, & Davis, 2018). The insect infectious status and

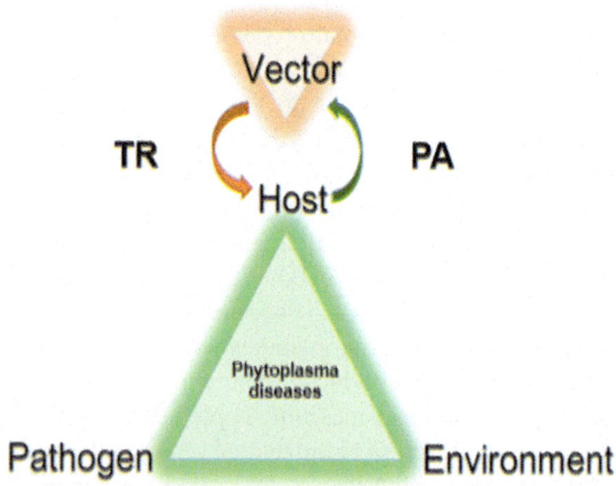

Fig. 8.2 Basic plant disease triangle linked with vector biology (*PA* Pathogen acquisition, *TR* Pathogen transmission)

transmission ability vary depending on the inoculum density. Vectors may have single or multiple hosts for feeding or reproduction, and recurrent insect populations may re-infect hosts. Multiple infections occurring over time, variation in virulence among different phytoplasma strains, and different phenological stages of the host, can produce fluctuations in the symptoms exhibited by infected trees (Schneider, Sule, Jelkmann, & Seemüller, 2014). The ecological interactions of vectors that transmit phytopathogens are complex. For example, the infectious status of the host can alter the vector settlement, feeding or reproduction biology (Eigenbrode et al., 2018). Aster yellow phytoplasma is known to increase the fitness of their insect vector by reducing the time needed from generation to generation (Beanland, Hoy, Miller, & Nault, 2000). In the following paragraphs, the host, environment and vector aspects of phytoplasma 'disease triangle' for perennial trees, are discussed and examples of control measures are included.

Trees represent a very important host for pathogens as they have long life spans, secondary growths, and different life cycle as compared to herbaceous plants (Eyles, Bonello, Ganley, & Mohammed, 2010). Vascular pathogens such as phytoplasma remain in perennial host for long periods of time, becoming a perdurable source of inoculum for other plants. During winter in temperate regions, the aerial phloem dies and the phytoplasma titter is reduced (Laimer et al., 2009). In contrast, tropical areas are characterized by warm temperatures displaying less variations and trees lack the dormancy period that can reduce phytoplasma titters. Tropical conditions may also favor large populations of insect vectors, since there are no overwintering periods. Ash trees grown experimentally in glasshouses have altered cycles of growth patterns (Sinclair & Griffiths, 2000). Indeed, it is expected that temperate trees growing in tropical conditions have different growth cycles, inducing different patterns in phytoplasma infections. Additionally, areas with greater biodiversity potentially offer more alternative host plants for insects. Urban trees frequently inhabit extensive areas and are not confined to a specific area, such as field crops, which represents an additional barrier for their management.

8.3.3 Host Resistance

As long-lasting plants, trees are exposed to several stress sources during their lives and possess constitutive and inducible defenses against pathogens, insects or predators to withstand dynamic challenges. The range of defenses depends on evolutionary adaptations in each species, their ecological niches, abiotic conditions, biotic interactions, and predispositions (Bostock, Pye, & Roubtsova, 2014; Eyles et al., 2010).

In plants, induced defense mechanisms are activated as a response to biotic and abiotic factors and include chemical, anatomical, ecological and compensatory defenses as well as specific proteins (Eyles et al., 2010). Defense signaling pathways act at the local or systemic level. At the systemic level, a broad spectrum defense involves conserved plant stress pathways that are interconnected with phytohormone signaling and production of reactive oxygen species (ROS) (Bostock et al., 2014). In response to abiotic and biotic stress, hormonal changes in abscisic acid (ABA) and cytokinins (CK) coincide with systemic increases of other phytohormones, hence generating a 'primed' plant that is more tolerant to future encounters with biotic stressors (Bostock et al., 2014; Eyles et al., 2010). Changes in gene expression of the salicylic acid (SA) hormone and of other protein intermediates activate the systemic acquired resistance (SAR) pathway. The alternative pathways, jasmonic acid (JA) and ethylene (ET) are associated with induced systemic resistance (ISR), which is typically activated as a response to bacterial colonization of the roots or to leaf-chewing insects (Bostock et al., 2014; Kallenbach, Bonaventure, Gilardoni, Wissgott, & Baldwin, 2012). While some phloem-feeding insects have several strategies to avoid effective plant defense during feeding or reproduction (Walling, 2008), insects such a whiteflies and aphids elicit SA mediated response (Walling, 2008; Zarate, Kempema, & Walling, 2007). Similarly, natural populations of *Empoasca* sp. Walsh (Hemiptera: Cicadellidae), are able to choose plant hosts with different capabilities of JA signaling and feed preferentially on plants with reduced JA signaling defense mechanisms (Kallenbach et al., 2012).

Interactions between phytoplasma and the tree's defense system are complex and not well understood. Phytoplasma infection up and down-regulate genes related to various pathways in a very coordinated way (Mardi, Farsad, Gharechahi, & Salekdeh, 2015). Using transcriptome profiling, Wang et al. (2018) observed that phytohormone biosynthesis, signal transduction, photosynthesis, and secondary metabolism played important roles in the interactions between '*Ca.* P. ziziphi' and its jujube tree host. Similarly, Zimmermann et al. (2015) observed a strong increase of SA accompanied with a decrease of JA in apple tree infected with '*Ca.* P mali' suggesting that phytohormones have an important role in tree defense signaling against phytoplasma infection.

In crop systems where chemical controls of pathogen and pest are limited, such as organic agriculture, urban areas, landscape trees or forests, the best long-term control measure is to harness the intrinsic plant host genetic tolerance or resistance (Eyles et al., 2010). Worldwide, there are very few examples of successful host resistance against bacterial phytopathogens transmitted by insects, as searching for natural resistance to phytoplasmas in trees is very time-consuming (Laimer et al., 2009). However, a few phenomenons of natural resistance have been described for woody plants such as recovery (Carraro, Ermacora, Loi, & Osler, 2004; Paolacci et al., 2017; Santi, De Marco, Polizzotto, Grisan, & Musetti, 2013), cross-protection (Osler et al., 2016; Schneider et al., 2014; Sinclair & Griffiths, 2000) and rootstock resistance (Seemüller, Moll, & Schneider, 2009).

The phenomenon of recovery is a resistance mechanism for which plants have biochemical and structural reactions that limit the phytoplasma propagation and which involves an increase in sucrose transport and defense signaling (Santi et al., 2013). The recovery process in fruit trees has been observed only in field conditions after decades of long-term documentation. For 12 years, Carraro et al. (2004) followed individual pear trees, characterized their symptoms and recorded the presence of phytoplasmas in roots and aerial parts. The authors demonstrated that phytoplasmas were detected in the roots of trees showing signs of recovery but not in branches or trunk. So far, most recovery descriptions occur in European stone fruits, pome fruits and grapes. Importantly, the virulence, infectivity and viability of phytoplasmas remain unchanged in other hosts, but in the recovered trees, no aerial symptoms were observed for several years in a row (Carraro et al., 2004). Apricot has been tested for resistance to European Stone Yellows Phytoplasma (ESFY) and no resistant genotypes have been found, but individual plants recover occasionally (Osler et al., 2016). Spontaneous recovery does not eliminate phytoplasma infection, but symptom expression is reduced and the probability of re-infection is low (Laimer et al., 2009). Recovery is a phenomenon that varies depending on host genotype and phytoplasma strain virulence. In grapevines, only one cultivar among many, showed recovery, after Flavescence Dorée infection (Osler et al., 2016). The recovery phenomenon is a promising strategy for the management of phytoplasmas in trees, but more knowledge of the biological mechanisms that trigger the response are needed in order to use it as a management tool.

The cross-protection effect has been observed in herbaceous and woody plant host as well as in leafhopper vectors (Schneider et al., 2014; Sinclair & Griffiths, 2000). Cross protection occurs when closely related phytoplasma strains have different virulence. A mild or weakly virulent strain (primary) that infect a host, elicits a process of protection. When a second virulent strain (challenger) infects the host, the mechanism associated with the primary strain or its effect on the host have antagonistic interference with multiplication or long-distance movement of the challenger strain (Sinclair & Griffiths, 2000). An example of cross-protection occurs in *Fraxinus* spp. trees in North-America, where several distinct natural occurring populations of '*Ca.* P. fraxini' were tested for aggressiveness on woody and herbaceous hosts (Sinclair & Griffiths, 2000). Similarly. using *C. roseus* and *Nicotiana occidentalis* as host, Schneider et al. (2014) demonstrated the presence of suppressive strains in '*Ca.* P. mali' from Europe. Documented cases of cross-protection in herbaceous hosts include '*Ca.* P. asteris', and potato witches' broom phytoplasmas among others (Osler et al., 2016). Describing the virulence of novel phytoplasmas strains requires the maintenance of live collections of phytoplasma strains, regular infection of various plant hosts, and the development of cost-effective genotyping tools for local phytoplasma strains.

Resistant genotypes, that are used as rootstocks, offer protection by limiting phytoplasma survival in the rootstock during the dormancy period of the plant, when root's vascular system is the only viable phloem, thus avoiding inoculum build-up year after year (Seemüller, Moll, & Schneider, 2008). A true resistant genotype is different from recovery or cross protection phenomena, as in the latter situations, the genotype background is susceptible and occasionally some individuals show reduced symptoms or infection (Osler et al., 2016). The recovery phenomenon is not well understood and the conditions conducive to this change in plant response, is hard to predict.

The underlying biochemical or defense pathways involved in resistance, recovery or cross-protection are not fully understood and they appear to be species or cultivar-specific. In a study of grapevine infected with Bois Noir (BN), key plant defense and metabolic enzymes such as callose synthase, invertase, and sugar transporters were evaluated during recovery and Santi et al. (2013) demonstrated an increase in expression of genes associated with sucrose transport. Studying the role of SA- and JA-mediated signaling in phytoplasma/plant interactions in BN infected grapevines, Paolacci et al. (2017) found that JA-responsive genes were up-regulated in resistant leaves. In apples affected by Apple proliferation phytoplasma (AP), the production of ROS was evaluated in healthy, symptomatic and recovered apples. Musetti, di Toppi, Ermacora, and Favali (2004) found that H_2O_2 concentration was higher in recovered leaves and associated recovery with sustained production of H_2O_2 in apple leaf tissue. In summary, while there may be promising examples of tolerance to phytoplasmas, the conditions are highly dependent on the pathosystem and more research is needed to integrate them in management schemes.

8.3.4 Environmental Conditions

Atmospheric and climatic conditions affect plant physiology and therefore, the ability of plants to respond to diseases (Eastburn, McElrone, & Bilgin, 2011; Garrett, Dendy, Frank, Rouse, & Travers, 2006). Frequently, urban trees are exposed to suboptimal growing conditions such as water stress, higher temperature due to urban heat islands, elevated concentrations of CO_2, NO_2 and O_3 and air particles PM2.5 and PM10, as compared to their original habitats (Dale & Frank, 2017; Richter, 2016; World Health Organization, 2014). The effect of these environmental conditions on the response of urban trees to pathogens, has been poorly studied. Studying climate change, Garrett et al. (2006) and Tubby and Webber (2010) demonstrated that physiological responses of plants change due to high concentrations of CO_2, NO_2 and O_3 in the air. For example, elevated concentrations of CO_2 and O_3 predispose *Populus tremuloides* Michx. (Salicaceae) to infection by the fungi *Melampsora medusae* Thüm. (Melampsoraceae) and affect stomata opening, the quantity and quality of leaf wax and photosynthetic activity of plants, therefore affecting the responses of the trees to pathogens (Eastburn et al., 2011; Karnosky et al., 2002). It has also been shown that herbivorous insects can be more abundant and damaging in urban environments. This can be attributed to less complex ecosystems in which the presence of biological controllers is reduced, the water-stressed trees are more attractive to insects and the dry and warm conditions of the city favor the pest fitness and abundance (Dale & Frank, 2017). Similarly, Canto, Aranda, and Fereres (2009) showed that the physiology, population dynamics and geographical range of hemipterans vectors in locations such as the Iberian peninsula, have changed as a direct consequence of climatic dynamics. Therefore, more research is needed to understand the effect of the urban climate in relation to phytoplasma diseases in urban trees.

8.3.5 Vector Management

Phytoplasmas are transmitted by insect vectors of the order Hemiptera, and vector species have been found in the Cicadellidae, Cixiidae, Psyllidae, Cercopidae, Delphacidae, Derbidae, Menoplidae and Flatidae families (Weintraub & Beanland, 2006). However, for most phytoplasma diseases, the insect vectors are unknown and only a few vectors have been studied within the context of their host crops (Foissac & Wilson, 2010). The geographical distribution of phytoplasma diseases depends on the range of their host plants and on the range of the plants needed by the insect vectors to breed and feed (Foissac & Wilson, 2010; Garnier et al., 2001; Weintraub & Beanland, 2006). Therefore, in order to propose management strategies, it is also important to know as much as possible of the life cycle, population dynamics and alternative host plants that insect vectors may have.

A typical management strategy is the application of insecticides to control insect vector populations, when the vector is monophagous and restricted to the affected crop (Garnier et al., 2001). Phytoplasma vectors are generally considered monophagous, but many polyphagous vectors have also been identified (Foissac & Wilson, 2010). The devastating effect of polyphagous insect vectors is observed in Bogota, where the occurrence of phytoplasmas has been detected by molecular methods at high prevalence, in nine urban tree species and symptoms have been observed in many other plant species (Perilla-Henao et al., 2012). The occurence of the polyphagous insects *Amplicephalus funzaensis* Linnavuori (Hemiptera: Cicadellidae) and *Exituanus atratus* Linnavuori (Hemiptera: Cicadellidae) has been reported in the area of Bogota (Perilla-Henao, Wilson & Franco-Lara, 2016). These insects complete their life cycles in grasses but also feed on *Apium graveolens* L. (Apiaceae), *Daucus carota* L. (Apiaceae), *Fragaria* x *ananassa* Duschene (Rosaceae), *Physalis vulgaris* Fleming (Balaenopteridae) and *Solanum tuberosum* L. (Solanaceae) among others, and have been recently collected from *Q. humboldtii* trees (unpublished results). Additionally, four other Deltocephalinae leafhopper species including a *Dalbulus* sp. DeLong (Hemiptera: Cicadellidae), are present in the ubiquitous grass cover found in parks and street separators of the city and its surroundings (unpublished results). *Dalbulus* genus is a known vector of the Maize Bushy Stunt phytoplasma in America, and is usually associated with *Zea* spp. L. (Poaceae) and *Tripsacum* spp. L. (Poaceae) grass species (Nault et al., 1984). These Deltocephalinae may also be vectors of phytoplasmas in this pathosystem.

Insect vector control using conventional insecticides has been attempted with variable results for the prevention of phytoplasma diseases, even when applied frequently (D'Amelio, Marzachì, & Bosco, 2010; Saracco, Marzachì, & Bosco, 2008; Wally et al., 2004; Weintraub, 2007). In the complex case of the trees in Bogota, insecticide spraying to control the insect vectors, would be impractical and expensive, as well as ineffective. Nets were used to cover papaya plants in Australia, and were considered more efficient than insecticide to protect the plants from the visit of insect vectors. This method seemed efficient to avoid the inoculation of phytoplasmas in the plants, but pollination was diminished (Walsh, Guthrie, & White, 2006). Insecticide spraying and tree netting are inapplicable for urban trees due to logistics and economic costs, as well as environmental and human health concerns for insecticides. Commercial and domestic devices are available to inject solutions of fertilizer and systemic insecticides directly in the phloem of the trees. Despite the lack of curative chemical treatment for phytoplasmas, the application of fertilizers to underfertilised and weak trees in combination with insecticides, has a positive effect on the health of the trees (L. Franco-Lara, personal communication). However, these treatments are expensive because of the chemicals, injecting device and manpower costs and produce only temporary improvement, as trees need to be treated regurlarly.

A developing alternative to traditional insecticides for vector control, is the reduction of vector transmission ability (Killiny, Rashed, & Almeida, 2012; Sinkins & Gould, 2006). In principle, understanding the bacterial mechanisms responsible for direct interactions between the phytoplasma and the insect host, opens opportunities to identify cellular targets for control. Interfering with any step during the bacterial recognition, adhesion, multiplication, survival or transmission process within the insect, would represent a breakthrough in dealing with emergent and unpredictable outbreaks of vector-borne phytopathogenic bacteria (Perilla-Henao & Casteel, 2016). One approach would be the identification of adhesion factors, which are extracellular proteins of the phytoplasma cell recognized to play a crucial role for the circulation of the phytoplasma within the vector. A potential control would be the use of molecules interfering with the adhesion factors within the insect, hence altering the biological cycle of the phytoplasma. For example, the Onion Yellow phytoplasma protein P38, plays a role in the adhesion of phytoplasmas to the insect vector and probably to the host plant, and could interfere with the transmission if blocked (Neriya et al., 2014). Using *in vitro* extracts containing antibodies that recognize the Aster Yellow phytoplasma Amp membrane protein, Rashidi et al. (2015) demonstrated that Amp could be another target for alternative control, as Amp is involved in the crossing of the midgut and salivary glands membranes by the phytoplasma. Other strategies such as RNA interference (RNAi) tested in plant virus transmission (Kanakala & Ghanim, 2016) or gene drive systems (Sinkins & Gould, 2006) are promissory, but are in the early stages of development.

Therefore, other strategies, such as the use of natural enemies, are needed to control insect vectors. For instance, in Central America *Dalbulus maidis* DeLong (Hemiptera: Cicadellidae), is parasitized by *Anagrus* sp. Haliday (Hymenoptera: Mymaridae) and its eggs are parasitized by *Paracentrobia* sp. Howard (Hymenoptera: Trichogrammatidae). In Mexico, *Gonatropus bartletti* Olmi (Hymenoptera: Dryinidae) and *Eudorus* sp. (Diptera: Pipunculidae) attack nymphs, and the fungi *Beauveria bassiana* (Bals.-Criv.) Vuill. (Cordycipitaceae) and *Metarhizium anisopliae* (Metchnikoff) Sorokin (Clavicipitaceae) infect nymphs and adults (Capinera,

2001). The eggs of *Empoasca fabae* Harris are parasitized by Hymenoptera Mymaridae and Dryinidae and generalist predators such as spiders, lacewings, nabids, lady beetles and ants destroy these leafhoppers. The fungus *Erynia radicans* Bref. (Entomophtoraceae) is also a natural enemy of *E. fabae*. In Bogota, *Eudorylas* sp. Aczél (Diptera: Pipunculidae) and *Tomosvaryella* sp. Aczél (Diptera: Pipunculidae) have been found parasitizing *Dalbulus* sp. and *E. atratus* respectively (Silva-Castaño, Brochero, Wilson, & Franco-Lara, 2017). However, so far, biological control schemes have not been developed for insect vectors of tree diseases.

8.4 Perspectives

Dealing with phytoplasma diseases in not an easy task. Their complex pathosystems involve a tritrophic interaction "plant-insect-phytoplasma", which depend on the susceptibility of the host plants and the insect vectors and is influenced by internal and external biotic and abiotic conditions. Moreover, dealing with urban trees is difficult since they are long lasting plants, often living in unfriendly environments where they are exposed to suboptimal conditions of temperature, water availability, high concentrations of gases and attack from herbivorous insects. Urban trees offer invaluable ecosystemic services to people living in the cities and are also important in the mitigation of climate changes. It is therefore necessary to find management strategies to decrease the impact of disease in urban trees and its dispersal to other plants.

Integrated management approaches require a deep knowledge of the complex relationships between all the components of the pathosystem. Potential integrated management can also be limited by the fact that urban trees frequently live in non-contained areas. For most of the urban tree pathosystems, more knowledge and research are needed to have a better overview of the interacting factors, to allow their manipulation in the future. Promising strategies such as the induction of the recovery phenomenon or SAR and ISR pathways, require more information about trees response to phytoplasmas and insect vector attacks. The search for genetic resistance to phytoplasmas is another avenue and biological control of insect vectors is an interesting approach that requires the identification of potential controllers, whether insects or entomopathogens, and the modelling of their behavior in open urban conditions. The effect of environmental variables on the disease and its transmission in urban trees has not been studied and should consider the differences between urban areas from temperate and tropical zones. Another promising approach would be to search in the microbiota of infected and healthy trees for possible competitors of phytoplasmas. In conclusion, there are many opportunities to study strategies of management for urban trees.

References

Agrios, G. (1997). *Plant pathology* (4th ed.). New York, NY: Academic.

Aldaghi, M., Massart, S., Dutrecq, O., Bertaccini, A., Jijakli, M., & Lepoivre, P. (2009). A simple and rapid protocol of crude DNA extraction from apple trees for PCR and real-time PCR detection of '*Candidatus* Phytoplasma mali'. *Journal of Virological Methods, 156*(1–2), 96–101.

Allen, C. D., Macalady, A. K., Chenchouni, H., Bachelet, D., McDowell, N., Vennetier, M., & Hogg, E. T. (2010). A global overview of drought and heat-induced tree mortality reveals emerging climate change risks for forests. *Forest Ecology and Management, 259*(4), 660–684.

Arismendi, N., Andrade, N., Riegel, R., & Carrillo, R. (2010). Presence of a phytoplasma associated with witches' broom disease in *Ugni molinae* Turcz. and *Gaultheria phillyreifolia* (Pers.) Sleumer determined by DAPI, PCR, and DNA sequencing. *Chilean Journal of Agricultural Research, 70*(1), 26–33.

Arneodo, J., Galdeano, E., Orrego, A., Stauffer, A., Nome, S., & Conci, L. (2005). Identification of two phytoplasmas detected in China-trees with decline symptoms in Paraguay. *Australasian Plant Pathology, 34*(4), 583–585.

Arneodo, J., Marini, D., Galdeano, E., Meneguzzi, N., Bacci, M., Jr., Domecq, C., & Conci, L. (2007). Diversity and geographical distribution of phytoplasmas infecting China-tree in Argentina. *Journal of Phytopathology, 155*(2), 70–75.

Ayres, M. P., & Lombardero, M. J. (2000). Assessing the consequences of global change for forest disturbance from herbivores and pathogens. *Science of the Total Environment, 262*(3), 263–286.

Beanland, L., Hoy, C., Miller, S., & Nault, L. (2000). Influence of aster yellows phytoplasma on the fitness of aster leafhopper (Homoptera: Cicadellidae). *Annals of the Entomological Society of America, 93*(2), 271–276.

Bebber, D. P. (2015). Range-expanding pests and pathogens in a warming world. *Annual Review of Phytopathology, 53*, 335–356.

Bertaccini, A., & Duduk, B. (2010). Phytoplasma and phytoplasma diseases: A review of recent research. *Phytopathologia Mediterranea, 48*(3), 355–378.

Bianco, P., Casati, P., & Marziliano, N. (2004). Detection of phytoplasmas associated with grapevine flavescence dorée disease using real-time PCR. *Journal of Plant Pathology, 86*(3), 257–261.

Bostock, R. M., Pye, M. F., & Roubtsova, T. V. (2014). Predisposition in plant disease: Exploiting the nexus in abiotic and biotic stress perception and response. *Annual Review of Phytopathology, 52*, 517–549.

Botti, S., & Bertaccini, A. (2006). *Phytoplasma infection through seed transmission: Further observations*. Paper presented at the 16th International Organization of Mycoplasmology Conference (July 9–14, 2006), Cambridge, UK.

Boyd, I., Freer-Smith, P., Gilligan, C., & Godfray, H. (2013). The consequence of tree pests and diseases for ecosystem services. *Science, 342*(6160), 1–8.

Bricker, J. S., & Stutz, J. C. (2004). Phytoplasmas associated with ash decline. *Journal of Arboriculture, 30*(3), 193–199.

Calari, A., Paltrinieri, S., Contaldo, N., Sakalieva, D., Mori, N., Duduk, B., & Bertaccini, A. (2011). Molecular evidence of phytoplasmas in winter oilseed rape, tomato and corn seedlings. *Bulletin of Insectology, 64*(Supplement), S157–S158.

Canto, T., Aranda, M. A., & Fereres, A. (2009). Climate change effects on physiology and population processes of hosts and vectors that influence the spread of hemipteran-borne plant viruses. *Global Change Biology, 15*(8), 1884–1894.

Capinera, J. (2001). Order Homoptera–aphids, leaf-and planthoppers, psyllids and whiteflies. In J. Capinera (Ed.), *Handbook of vegetable pests*. (Vol. 1, pp. 279–282): San Diego, CA: Academic.

Carraro, L., Ermacora, P., Loi, N., & Osler, R. (2004). The recovery phenomenon in apple proliferation-infected apple trees. *Journal of Plant Pathology, 86*, 141–146.

Chalak, L., Dekwaneh, B., Elbitar, A., Mourad, N., Mortada, C., & Choueiri, E. (2013). Elimination of grapevine bois noir Phytoplasma by tissue culture coupled or not with heat therapy or hot water treatment. *Advances in Crop Science and Technology, 1*(2), 107.

Chalak, L., Elbitar, A., Rizk, R., Choueiri, E., Salar, P., & Bové, J. (2005). Attempts to eliminate '*Candidatus* phytoplasma phoenicium' from infected Lebanese almond varieties by tissue culture techniques combined or not with thermotherapy. *European Journal of Plant Pathology, 112*(1), 85–89.

D'Amelio, R., Marzachì, C., & Bosco, D. (2010). Activity of benzothiadiazole on chrysanthemum yellows phytoplasma ('*Candidatus* Phytoplasma asteris') infection in daisy plants. *Crop Protection, 29*(10), 1094–1099.

Dai, Q., He, F. T., & Liu, P. Y. (1997). Elimination of phytoplasma by stem culture from mulberry plants (*Morus alba*) with dwarf disease. *Plant Pathology, 46*(1), 56–61.

Dale, A. G., & Frank, S. D. (2017). Warming and drought combine to increase pest insect fitness on urban trees. *PLoS One, 12*(3), e0173844.

Davey, M. (2017). Regeneration of fruit and ornamental trees via cell and tissue culture. In B. Thomas, B. Murray, & D. murphy (Eds.), *Encyclopedia of applied sciences. Second Edition.* (pp. 448–454). Dordrecht, The Netherlands: Academic..

Davies, D., & Clark, M. (1994). Maintenance of mycoplasma-like organisms occurring in *Pyrus* species by micropropagation and their elimination by tetracycline therapy. *Plant Pathology, 43*(5), 819–823.

De Jonghe, K., De Roo, I., & Maes, M. (2017). Fast and sensitive on-site isothermal assay (LAMP) for diagnosis and detection of three fruit tree phytoplasmas. *European Journal of Plant Pathology, 147*(4), 749–759.

Demuzere, M., Orru, K., Heidrich, O., Olazabal, E., Geneletti, D., Orru, H., & Faehnle, M. (2014). Mitigating and adapting to climate change: Multi-functional and multi-scale assessment of green urban infrastructure. *Journal of Environmental Management, 146*, 107–115.

Dickinson, M. (2015). Loop-mediated isothermal amplification (LAMP) for detection of phytoplasmas in the field. In C. Lacomme (Ed.), *Plant pathology; methods in molecular biology* (pp. 99–111). New York, NY: Humana Press/Springer.

Doi, Y., Teranaka, M., Yona, K., & Asuyama, H. (1967). Elements of plants infected with mulberry dwarf, potato witches' broom, aster yellows or Paulownia witches' broom. *Ann Phytopath. Soc. Japan, 33*, 259–266.

Eastburn, D., McElrone, A., & Bilgin, D. (2011). Influence of atmospheric and climatic change on plant–pathogen interactions. *Plant Pathology, 60*(1), 54–69.

Eigenbrode, S. D., Bosque-Pérez, N. A., & Davis, T. S. (2018). Insect-borne plant pathogens and their vectors: Ecology, evolution, and complex interactions. *Annual Review of Entomology, 63*, 169–191.

Elevitch, C. R. (2004). *The overstory book: Cultivating connections with trees*. Holualoa, HI: Permanent Agricultural Ressources.

Escobedo, F. J., Kroeger, T., & Wagner, J. E. (2011). Urban forests and pollution mitigation: Analyzing ecosystem services and disservices. *Environmental Pollution, 159*(8–9), 2078–2087.

Eyles, A., Bonello, P., Ganley, R., & Mohammed, C. (2010). Induced resistance to pests and pathogens in trees. *New Phytologist, 185*(4), 893–908.

Feng, C., Yin, Z., Ma, Y., Zhang, Z., Chen, L., Wang, B., & Wang, Q. (2011). Cryopreservation of sweetpotato (*Ipomoea batatas*) and its pathogen eradication by cryotherapy. *Biotechnology Advances, 29*(1), 84–93.

Foissac, X., & Wilson, M. R. (2010). Current and possible future distributions of phytoplasma diseases and their vectors. In P. G. Weintraub & P. Jones (Eds.), *Phytoplasmas: Genomes, plant hosts and vectors* (pp. 309–324). Oxfordshire, UK: CABI.

Franco-Lara, L., Contaldo, N., Mejia, J. F., Paltrinieri, S., Duduk, B., & Bertaccini, A. (2017). Detection and identification of phytoplasmas associated with declining Liquidambar styraciflua trees in Colombia. *Tropical Plant Pathology, 42*(5), 352–361.

Franco-Lara, L., & Perilla-Henao, L. (2014). Phytoplasma diseases in trees of Bogotá, Colombia: A serious risk for urban trees and crops. In A. Bertaccini (Ed.), *Phytoplasmas and phytoplasma disease management: How to reduce their economic impact* (pp. 90–100). Bologna, Italy: IPWG.

Galdeano, E., Torres, L., Meneguzzi, N., Guzmán, F., Gomez, G., Docampo, D., & Conci, L. (2004). Molecular characterization of 16S ribosomal DNA and phylogenetic analysis of two X-disease group phytoplasmas affecting China-tree (*Melia azedarach* L.) and garlic (*Allium sativum* L.) in Argentina. *Journal of Phytopathology, 152*(3), 174–181.

Garnier, M., Foissac, X., Gaurivaud, P., Laigret, F., Renaudin, J., Saillard, C., & Bové, J. (2001). Mycoplasmas, plants, insect vectors: A matrimonial triangle. *Comptes Rendus de l'Académie des Sciences-Series III-Sciences de la Vie, 324*(10), 923–928.

Garrett, K. A., Dendy, S. P., Frank, E. E., Rouse, M. N., & Travers, S. E. (2006). Climate change effects on plant disease: Genomes to ecosystems. *Annual Review of Phytopathology, 44*, 489–509.

Griffiths, H., Sinclair, W., Smart, C. D., & Davis, R. E. (1999). The phytoplasma associated with ash yellows and lilac witches'-broom: '*Candidatus* Phytoplasma fraxini'. *International Journal of Systematic and Evolutionary Microbiology, 49*(4), 1605–1614.

Harrison, N., Boa, E., & Carpio, M. (2003). Characterization of phytoplasmas detected in chinaberry trees with symptoms of leaf yellowing and decline in Bolivia. *Plant Pathology, 52*(2), 147–157.

IRPCM Phytoplasma/Spiroplasma Working Team–Phytoplasma Taxonomy Group. (2004). 'Candidatus Phytoplasma', a taxon for the wall-less, non-helical prokaryotes that colonize plant phloem and insects. *International Journal of Systematic and Evolutionary Microbiology, 54*(4), 1243–1255.

Jacobs, K., Lee, I.-M., Griffiths, H., Miller, F., Jr., & Bottner, K. (2003). A new member of the clover proliferation phytoplasma group (16SrVI) associated with elm yellows in Illinois. *Plant Disease, 87*(3), 241–246.

Kallenbach, M., Bonaventure, G., Gilardoni, P. A., Wissgott, A., & Baldwin, I. T. (2012). *Empoasca* leafhoppers attack wild tobacco plants in a jasmonate-dependent manner and identify jasmonate mutants in natural populations. *Proceedings of the National Academy of Sciences, 109*(24), E1548–E1557.

Kanakala, S., & Ghanim, M. (2016). RNA interference in insect vectors for plant viruses. *Viruses, 8*(12), 329.

Karnosky, D., Percy, K. E., Xiang, B., Callan, B., Noormets, A., Mankovska, B., & Dickson, R. (2002). Interacting elevated CO2 and tropospheric O3 predisposes aspen (*Populus tremuloides* Michx.) to infection by rust (*Melampsora medusae* f. sp. *tremuloidae*). *Global Change Biology, 8*(4), 329–338.

Killiny, N., Rashed, A., & Almeida, R. P. (2012). Disrupting the transmission of a vector-borne plant pathogen. *Applied and Environmental Microbiology, 78*(3), 638–643.

Laimer, M., Lemaire, O., Herrbach, E., Goldschmidt, V., Minafra, A., Bianco, P., & Wetzel, T. (2009). Resistance to viruses, phytoplasmas and their vectors in the grapevine in Europe: A review. *Journal of Plant Pathology, 91*(1), 7–23.

Lee, I.-M., Bertaccini, A., Vibio, M., Gundersen, D., Davis, R. E., Mittempergher, L., & Gennari, F. (1995). Detection and characterization of phytoplasmas associated with disease in *Ulmus* and *Rubus* in northern and Central Italy. *Phytopathologia Mediterranea, 34*(3), 174–183.

Lee, I.-M., Davis, R. E., & Gundersen-Rindal, D. E. (2000). Phytoplasma: Phytopathogenic mollicutes. *Annual Reviews in Microbiology, 54*(1), 221–255.

Lee, I.-M., Gundersen-Rindal, D. E., Davis, R. E., & Bartoszyk, I. M. (1998). Revised classification scheme of phytoplasmas based on RFLP analyses of 16S rRNA and ribosomal protein gene sequences. *International Journal of Systematic and Evolutionary Microbiology, 48*(4), 1153–1169.

Lee, I.-M., Gundersen-Rindal, D. E., Davis, R. E., Bottner, K., Marcone, C., & Seemüller, E. (2004). 'Candidatus Phytoplasma asteris', a novel phytoplasma taxon associated with aster yellows and related diseases. *International Journal of Systematic and Evolutionary Microbiology, 54*(4), 1037–1048.

Luque, S., & Iverson, L. (2016). Forest-related ecosystem services. In M. Potschin, R. Haines-Young, & K. Turnes (Eds.), *Routledge handbook of ecosystem services* (pp. 383–393). New York, NY: Routledge.

Marcone, C. (2015). Current status of phytoplasma diseases of forest and landscape trees and shrubs. *Journal of Plant Pathology, 97*(1), 9–36.

Marcone, C., Ragozzino, A., Cousin, M., Berges, R., & Seemüller, E. (1999). Phytoplasma diseases of trees and shrubs of urban areas in Europe. *Acta Horticulturae, 496*, 69–76.

Marcone, C., & Rao, G. P. (2016). Elm and alder yellows: Etiology, symptomatology and epidemiology. *Phytopathogenic Mollicutes, 6*(2), 63–70.

Mardi, M., Farsad, L. K., Gharechahi, J., & Salekdeh, G. H. (2015). In-depth transcriptome sequencing of Mexican lime trees infected with '*Candidatus* Phytoplasma aurantifolia'. *PLoS One, 10*(7), e0130425.

Musetti, R., di Toppi, L. S., Ermacora, P., & Favali, M. (2004). Recovery in apple trees infected with the apple proliferation phytoplasma: An ultrastructural and biochemical study. *Phytopathology, 94*(2), 203–208.

Nault, L., Madden, L., Styer, W., Triplehorn, B., Shambaugh, G., & Heady, S. (1984). Pathogenicity of corn stunt spiroplasma and maize bushy stunt mycoplasma to their vector, *Dalbubus longulus* [survival analyses, Mexico]. *Phytopathology, 74*, 977–979.

Neriya, Y., Maejima, K., Nijo, T., Tomomitsu, T., Yusa, A., Himeno, M., & Namba, S. (2014). Onion yellow phytoplasma P38 protein plays a role in adhesion to the hosts. *FEMS Microbiology Letters, 361*(2), 115–122.

Nowak, D. J., Greenfield, E. J., Hoehn, R. E., & Lapoint, E. (2013). Carbon storage and sequestration by trees in urban and community areas of the United States. *Environmental Pollution, 178*, 229–236.

Oropeza, C., Cordova, I., Puch-Hau, C., Castillo, R., Chan, J., & Sáenz, L. (2017). Detection of lethal yellowing phytoplasma in coconut plantlets obtained through *in vitro* germination of zygotic embryos from the seeds of infected palms. *Annals of Applied Biology, 171*(1), 28–36.

Osler, R., Borselli, S., Ermacora, P., Ferrini, F., Loschi, A., Martini, M., & Serra, S. (2016). Transmissible tolerance to European stone fruit yellows (ESFY) in apricot: Cross-protection or a plant mediated process? *Phytoparasitica, 44*(2), 203–211.

Paolacci, A. R., Catarcione, G., Ederli, L., Zadra, C., Pasqualini, S., Badiani, M., & Ciaffi, M. (2017). Jasmonate-mediated defence responses, unlike salicylate-mediated responses, are involved in the recovery of grapevine from bois noir disease. *BMC Plant Biology, 17*(1), 118.

Parmessur, Y., Aljanabi, S., Saumtally, S., & Dookun-Saumtally, A. (2002). Sugarcane yellow leaf virus and sugarcane yellows phytoplasma: Elimination by tissue culture. *Plant Pathology, 51*(5), 561–566.

Perilla-Henao, L., & Casteel, C. L. (2016). Vector-borne bacterial plant pathogens: Interactions with hemipteran insects and plants. *Frontiers in Plant Science, 7*, 1163.

Perilla-Henao, L., Dickinson, M., & Franco-Lara, L. (2012). First report of '*Candidatus* Phytoplasma asteris' affecting woody hosts (*Fraxinus uhdei, Populus nigra, Pittosporum undulatum*, and *Croton* spp.) in Colombia. *Plant Disease, 96*(9), 1372–1372.

Perilla-Henao, L., & Franco-Lara, L. (2012). *Phytoplasmas of group 16SrI associated with strawberry (Fragaria x Ananassa) in Colombia.* Paper presented at the 22nd international conference on virus and other transmissible diseases of fruit crops (June 3–8, 2012), Roma, Italy.

Perilla-Henao, L., Wilson, M. R., & Franco-Lara, L. (2016) Leafhoppers *Exitianus atratus* and *Amplicephalus funzaensis* transmit phytoplasmas of groups 16SrI and 16SrVII in Colombia. *Plant Pathology, 65*(7), 1200–1209.

Pimentel, D., Zuniga, R., & Morrison, D. (2005). Update on the environmental and economic costs associated with alien-invasive species in the United States. *Ecological Economics, 52*(3), 273–288.

Rashidi, M., Galetto, L., Bosco, D., Bulgarelli, A., Vallino, M., Veratti, F., & Marzachì, C. (2015). Role of the major antigenic membrane protein in phytoplasma transmission by two insect vector species. *BMC Microbiology, 15*(1), 193.

Reed, J., van Vianen, J., Foli, S., Clendenning, J., Yang, K., MacDonald, M., & Sunderland, T. (2017). Trees for life: The ecosystem service contribution of trees to food production and livelihoods in the tropics. *Forest Policy and Economics, 84*, 62–71.

Richter, M. (2016). Urban climate change-related effects on extreme heat events in Rostock, Germany. *Urban Ecosystem, 19*(2), 849–866.

Santi, S., De Marco, F., Polizzotto, R., Grisan, S., & Musetti, R. (2013). Recovery from stolbur disease in grapevine involves changes in sugar transport and metabolism. *Frontiers in Plant Science, 4*, 171.

Saracco, P., Marzachì, C., & Bosco, D. (2008). Activity of some insecticides in preventing transmission of chrysanthemum yellows phytoplasma ('*Candidatus* Phytoplasma asteris') by the leafhopper *Macrosteles quadripunctulatus* Kirschbaum. *Crop Protection, 27*(1), 130–136.

Schneider, B., Sule, S., Jelkmann, W., & Seemüller, E. (2014). Suppression of aggressive strains of '*Candidatus* Phytoplasma mali'by mild strains in *Catharanthus roseus* and *Nicotiana occidentalis* and indication of similar action in apple trees. *Phytopathology, 104*(5), 453–461.

Seemüller, E., & Lederer, W. (1988). MLO-associated decline of *Alnus glutinosa*, Populus tremula and Crataegus monogyna. *Journal of Phytopathology, 121*(1), 33–39.

Seemüller, E., Moll, E., & Schneider, B. (2008). Pear decline resistance in progenies of Pyrus taxa used as rootstocks. *European Journal of Plant Pathology, 123*(2), 217–223. https://doi.org/10.1007/s10658-008-9360-8

Seemüller, E., Moll, E., & Schneider, B. (2009). Pear decline resistance in progenies of Pyrus taxa used as rootstocks. *European Journal of Plant Pathology, 123*(2), 217–223.

Seemuller, E., Schneider, B., Maurer, R., Ahrens, U., Daire, X., Kison, H., & Sears, B. B. (1994). Phylogenetic classification of phytopathogenic mollicutes by sequence analysis of 16S ribosomal DNA. *International Journal of Systematic and Evolutionary Microbiology, 44*(3), 440–446.

Šeruga, M., Škorić, D., Botti, S., Paltrinieri, S., Juretić, N., & Bertaccini, A. (2003). Molecular characterization of a phytoplasma from the aster yellows (16SrI) group naturally infecting *Populus nigra* L.'Italica'trees in Croatia. *Forest Pathology, 33*(2), 113–125.

Sfalanga, A., Martini, M., Surico, G., & Bertaccini, A. (2002). Involvement of phytoplasmas in a decline of *Ulmus chenmoui* in Central Italy. *Forest Pathology, 32*(4–5), 265–275.

Silva-Castaño, A., Brochero, H., Wilson, M. R., & Franco-Lara, L. (2017, June 4–8). *Composition, richness and abundance of Hemiptera: Cicadellidae in two areas of grassland in the Sabana de Bogota: preliminary study.* Paper presented at the 3rd hemipteran-plant interactions symposium, Madrid, Spain.

Sinclair, W., & Griffiths, H. (1994). Ash yellows and its relationship to dieback and decline of ash. *Annual Review of Phytopathology, 32*(1), 49–60.

Sinclair, W., & Griffiths, H. (2000). Variation in aggressiveness of ash yellows phytoplasmas. *Plant Disease, 84*(3), 282–288.

Sinclair, W., Griffiths, H., & Davis, R. E. (1996). Ash yellows and lilac witches'-broom: Phytoplasmal diseases of concern in forestry and horticulture. *Plant Disease, 80*(5), 468–475.

Sinkins, S. P., & Gould, F. (2006). Gene drive systems for insect disease vectors. *Nature Reviews Genetics, 7*(6), 427–435.

Smart, C., Schneider, B., Blomquist, C., Guerra, L., Harrison, N., Ahrens, U., & Kirkpatrick, B. (1996). Phytoplasma-specific PCR primers based on sequences of the 16S-23S rRNA spacer region. *Applied and Environmental Microbiology, 62*(8), 2988–2993.

Tubby, K., & Webber, J. (2010). Pests and diseases threatening urban trees under a changing climate. *Forestry: An International Journal of Forest Research, 83*(4), 451–459.

United Nations DESA/Population Division. (2015). *World urbanization prospects: The 2014 revision, highlights.* Department of economic and social affairs. Retrieved from New York, NY.

Varela, C., Quintero, J., Delgado, L., Arevalo, E., & Franco-Lara, L. (2017). *Evidencia de fitoplasmas asociados a una nueva enfermedad de papa en Cundinamarca.* Paper presented at the Simposio Internacional sobre Solanáceas cultivadas en el Trópico y VII Congreso Colombiano de Horticultura, Cajica, Colombia.

Walling, L. L. (2008). Avoiding effective defenses: Strategies employed by phloem-feeding insects. *Plant Physiology, 146*(3), 859–866.

Wally, O., Daayf, F., Iranpour, M., Adam, L., Elliott, B., Shinners-Carnelley, T., … Khadhair, A. (2004). Incidence and molecular detection of yellows-type disease in carrots, associated with leafhoppers in southern Manitoba, Canada. *Canadian Journal of Plant Pathology, 26*(4), 498–505.

Walsh, K. B., Guthrie, J. N., & White, D. (2006). Control of phytoplasma diseases of papaya in Australia using netting. *Australasian Plant Pathology, 35*(1), 49–54.

Wang, H., Ye, X., Li, J., Tan, B., Chen, P., Cheng, J., … Feng, J. (2018). Transcriptome profiling analysis revealed co-regulation of multiple pathways in jujube during infection by '*Candidatus* Phytoplasma ziziphi'. *Gene, 665*, 82–95.

Weintraub, P. G. (2007). Insect vectors of phytoplasmas and their control-an update. *Bulletin of Insectology, 60*(2), 169.

Weintraub, P. G., & Beanland, L. (2006). Insect vectors of phytoplasmas. *Annual Review of Entomology, 51*, 91–111.

Winkelmann, T. (2013). Recent advances in propagation of woody plants. *Acta Horticulturae, 990*, 375–381.

Wongkaew, P., & Fletcher, J. (2004). Sugarcane white leaf phytoplasma in tissue culture: Long-term maintenance, transmission, and oxytetracycline remission. *Plant Cell Reports, 23*(6), 426–434.

World Health Organization. (2014). *Ambient (Outdoor) air pollution in cities database 2014.* Retrieved August 2, 2017 https://www.who.int/phe/health_topics/outdoorair/databases/cities-2014/en/

Zarate, S. I., Kempema, L. A., & Walling, L. L. (2007). Silverleaf whitefly induces salicylic acid defenses and suppresses effectual jasmonic acid defenses. *Plant Physiology, 143*(2), 866–875.

Zimmermann, M. R., Schneider, B., Mithöfer, A., Reichelt, M., Seemüller, E., & Furch, A. C. (2015). Implications of *Candidatus* Phytoplasma mali infection on phloem function of apple trees. *Endocytobiosis and Cell Research: Journal of the International Society of Endocytobiology, 26*, 67–75.

Chapter 9
Biological Control of the Leafhopper *Dalbulus maidis* in Corn Throughout the Americas: Interaction Among Phytoplasma- Insect Vector- Parasitoids

Gustavo Moya-Raygoza

Abstract Little is known about the biological control of insect pests that transmit plant pathogenic bacteria such as phytoplasmas. This chapter is a review of the status of the biological control of the corn leafhopper *Dalbulus maidis* DeLong (Hemiptera: Cicadellidae), an efficient vector of maize (*Zea mays* ssp. *mays* L., Poaceae) pathogens in the Neotropic region. The chapter reviews the biodiversity of parasitoids that attack the corn leafhopper in the Americas and relates the efforts conducted to understand the interactions among *D. maidis*- phytoplasmas-parasitoids-environment. Nymphs and adults of the corn leafhopper are parasitized by hymenopteran, dipteran, and strepsipteran species, with *Gonatopus bartletti* Olmi (Hymenoptera: Dryinidae) being the most common species, whereas eggs of the corn leafhopper are parasitized by hymenopterans, with *Anagrus incarnatus* Haliday (Hymenoptera: Mymaridae) and *Paracentrobia* sp. Howard (Hymenoptera: Trichogrammatidae) being the most common species. Knowledge of the transmission biology of phytoplasma is a key factor to identify parasitoid species that could be used as biological agents and to determine the best environmental conditions that will favor parasitoid population and their use as biological control. Maize bushy stunt phytoplasma (MBSP (*Ca.* Phytoplasma *asteris*)) has a long latent period in *D. maidis*, meaning that the most common parasitoids are able to complete their life cycle before the vector transmits the plant pathogen and can therefore be used as biological control. As well, maize edges growing during winter harbor a high diversity of leafhoppers and none of the leafhopper species sampled in those edges have been found to be positive for MBSP. These edges have a high diversity of nymph-adult parasitoids and egg parasitoids that attack the corn leafhopper, not only during winter but also during the maize growing season, meaning that habitat management such as field edges, is an important part of the biological control of corn leafhopper.

G. Moya-Raygoza (✉)
Departamento de Botánica y Zoología, CUCBA, Universidad de Guadalajara, Zapopan, Jalisco, Mexico

C. Y. Olivier et al. (eds.), *Sustainable Management of Phytoplasma Diseases in Crops Grown in the Tropical Belt*, Sustainability in Plant and Crop Protection 12, https://doi.org/10.1007/978-3-030-29650-6_9

Keywords Biological control · insect vector · maize bushy stunt phytoplasma · *Dalbulus maidis* · corn · parasitoids

9.1 Introduction: General Aspects of Hemiptera-Phytoplasma-Host Plant Interactions

Phytoplasmas are transmitted by sap-feeding Hemiptera belonging to the Cercopidae, Cicadellidae, Cixiidae, Delphacidae, Derbidae, and Psyllidae families. The Cicadellidae family contains the largest number of vector species, with more than 75% of the known vector species of phytoplasmas belonging to the Deltocephalinae subfamily (Weintraub & Beanland, 2006). Phytoplasmas affect wild plants as well as economically important crops such as coconut, grapevine, peach, onion, peanut, sweet potato, papaya, alfalfa, sugarcane, strawberry, and maize (Marzachi, Milne, & Bosco, 2004). Phytoplasmas are transmitted by insect vectors under a persistent circulative propagative mechanism (Bosco & Marzachì, 2016).

Most studies on phytoplasmas focused on the phytoplasma-vector-host plant interactions, with special attention to the effects of phytoplasma on the insect vector and the host plant, as well as on detection and identification of new hemipteran vectors. Over the past few years, the use of and advances in molecular technologies allowed increasingly rapid detection and identification of new phytoplasmas strains, species, and isolates (Orlovskis et al., 2017). Full genome sequencing has shed light on key roles of mechanisms in phytoplasma evolution and pathogenicity (Orlovskis et al., 2017; Zhao, Davis, Wei, Shao, & Jomantiene, 2014). However, little is known about other aspects of phytoplasma epidemiology, such as information on the natural enemies of the vectors and the interactions between natural enemies, phytoplasma, insect vector and habitat.

Being phloem-restricted, phytoplasmas are cryptic and difficult to control. Elimination of phytoplasmas from plants might be possible using antibiotics (e.g. tetracycline) which are not usually permitted in crops (Bertaccini & Duduk, 2010). Meristem tip culture and hot water treatments are other preventive methods used to eliminate phytoplasmas from plant tissues before plant transplantation, and generally used for high value horticultural or forestry plants (Laimer & Barba, 2011). Managing the vector populations with repeated insecticide treatments is the main method used by growers to control phytoplasma diseases in crops (Weintraub, 2007). However, pesticides are not environmentally friendly and other methods such as the use of parasitoids to control the populations of leafhopper vectors are being researched.

In general, leafhopper nymphs and adults are parasitized by Dyinidae (Hymenoptera), Pipunculidae (Diptera) and Halictophagidae (Strepsiptera) (Freytag, 1985), whereas leafhopper eggs are parasitized by micro-hymenoptera of the families Mymaridae and Trichogrammatidae. Parasitoids can be used as biological control agents to reduce the population of insect pests such as leafhopper vectors (Gillespie, Gurr, & Wratten, 2016; Mills, 2009). However, an extended knowledge of the epidemiology of the disease and biology of the leafhopper vectors is paramount to develop efficient and sustainable management methods using parasitoids. For this reason, few attempts of biological control for leafhopper vectors of phytoplasmas using parasitoids were undertaken, and those were not highly successful. For example, the biological control of *Scaphoideus titanus* Ball (Hemiptera: Cicadellidae), the vector of flavescence doreé in grape, has been attempted in France by releasing the natural enemy *Gonatopus flavipes* Olmi (Hymenoptera: Dryinidae) or by increasing local natural enemy populations. However, these two strategies have produced poor results, with a low parasitism of <0.4% (Chuche & Thiéry, 2014).

The Deltocephalinae species *Dalbulus maidis* DeLong (Hemiptera: Cicadellidae) (Fig. 9.1a) efficiently transmits maize bushy stunt phytoplasma (MBSP ('*Ca.* Phytoplasma asteris')) to maize (*Zea mays* ssp. *mays* L., Poaceae). This phytoplasma belongs to the 16SrI-B cluster that includes aster yellows phytoplasmas and periwinkle yellows phytoplasmas, making *D. maidis* one of the most important pests of maize in Neotropics (DeOliveira, DeSousa, & Landau, 2011; Nault, 1990). *Dalbulus elimatus* Ball (Hemiptera: Cicadellidae) is another efficient vector of '*Ca.* Phytoplasma asteris', and a pest of maize (Nault, 1980). *D. elimatus* is distributed only in Mexico and the United States of America (USA), whereas *D. maidis* is broadly distributed in the Americas, mostly in Latin America, from Argentina to southern USA (Triplehorn & Nault, 1985). Isolates of '*Ca* phytoplasma asteris' from maize plants have been found in USA, Mexico, Costa Rica and Brazil;

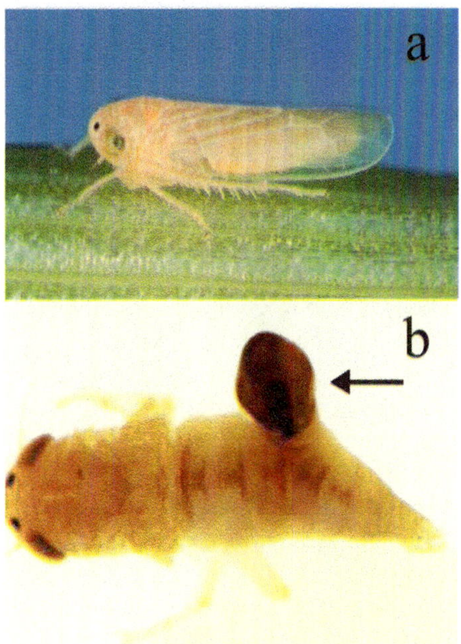

Fig. 9.1 (**a**) *Dalbulus maidis* adult, (**b**) *Dalbulus maidis* adult with evidence of parasitism by the larval sac formed by *Gonatopus barletti* (arrow)

however, the role of MBSP in limiting corn production has not been determined (Bedendo, Davis, & Dally, 2000).

D. maidis is also one of the most studied hemipteran vectors and its behavior, ecology, evolution, migration, and interaction with phytoplasma and other plant pathogens are well known (Nault, 1990). As well, interactions between *D. maidis* and its parasitoid species in maize crops have been extensively studied. The focus of this chapter is to review the status of the biological control for *D. maidis* using parasitoids.

9.2 *Dalbulus maidis*- MBSP- *Zea* spp. Interactions

The genus *Dalbulus* DeLong is composed of 13 species organized in three monophyletic groups: (1) the *maidis* group, which includes the corn leafhopper *D. maidis*, (2) the *elimatus* group, which includes the Mexican corn leafhopper, *D. elimatus*, and (3) the *quinquenotatus* group, which contains the five spotted gamagrass leafhopper *D. quinquenotatus* DeLong & Nault (Nault & Styer, 1994; Triplehorn & Nault, 1985). *D. maidis* is a phloem-feeding pest that only feeds and reproduces on maize, annual teosintes (i.e., *Zea mays* ssp. *parviglumis* Iltis & Doebley), and perennial teosintes (i.e., *Zea perennis* (Hitchc.) Reeves & Mangelsd.) (Nault & DeLong, 1980). *D. maidis* has evolved with the teosinte relatives of maize and the closely related gamagrass, became very well adapted to maize and achieved the status of maize pest (Nault, 1990). In addition, MBSP has been detected in teosintes, indicating that the *D. maidis*-MBSP association probably started before maize domestication. Maize evolved from *Z. m.* ssp. *parviglumis* about 9000 years ago (Matsuoka et al., 2002). The highest diversity of *Dalbulus* species and their host plants as well as *D. maidis* parasitoids is found in Mexico.

MBSP is transmitted by *D. maidis* in a persistent-propagative mode. After a 48 hrs acquisition access period, phytoplasmas spend a long latent period into the insects before being transmitted to plants by their insect vectors. For example, the mean latent period for 3 MBSP Mexican isolates (Tlatizapan, Poza Rica, and El Batan) (Fig. 9.2) is roughly 20 days at 25 °C in nymphs or adults. Similarly, the incubation period into the maize (Moya-Raygoza & Nault, 1998) is long, as a 20-day period is required before infected plants express the first symptoms. Maize infected with any of the 3 isolates first showed reddening in the margins of the old leaves. Three to 4 days later, symptoms of chlorosis showed up followed by leaf senescence. Often, stunting and development of basal tillers occur simultaneously with the reddening and chlorosis symptoms (Moya-Raygoza & Nault, 1998).

In the Americas, maize is cultivated under two systems; a) all year continual planting with water supply provided by irrigation during the dry season and rainfall during the wet season, and b) "seasonal planting", i.e., planting during the wet season. In Mexico, 97.3% of the maize surface is planted during the wet season and the remaining 2.7% are planted during the dry season (Moya-Raygoza, Hogenhout, & Nault, 2007). In most countries from Central America and South America, maize is

Fig. 9.2 Maize plants showing symptoms caused by three maize bushy stunt phytoplasma isolates from Mexican origin (Tlaltizapan (Morelos state), Poza Rica (Veracruz state) and El Batan, (Mexico state))

planted by growers at the beginning of the rainfall season. The cropping system has direct impacts on *D. maidis* populations, their dispersal and migration. In Mexico, high populations of corn leafhoppers (400 corn leafhopper adults per 1110 sweeps) are found during the dry winter season in areas where maize is planted continuously as compared to low population (<100 corn leafhopper adults per 11,110 sweeps) in areas where maize is grown in wet season only. This means that 4 times more *D. maidis* adults are collected in the large maize fields grown in the Sinaloa state as compared to regions practicing seasonal maize cropping (Moya-Raygoza et al., 2007). In areas practicing seasonal cropping in Mexico, part of the population of *D. maidis* survive the dry season in local habitats (Larsen, Nault, & Moya-Raygoza, 1992). However, when the wet season starts and new maize crops are seeded in seasonal cropping areas, populations of *D. maidis* consisting of large-sized migrant females that lay large numbers of eggs is observed in young maize crops, along with the overwintering *D. maidis*. Migrant females that arrive at the beginning of the wet season in July are very important, as they can establish new vigorous population in young corn crops. These migrant females might come from areas where maize is planted all year long (Moya-Raygoza, Palomera-Avalos, & Galaviz-Mejia, 2007). These larger females are adapted to colonize newly planted maize seedlings because their larger body size and resultant greater energy storage capacity promotes longer survival than that of smaller females (Larsen & Nault, 1994). Therefore, migrant *D. maidis* females are facilitating the re-occurrence and increase of *D. maidis* populations in areas with a low *D. maidis* population due to the dry seasons. Similarly, female delphacids with longer body size show high migratory ability (Denno, 1994) and higher fecundity than smaller females (Denno & McCloud, 1985).

In Mexico and throughout tropical countries in Central America and South America, maize fields are surrounded by grassy edges. Molecular tests on leafhoppers sampled in the grassy edges of maize fields grown in central Mexico showed the absence of MBSP in *D. maidis* and 26 other species of leafhoppers, suggesting that MBSP does not overwinter in leafhoppers living in the grassy edges of the fields (Torres-Moreno, Moya-Raygoza, & Pérez-López, 2015).

9.3 Biological Control of *Dalbulus maidis*

In the Netropical region the corn leafhopper nymphs, adults and eggs are parasitized by endoparasitoids. Nymphs and adults of *D. maidis* are parasitized by the wasps *Gonatopus bartletti* (Figs. 9.1b and 9.3a), *G. desantisi* Olmi & Virla, *G. contartus* Olmi, *G. moyaraygozai* Olmi, *G. flavipes* Olmi and *Anteon pilicorne* Ogloblin (Hymenoptera: Dryinidae). The flies *Eudorylas spinosus* Hardy (Diptera: Pipunculidae), *E. subopacus* Loew and *E. schreiteri* Shannon, and the strepsipteran *Halictophagus naulti* Kathirithamby and Moya-Raygoza (Strepsiptera: Halictophagidae) (Table 9.1). *Gonatopus bartletti* is very important numerically because it is broadly distributed in the Neotropics and is found on maize planted seasonally and continuously in Mexico, Central and South America. In Mexico,

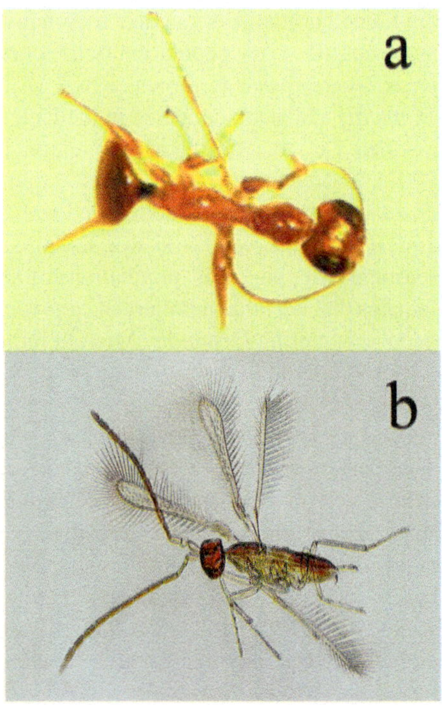

Fig. 9.3 (**a**) *Gonatopus bartletti* female, (**b**) *Anagrus incarnatus* male

Table 9.1 Checklist of nymph and adult parasitoids of *Dalbulus maidis* in Latin America at different elevations (m above sea level) on maize (*Zea mays* ssp. *mays*) planted seasonally or continually

Nymph and adult parasitoids	Country (State; site-elevation in m) *Maize cropping system*[a]	References
Hymenoptera Dryinidae		
Gonatopus bartletti	**Mexico** (Jalisco; Cuzalapa-680 m, Ayuquila-880 m, El Chante-925 m, Ahuacapan-1000 m, Corralitos-1760 m). *Seasonally*	(Moya-Raygoza & Trujillo-Arriaga, 1993)
	Mexico (Guanajuato; Celaya. Mexico; Montecillo. Veracruz)	(Vega & Barbosa, 1990)
	Mexico (Jalisco; El Grullo-880 m, Tecomates-280 m, Las Palmas-2 m. Colima; Colima-495 m. Nayarit; Ixtapan-63 m. Sinaloa; Guasave-21 m, Zapotillo-29 m. Chiapas; Nueva Esperanza-37 m. Morelos; Yautepec-1203 m. Oaxaca; Copalita-28 m, El Coco-26 m, El Cacahuete-22 m. Tabasco; Cucuyalapa-22 m, Villa Chable-14 m. Veracruz; Cardel-40 m. Zacatecas; Apozol-771 m. *Continually*	(Moya-Raygoza et al., 2004)
	USA, Puerto Rico	(Virla & Olmi, 1998)
	Nicaragua	(Guglielmino, Olmi, & Bückle, 2013; Olmi & Virla, 2014)
	Venezuela	(Guglielmino et al., 2013)
Gonatopus desantisi	**Argentina** (Tucuman-387 m)	(Olmi & Virla, 2014)
Gonatopus contartus	**Argentina** (Tucuman-387 m)	(Virla & Olmi, 2007)
Gonatopus moyaraygozai	**Argentina** (Tucuman-387 m).	(Virla & Olmi, 2007)
Gonatopus flavipes	**Brazil** (Teresina-PI).	(Olmi & Virla, 2014)
Anteon pilicorne	**Argentina** (Tucuman; Los Nogales-585 m).	(Virla, Espinosa, & Moya-Raygoza, 2011)
Diptera Pipunculidae		
Eudorylas spinosus	**Mexico** (Nayarit; Acaponeta-22 m. Oaxaca; Tehuantepec-26 m. Puebla; Tehuacan-1013 m. Sinaloa; Villa Union-15 m. Veracruz; Paso del Toro-20 m). *Continually*	(Moya-Raygoza et al., 2004)
Eudorylas subopacus	**Mexico** (Jalisco; Sayula-1366 m). *Continually*	(Moya-Raygoza, 2007)
	Mexico (Texcoco-2545 m). *Seasonally*	(Vega, Barbosa, & Panduro, 1991)

(continued)

Table 9.1 (continued)

Nymph and adult parasitoids	Country (State; site-elevation in m) *Maize cropping system*[a]	References
Eudorylas schreiteri	**Argentina** (Tucuman; El Manantial-495 m, Los Nogales-585 m).	(Virla, Moya-Raygoza, & Rafael, 2009)
Strepsiptera Halictophagidae		
Halictophagus naulti	**Mexico** (Morelos; Tlaltizapán-940 m). *Seasonally*	(Kathirithamby & Moya-Raygoza, 2000)
	Mexico (Chiapas; Nueva Esperanza-37 m.Nayarit; Tuxpan-8 m. Oaxaca; El Coco-26 m, Huitzo-1036 m). *Continually*	(Moya-Raygoza et al., 2004)

[a]Elevation and cropping system indicated when mentionned in the references

G. bartletti was found in the states of Chiapas, Colima, Jalisco, Morelos, Nayarit, Oaxaca, Sinaloa, Tabasco, Veracruz, and Zacatecas, and was, therefore, the most common nymph-adult parasitoid species. *G. bartletti* also reached the highest parasitism rate with 24% of parasitism in Colima in January 2001 when maize was cultivated with irrigation (Moya-Raygoza, Kathirithamby, & Larsen, 2004). An investigation of the biological cycle of *G. bartletti* under controlled conditions at 25 °C showed that the parasitoid needs 11.9 days to develop from egg to the last larval instar and that female adults lived 21.6 days (Rios Reyes & Moya-Raygoza, 2004). This means that the development of *G. barletti* in *D. maidis* is faster (11.9 days) than the mean latent period (20 days) of MBSP into *D. maidis* at 25 °C (Nault, 1980). By definition, a parasitoid develops in the host and causes its death after emergence. Parasitoids that develop faster than the MBSP latent period in *D. maidis* are appropriate candidates as biological control agents to reduce the population of MBSP leafhopper vectors. Therefore, *G. barletti* is a good candidate to control the populations of *D. maidis.*

The effects of MBSP on *G. barletti* larvae developing in MBSP-infected *D. maidis* are not known. However, it was found that *G. barletti* reduces the presence of *Spiroplasma kunkelii* within *D. maidis* (Moya-Raygoza, Palomera-Avalos, Chacón-Torres, & Becerra-Chiron, 2006). These authors found that parasitoid development was successfully completed when leafhoppers acquired *S. kunkelii* and showed that the survival of *S. kunkelii* in the leafhopper was negatively affected by the parasitoid larva.

Several micro-hymenopteran species have been reported to attack the eggs of *D. maidis* in maize agroecosystems in Mexico, Central America, and South America (Table 9.2). *Anagrus breviphragma* Soyka which was recently synonymized under *A. incarnatus* (Triapitsyn, 2015a) (Fig. 9.3b) and *Paracentrobia* sp. Howard (Hymenoptera: Trichogrammatidae) are the most abundant and common egg parasitoids of *D. maidis* on maize throughout Latin America (Gladstone, de la Llana,

Table 9.2 Egg parasitoids of *Dalbulus maidis* in Latin America on maize (*Zea mays* ssp. *mays*) at different elevations (m above sea level)

Egg parasitoid taxa	Country (site)	Elevation (m), abundance	References
Hymenoptera Mimaridae			
Anagrus incarnatus	Mexico (El Grullo)	(869 m), ↓,+	(Moya-Raygoza et al., 2012; Moya-Raygoza et al., 2014; Virla, Luft Albarracin, et al., 2009)
	Mexico (Zapopan)	(1661 m), ↑, +	(Moya-Raygoza et al., 2012; Moya-Raygoza et al., 2014; Virla, Luft Albarracin, et al., 2009)
	Argentina (Los Nagoles)	(588 m), ↑, +	(Moya-Raygoza et al., 2012)
	Argentina (V de María del Río)	(353 m), ↓,*	(Virla et al., 2013)
	Argentina (Sinsacate)	(490 m), ↓,*	(Virla et al., 2013)
	Argentina (Jesús María)	(544 m), ↑, +	(Virla et al., 2013)
	Argentina (Elena)	(613 m), ↓,*	(Virla et al., 2013)
	Argentina (El Mollar)	(1945 m), ↑, +	(Moya-Raygoza et al., 2012)
	Brazil (Piracicaba)	(530 m), ↓, +	(Virla et al., 2013)
Anagrus columbi	Mexico. (Zapopan)	(1661 m), ↓, +	(Moya-Raygoza & Becerra-Chiron, 2014)
Anagrus flaveolus	Peru (La Molina)	(nm), ↑, *	(Marín, 1987)
	Argentina (Jesús María)	(544 m), ↑, +	(Virla et al., 2013)
	Argentina (Los Nogales)	(588 m), ↓, +	(Moya-Raygoza et al., 2012)
Anagrus miramae	Argentina (El Mollar)	(1945 m), ↓, +	(Moya-Raygoza et al., 2012)
Anagrus nigriventris	Argentina (El Manantial)	(495 m), ↓, +	(Luft Albarracin, Virla, & Triapitsyn, 2006)
Anagrus sp.	Nicaragua (Managua)	(60 m), ↑, +	(Gladstone et al., 1994)
Polynema saga	Mexico (El Grullo)	(869 m), ↓, +	(Moya-Raygoza et al., 2012)
Polynema sp.	Argentina (Los Nogales)	(588 m), ↓, +	(Moya-Raygoza et al., 2012)
	Argentina (El Mollar)	(1945 m), ↓, +	(Moya-Raygoza et al., 2012)

(continued)

Table 9.2 (continued)

Egg parasitoid taxa	Country (site)	Elevation (m), abundance	References
Trichogrammatidae			
Paracentrobia tapajosae	Mexico (El Grullo)	(869 m), ↑, +	(Moya-Raygoza et al., 2012; Moya-Raygoza et al., 2014; Virla, Luft Albarracin, et al., 2009)
	Mexico (Zapopan)	(1661 m), ↓, +	(Moya-Raygoza et al., 2012; Moya-Raygoza et al., 2014; Virla, Luft Albarracin, et al., 2009)
	Argentina. (Los Nogales)	(588 m), ↓, +	(Moya-Raygoza et al., 2012)
Paracentrobia subflava	Argentina (La Plata)	Lab. Conditions	(Virla, 1999)
Paracentrobia sp.	Nicaragua (Managua)	(60 m), ↓, +	(Gladstone et al., 1994)
Pseudoligosita longifrangiata	Mexico (El Grullo)	(869 m), ↓, +	(Moya-Raygoza et al., 2012; Moya-Raygoza et al., 2014; Virla, Luft Albarracin, et al., 2009)
	Mexico (Zapopan)	(1661 m), ↓, +	(Moya-Raygoza et al., 2014)
	Argentina (Jesús María)	(544 m), ↑, +	(Virla et al., 2013)
	Argentina (Los Nogales)	(588 m), ↓, +	(Moya-Raygoza et al., 2014)
	Argentina (El Mollar)	(1945 m), ↓, +	(Moya-Raygoza et al., 2014)
Aphelinoidea semifuscipenis	Mexico (El Grullo)	(869 m), ↓, +	(Moya-Raygoza et al., 2014; Virla, Luft Albarracin, et al., 2009)
	Mexico (Zapopan)	(1661 m), ↓, +	(Moya-Raygoza et al., 2014; Virla, Luft Albarracin, et al., 2009)
Oligosita clarimaculosa	Mexico (El Grullo)	(869 m), ↓, +	(Moya-Raygoza et al., 2012)
	Mexico (Zapopan)	(1661 m), ↓, +	(Moya-Raygoza et al., 2012)
Oligosita desantisi	Mexico (El Grullo)	(869 m), ↓, +	(Moya-Raygoza et al., 2014)
	Argentina (El Mollar)	(1945 m), ↓, +	(Moya-Raygoza et al., 2012)
Oligosita giraulti	Argentina (Los Nogales)	(588 m), ↓, +	(Moya-Raygoza et al., 2012)
Oligosita sp.	Mexico (Zapopan)	(1661 m), ↓, +	(Moya-Raygoza & Becerra-Chiron, 2014)
Oligosita sp.	Brazil (Piracicaba)	(530 m), ↑, +	(Oliveira & Spotti Lopez, 2000)
Zagella nanula	Argentina (Los Nogales)	(588 m), ↓, +	(Moya-Raygoza et al., 2012)

↑ = High abundance species in maize agroecosystems, ↓ = Low abundance species in maize agroecosystems, + = found with other egg parasitoid species, ∗ = single parasitoids species

Rios, & Lopez, 1994; Moya-Raygoza, Renteria, Luft Albarracin, & Virla, 2014; Oliveira & Spotti Lopez, 2000; Virla, Moya-Raygoza, & Luft Albarracin, 2013). In central Mexico, *A. incarnatus* is very common in maize fields grown in high elevations, but it has also been found in fields grown in low elevations, whereas *Paracentrobia* sp. is most common at low elevations (Moya-Raygoza, Luft Albarracin, & Virla, 2012; Moya-Raygoza et al., 2014). *Pseudoligosita* sp. Girault (Hymenoptera: Trichogrammatidae) has been observed attacking *D. maidis* eggs at both low and high elevations throughout tropical America, but this parasitoid is less abundant and common as compared to *A. incarnatus* (Moya-Raygoza et al., 2012; Moya-Raygoza et al., 2014; Virla, Luft Albarracin, & Moya-Raygoza, 2009).

These egg parasitoids are generalists and have been found parasitizing *D. maidis* eggs as well as eggs of other insects throughout Latin America. For example, *A. incarnatus* parasitizes the eggs of *D. maidis* in Argentina, Brazil, and Mexico, and *D. elimatus* in Mexico (Moya-Raygoza et al., 2012, 2014). *Anagrus incarnatus* also parasitizes the eggs of two types of Delphacidae, *Peregrinus maidis* Ashmead (Hemiptera: Delphacidae) and *Delphacodes kuscheli* Fennah (Hemiptera: Delphacidae) in South America (Triapitsyn, 2015b). *Paracentrobia* sp. attacks the eggs of *D. maidis* in Argentina and Mexico (Moya-Raygoza et al., 2012) and *D. elimatus* in Mexico (Moya-Raygoza et al., 2014). *Pseudoligosita* sp. attacks the eggs of *D. maidis* in Argentina and Mexico and *D. elimatus* in Mexico (Moya-Raygoza et al., 2012, 2014). Generalist parasitoids show more plastic behavior and a strong ability to use a wide diversity of host resources in disturbed environments such as agroecosystems where host availability changes through time (Raymond, Plantegenest, Gagic, Navasse, & Lavandero, 2016). In addition, the populations of *D. maidis* follows the r reproductive strategy, a survival strategy characterized by short generations, high numbers of descendants, greater mobility, and adaptation to systems with a higher incidence of disturbance (Nault & Madden, 1985), meaning that, in maize agroecosystems, the populations of *D. maidis* are unpredictable in time and space (Nault, 1990). In natural conditions, finding two or more egg parasitoids in the same maize fields is very common (Table 9.2) and the percentage of egg parasitism can reach 60%. For instance, in Mexico and Argentina, the percentage of egg parasitism by the two most abundant parasitoids *A. incarnatus* and *Paracentrobia* sp. reaches 60.9% and 44.8%, respectively (Moya-Raygoza et al., 2012).

The time from oviposition to egg-hatch at 25 °C is approximately 14–15 days for *D. maidis* (Nault, 1990). The development time from oviposition to adult emergence of *A. incarnatus* is around 16.5 days at 25 °C in *D. maidis* (Virla, 2001), and around 16 days at 23–24 °C for *Paracentrobia* sp. (Virla, 1999). Because the development time of these two parasitoids is short, both species could also be used as biological control agents to reduce *D. maidis* populations.

Host-parasitoid interactions also occur during winter on grassy maize edges. Dryinids (Hymenoptera) parasitized nymphs and adults of several leafhopper species, including *D. maidis* (Becerra-Chiron, Moya-Raygoza, & Muñoz-Urias, 2017). Similarly, *D. maidis* eggs are parasitized by egg parasitoids in the hymenopteran families Mymaridae and Trichogrammatidae (Moya-Raygoza & Becerra-Chiron,

2014). Therefore, the presence of parasitoids in winter edges highlights the importance of conserving maize edges for biological control.

Surveys done at the beginning of the wet season demonstrated that *D. maidis* as well as parasitoids are migrating in the newly planted maize fields, raising the question of the origin of the parasitoids. Recently, Hymenoptera dryinids species such as *G. huggerti* Olmi, *G. mimoides* Perkins, *G. breviforceps* Kieffer, and *G. caraibicus* Olmi were found attacking leafhoppers species such as *D. maidis*, *Balclutha incisa* Matsumura (Hemiptera: Cicadelliade), *Exitianus picatus* Gibson (Hemiptera: Cicadellidae), and *Stirellus bicolor* Van Duzee (Hemiptera: Cicadellidae) on the edges of maize crops during winter (Becerra-Chiron et al., 2017). Similarly surveys performed on the edges of maize crops show that during the dry season, several leafhopper species were parasitized by strepsipterans and dryinids parasitoids (Moya-Raygoza & Becerra-Chiron, 2014). Therefore, parasitoids inhabiting grassy edges (Fig. 9.4) during the dry season, when maize is absent, are the source of vector parasitism in maize fields planted in the wet season.

During this dry season, parasitoids attack leafhoppers species other than *D. maidis*. At the beginning of the wet season, maize crops are planted and parasitoids migrate from the grassy edges and colonize and parasitize the migrant *D. maidis* (Moya-Raygoza et al., 2014). At the end of the dry winter season, *D. maidis* eggs located on perennial grasses grown in grassy edges, were found parasitized by *Oligosita* sp. and *Anagrus columbi* Perkins (Moya-Raygoza & Becerra-Chiron, 2014). These results suggest that grasses grown at the field's edges are winter refuges for egg parasitoids of *D. maidis* as well as for other parasitoids such dryinids

Fig. 9.4 Edge grasses growing during the dry and wet seasons in maize agroecosystems cultivated under the "seasonal planting" system in Mexico. Field edges serve as winter reservoir for parasitoids that attack leafhoppers feeding on maize during the rainy summer season

and strepsipterans known to attack nymphs and adults of leafhoppers. Grassy edges also occur in Central America and South America and should be maintain in order to conserve habitats promoting diversity and abundance of parasitoids of *D. maidis* throughout the Neotropic region. Habitat management, particularly of field edges, is an important alternative for the control of insect pests because edges provide shelter, food (nectar and pollen) and alternative hosts (Gurr, Wratten, Landis, & You, 2017). In addition, the use of mass release of species such *G. bartletti*, *A. incarnatus* and *Paracentrobia* sp. could be a good alternative in corn fields where their abundance is limited. These parasitoid species are easy to mass rear and native to the Americas.

9.4 Conclusion

Biological control of other Delthocepaline leafhoppers that transmit phytoplasmas is generally limited to species identification of both vectors and parasitoids, as successful sustainable biological control can only be achieved when the hemipteran vector and its parasitoids have been identified and the interactions between phytoplasmas, vectors, crops and crop edges, agronomic practices and parasitoids have been understood. *D. maidis* is an important pest on maize and its biology, interactions with maize and vectoring ability have been extensively studied. As well, the common parasitoid species that attack nymphs, adults and eggs of *D. maidis* and their biology have also been extensively investigated. The positive effect of maize edges in the control of the corn leafhopper populations has been demonstrated, although more research is needed, especially to evaluate the effects of habitat management on the parasitoids, vectors and phytoplasmas. The functional and numerical responses of parasitoids such as *G. bartletti*, *A. incarnatus* and *Paracentrobia* sp. should be investigated before they could be used as a large biological control program. In addition, more research is required to know the effects of MBSP on parasitoid development. Finally, the study of *D. maidis* parasitoids in the Neotropics can serve as a basis to investigate the use of biological control programs for other leafhopper vectors throughout the world.

References

Becerra-Chiron, I. M., Moya-Raygoza, G., & Muñoz-Urias, A. (2017). Host-Dryinidae (Hymenoptera) interactions on edge grasses of maize agroecosystem throughout winter in Mexico. *Journal of Hymenoptera Research, 57*, 155.

Bedendo, I. P., Davis, R. E., & Dally, E. L. (2000). Detection and identification of the maize bushy stunt phytoplasma in corn plants in Brazil using PCR and RFLP. *International Journal of Pest Management, 46*(1), 73–76.

Bertaccini, A., & Duduk, B. (2010). Phytoplasma and phytoplasma diseases: A review of recent research. *Phytopathologia Mediterranea, 48*(3), 355–378.

Bosco, D., & Marzachì, C. (2016). Insect transmission of phytoplasmas. In J. Brown (Ed.), *Vector-Mediated Transmission of Plant Pathogens* (pp. 319–327). St. Paul, MI: APS Press.

Chuche, J., & Thiéry, D. (2014). Biology and ecology of the Flavescence dorée vector *Scaphoideus titanus:* a review. *Agronomy for Sustainable Development, 34*(2), 381–403.

Denno, R. F. (1994). The evolution of dispersal polymorphisms in insects: The influence of habitats, host plants and mates. *Researches on Population Ecology, 36*(2), 127.

Denno, R. F., & McCloud, E. S. (1985). Predicting fecundity from body size in the planthopper, *Prokelisia marginata* (Homoptera: Delphacidae). *Environmental Entomology, 14*(6), 846–849.

DeOliveira, E., DeSousa, S. M., & Landau, E. (2011). Transmission of maize bushy stunt phytoplasma by *Dalbulus maidis* leafhopper. *Bulletin of Insectology, 64*, S153–S154.

Freytag, P. (1985). The insect parasites of leafhoppers, and related groups. In L. R. Nault & J. G. Rodriquez (Eds.), *The leafhoppers and planthoppers* (pp. 423–467). New York: Wiley.

Gillespie, M. A., Gurr, G. M., & Wratten, S. D. (2016). Beyond nectar provision: the other resource requirements of parasitoid biological control agents. *Entomologia Experimentalis et Applicata, 159*(2), 207–221.

Gladstone, S., de la Llana, A., Rios, R., & Lopez, L. (1994). Egg parasitoids of the corn leafhopper, *Dalbulus maidis* (DeLong and Wolcott) (Homoptera: Cicadellidae) in Nicaraguan maize. *Proceedings of the Entomological Society of Washington, 96*(1), 143–146.

Guglielmino, A., Olmi, M., & Bückle, C. (2013). An updated host-parasite catalogue of world Dryinidae (Hymenoptera: Chrysidoidea). *Zootaxa, 3740*(1), 1–113.

Gurr, G. M., Wratten, S. D., Landis, D. A., & You, M. (2017). Habitat management to suppress pest populations: progress and prospects. *Annual Review of Entomology, 62*, 91–109.

Kathirithamby, J., & Moya-Raygoza, G. (2000). *Halictophagus naulti* sp. n.(Strepsiptera: Halictophagidae), a new species parasitic in the corn leafhopper (Homoptera: Cicadellidae) from Mexico. *Annals of the Entomological Society of America, 93*(5), 1039–1044.

Laimer, M., & Barba, M. (2011). Elimination of systemic pathogens by thermotherapy, tissue culture, or *in vitro* micrografting. In A. Hadidi, M. Barba, T. Candresse, & W. Jelkmann (Eds.), *Virus and virus-like diseases of pome and stone fruits* (pp. 389–393). St. Paul, MI: APS Press.

Larsen, K. J., & Nault, L. (1994). Seasonal polyphenism of adult *Dalbulus* leafhoppers (Homoptera: Cicadellidae). *Annals of the Entomological Society of America, 87*(3), 355–362.

Larsen, K. J., Nault, L., & Moya-Raygoza, G. (1992). Overwintering biology of *Dalbulus* leafhoppers (Homoptera: Cicadellidae): Adult populations and drought hardiness. *Environmental Entomology, 21*(3), 566–577.

Luft Albarracin, E., Virla, E. G., & Triapitsyn, S. V. (2006). A new host record for the egg parasitoid *Anagrus nigriventris* (Hymenoptera: Mymaridae) of the corn leafhopper, *Dalbulus maidis* (Hemiptera: Cicadellidae). *Florida Entomologist, 89*(2), 284–285.

Marín, L. (1987). Biologia y comportamiento de *Dalbulus maides* (Homoptera: Cicadellidae). *Revista peruana de entomologia, 30*, 113–117.

Marzachi, C., Milne, R. G., & Bosco, D. (2004). Phytoplasma plant-vector relationships. In S. G. Pandalai & A. Gayathri (Eds.), *Recent research development in plant pathology* (Vol. 3, pp. 211–241). Trivandrum, India: Research Signpost.

Matsuoka, Y., Vigouroux, Y., Goodman, M. M., Sanchez, J., Buckler, E., & Doebley, J. (2002). A single domestication for maize shown by multilocus microsatellite genotyping. *Proceedings of the National Academy of Sciences, 99*(9), 6080–6084.

Mills, N. (2009). Egg parasitoids in biological control and integrated pest management. In F. Consoli, J. Parra, & R. Zucchi (Eds.), *Egg parasitoids in agroecosystems with emphasis on Trichogramma* (Vol. 9, pp. 389–411). New York, NY: Springer.

Moya-Raygoza, G. (2007). Native parasitoids of *Dalbulus elimatus* and *Dalbulus maidis* (Hemiptera: Cicadellidae) in winter crops of central Mexico. *The Canadian Entomologist, 139*(5), 722–727.

Moya-Raygoza, G., & Becerra-Chiron, I. M. (2014). Overwintering biology of egg parasitoids of *Dalbulus maidis* (Hemiptera: Cicadellidae) on perennial grasses, volunteer maize, stubble, and drip-irrigated maize. *Annals of the Entomological Society of America, 107*(5), 926–932.

Moya-Raygoza, G., Hogenhout, S. A., & Nault, L. (2007). Habitat of the corn leafhopper (Hemiptera: Cicadellidae) during the dry (winter) season in Mexico. *Environmental Entomology, 36*(5), 1066–1072.

Moya-Raygoza, G., Kathirithamby, J., & Larsen, K. J. (2004). Dry season parasitoids of adult corn leafhoppers (Hemiptera: Cicadellidae) on irrigated maize in Mexico. *The Canadian Entomologist, 136*(1), 119–127.

Moya-Raygoza, G., Luft Albarracin, E., & Virla, E. G. (2012). Diversity of egg parasitoids attacking *Dalbulus maidis* (Hemiptera: Cicadellidae) populations at low and high elevation sites in Mexico and Argentina. *Florida Entomologist, 95*(1), 105–112.

Moya-Raygoza, G., & Nault, L. (1998). Transmission biology of maize bushy stunt phytoplasma by the corn leafhopper (Homoptera: Cicadellidae). *Annals of the Entomological Society of America, 91*(5), 668–676.

Moya-Raygoza, G., Palomera-Avalos, V., Chacón-Torres, N. M., & Becerra-Chiron, I. M. (2006). The parasitoid *Gonatopus bartletti* reduces presence of plant-pathogenic *Spiroplasma kunkelii* within the leafhopper vector *Dalbulus maidis*. *Entomologia Experimentalis et Applicata, 119*(3), 189–196.

Moya-Raygoza, G., Palomera-Avalos, V., & Galaviz-Mejia, C. (2007). Field overwintering biology of *Spiroplasma kunkelii* (Mycoplasmatales: Spiroplasmataceae) and its vector *Dalbulus maidis* (Hemiptera: Cicadellidae). *Annals of Applied Biology, 151*(3), 373–379.

Moya-Raygoza, G., Renteria, I., Luft Albarracin, E., & Virla, E. G. (2014). Egg parasitoids of the leafhoppers *Dalbulus maidis* and *Dalbulus elimatus* (Hemiptera: Cicadellidae) in two maize habitats. *Florida Entomologist, 97*(1), 309–312.

Moya-Raygoza, G., & Trujillo-Arriaga, J. (1993). Dryinid (Hym.: Dryinidae) parasitoids of *Dalbulus* leafhopper (Hom.: Cicadellidae) in Mexico. *Entomophaga, 38*(1), 41–49.

Nault, L. (1980). Maize bushy stunt and corn stunt: a comparison of disease symptoms, pathogen host ranges, and vectors. *Phytopathology, 70*(7), 659–662.

Nault, L. (1990). Evolution of an insect pest: maize and the corn leafhopper, a case study. *Maydica, 35*(2), 165–175.

Nault, L., & DeLong, D. (1980). Evidence for co-evolution of leafhoppers in the genus *Dalbulus* (Cicadellidae: Homoptera) with maize and its ancestors. *Annals of the Entomological Society of America, 73*(4), 349–353.

Nault, L., & Madden, L. (1985). Ecological strategies of *Dalbulus* leafhoppers. *Ecological Entomology, 10*(1), 57–63.

Nault, L., & Styer, W. (1994). High elevation gamagrass is host to two new endemic *Dalbulus* species in Mexico (Homoptera: Cicadellidae: Deltocephalinae). *Annals of the Entomological Society of America, 87*(3), 257–262.

Oliveira, C., & Spotti Lopez, J. (2000). Parasitoides de ovos da cigarrinha-do-milho, *Dalbulus maidis* (DeLong & Wolcott)(Hemiptera: Cicadellidae), em Piracicaba. *Agricultural Reviews, 75*, 263–270.

Olmi, M., & Virla, E. G. (2014). Dryinidae of the Neotropical region (Hymenoptera: Chrysidoidea). *Zootaxa, 3792*(1), 1–534.

Orlovskis, Z., Canale, M. C., Haryono, M., Lopes, J. R. S., Kuo, C.-H., & Hogenhout, S. A. (2017). A few sequence polymorphisms among isolates of Maize bushy stunt phytoplasma associate with organ proliferation symptoms of infected maize plants. *Annals of Botany, 119*(5), 869–884.

Raymond, L., Plantegenest, M., Gagic, V., Navasse, Y., & Lavandero, B. (2016). Aphid parasitoid generalism: development, assessment, and implications for biocontrol. *Journal of Pest Science, 89*(1), 7–20.

Rios Reyes, A. V., & Moya-Raygoza, G. (2004). Ciclo bilógico, parasitismo y depredación de *Gonatopus bartletti* Olmi, 1984 (Hymenoptera: Dryinidae), un enemigo natural de la chicharrita *Dalbulus maidis* (Delong y Wolcott, 1923). *Folia Entomológica Mexicana, 43*(3).

Torres-Moreno, R., Moya-Raygoza, G., & Pérez-López, E. (2015). Absence of corn stunt spiroplasma and maize bushy stunt phytoplasma in leafhoppers (Hemiptera: Cicadellidae) that inhabit edge grasses throughout winter in Jalisco, Mexico. *Florida Entomologist, 98*(3), 967–969.

Triapitsyn, S. V. (2015a). Taxonomic notes on *Anagrus incarnatus* Haliday and some other fairy-flies (Insecta: Hymenoptera: Mymaridae) from the AH Haliday collection in the National Museum of Ireland. *Bulletin of the Irish Biogeographical Society, 39*, 215–221.

Triapitsyn, S. V. (2015b). Taxonomy of the genus *Anagrus* Haliday (Hymenoptera: Mymaridae) of the world: an annotated key to the described species, discussion of the remaining problems, and a checklist. *Acta Zoológica Lilloana, 59*(1–2), 3–50.

Triplehorn, B., & Nault, L. (1985). Phylogenetic classification of the genus *Dalbulus* (Homoptera: Cicadellidae), and notes on the phylogeny of the Macrostelini. *Annals of the Entomological Society of America, 78*(3), 291–315.

Vega, F. E., & Barbosa, P. (1990). *Gonatopus bartletti* Olmi (Hymenoptera: Dryinidae) in Mexico: a previously unreported parasitoid of the corn leafhopper *Dalbulus maidis* (Delong & Wolcott) and the Mexican corn leafhopper *Dalbulus elimatus* (Ball)(Homoptera: Cicadellidae). *Proceedings of the Entomological Society of Washington, 92*(3), 461–464.

Vega, F. E., Barbosa, P., & Panduro, A. P. (1991). *Eudorylas* (Metadorylas) sp.(Diptera: Pipunculidae): a previously unreported parasitoid of *Dalbulus maidis* (DeLong and Wolcott) and *Dalbulus elimatus* (Ball)(Homoptera: Cicadellidae). *The Canadian Entomologist, 123*(1), 241–242.

Virla, E. G. (1999). Aportes acerca de la bionomía de *Paracentrobia subflava* (Hymenoptera: Trichogrammatidae), parasitoide de Hemipteros Cicadeloideos argentinos. *Revista de la Sociedad Entomológica Argentina, 58*(3–4), 17–22.

Virla, E. G. (2001). Notes on the biology of *Anagrus breviphragma* (Hymenoptera: Mymaridae), natural enemy of the corn leafhopper *Dalbulus maidis* (Hemiptera: Cicadellidae) and other plant diseases vectors in South America. *Boletín de Sanidad Vegetal Plagas, 27*(2), 239–247.

Virla, E. G., Espinosa, M., & Moya-Raygoza, G. (2011). First host record for *Anteon pilicorne* (Ogloblin)(Hymenoptera: Dryinidae), a parasitoid of Cicadellidae, including the corn leafhopper (Hemiptera: Cicadellidae). *Neotropical Entomology, 40*(2), 285–287.

Virla, E. G., Luft Albarracin, E., & Moya-Raygoza, G. (2009). Egg Parasitoids of *Dalbulus maidis* (Hemiptera: Cicadellidae) in Jalisco State, Mexico. *Florida Entomologist, 92*(3), 508–510.

Virla, E. G., Moya-Raygoza, G., & Luft Albarracin, E. (2013). Egg parasitoids of the corn leafhopper, *Dalbulus maidis*, in the southernmost area of its distribution range. *Journal of Insect Science, 13*(1).

Virla, E. G., Moya-Raygoza, G., & Rafael, J. A. (2009). First record of *Eudorylas schreiteri* (Shannon)(Diptera: Pipunculidae) as a parasitoid of the corn leafhopper (Hemiptera: Cicadellidae) in Argentina, with a table of pipunculid-host associations in the neotropical region. *Neotropical Entomology, 38*(1), 152–154.

Virla, E. G., & Olmi, M. (1998). The Dryinidae of Argentina:(Hymenoptera-Chrysidoidea). *Acta Entomologica Chilena 22*, 19–35.

Virla, E. G., & Olmi, M. (2007). Dryinidae (Hymenoptera: Chrysidoidea) parasitoides de la chicharrita del maíz, *Dalbulus maidis* (Delong & Wolcott)(Hemiptera: Cicadellidae), en Argentina, y descripción del macho de *Gonatopus moyaraygozai* Olmi. *Interciencia, 32*(12), 847–849.

Weintraub, P. G. (2007). Insect vectors of phytoplasmas and their control-an update. *Bulletin of Insectology, 60*(2), 169.

Weintraub, P. G., & Beanland, L. (2006). Insect vectors of phytoplasmas. *Annual Review of Entomology, 51*, 91–111.

Zhao, Y., Davis, R. E., Wei, W., Shao, J., & Jomantiene, R. (2014). Phytoplasma genomes: evolution through mutually complementary mechanisms, gene loss and horizontal acquisition. In D. C. Gross, A. Lichens-Park, & C. Kole (Eds.), Genomics of plant-associated bacteria (pp. 235–271). Berlin/Heidelberg, Germany: Springer Berlin Heidelberg.

Chapter 10
The Resistance of Jujube Trees to Jujube Witches' Broom Disease in China

Jin Zhao, Zhiguo Liu, and Mengjun Liu

Abstract Chinese jujube is an important multi-purpose fruit tree in Asia. Jujube Witches' Broom Disease (JWB), caused by phytoplasma, is the most destructive disease of Chinese jujube. Typical symptoms include witches' broom, small leaves, phyllody and dense root suckers. This chapter focuses on the disease management using antibiotics and resistant germplasms. Two effective antibiotic combinations (3 g/L Oxytetracycline hydrochloride +1% $MgSO_4$ + 2% Citric acid; 3 g/L Tetracycline hydrochloride +1% MgSO4) were selected for their curing efficiency. Curing rate and effective rate of one trunk injection before blooming were 85% and 100%, respectively. In our breeding program, four lines were selected for their resistance to JWB and different lines of the same cultivar were found to exhibit differential level of resistance to JWB, probably due to genetic variation. We demonstrated that a subset of *ZjMAPKs* and *ZjWRKYs* was activated in response to phytoplasma infection and that photosynthetic response patterns differed strongly between susceptible and resistant cultivars.

Keywords Chinese jujube · Jujube witches' broom disease (JWB) · Symptoms · Trunk injection · Resistance · Photosynthetic activity

J. Zhao (✉)
College of Life Science, Hebei Agricultural University, Baoding, Hebei, China

Key Laboratory of Hebei Province for Plant Physiology and Molecular Pathology, Hebei Agricultural University, Baoding, Hebei, China
e-mail: zhaojinbd@126.com

Z. Liu · M. Liu
Research Center of Chinese Jujube, Hebei Agricultural University, Baoding, Hebei, China

C. Y. Olivier et al. (eds.), *Sustainable Management of Phytoplasma Diseases in Crops Grown in the Tropical Belt*, Sustainability in Plant and Crop Protection 12,
https://doi.org/10.1007/978-3-030-29650-6_10

10.1 Introduction

Chinese jujube (*Ziziphus jujuba* Mill.) belongs to the Rhamnaceae family (Order Rosales) and is also called Chinese date, red date or Zao or Hongzao in Chinese. Chinese jujube is native to China and has been cultivated in China for over 4000 years (M. Liu, 2006). Jujube is an important multi-purpose fruit tree in Asia, especially in China, Japan and Korea, where fruits are consumed fresh, dried or processed and also used in Chinese herbal medicine (Gao, Wu, & Wang, 2013). Jujube fruit is rich in antioxidants, such as vitamin C, vitamin B, and various organic compounds and acids, which are beneficial to the human immune system. It is also a good source of iron and phosphorous, both elements being key ingredients in red blood cells. In traditional medicine, consuming jujube seed is considered to be a good remedy for insomnia or restlessness.

China is the main producer of jujube with an annual production of over 7.34 million tons in 2014 (L. Wang & Hu, 2016; Zhang, 2015). Jujube trees are very popular not only for the fruit production, but also for its numerous agronomic advantages, such as early fruit production (fruits are produced the same year of the planting), long flowering season and high tolerance to various biotic and abiotic stresses such as drought and salinity, making jujube trees, ideal cash crops for arid and semi-arid areas (Fig. 10.1A, B) (M. Liu, 2006). As well, jujube trees can be successfully grown in long-term intercropping systems which have proven to be profitable and environmentally friendly (Yang, Ding, Liu, Li, & Eneji, 2016). Jujube trees have been introduced into more than 40 countries, spreading to neighbouring countries such as Japan and Korea or to Europe via the Silk Road. However, jujube production is threatened by several devastating diseases, including phytoplasma diseases such as Jujube Witches' Broom Disease (JWB).

JWB was first reported in the 1940s, and the pathogen was confirmed to be phytoplasma (formerly called mycoplasma like organism or MLO) in 1973 (Yi & La, 1973). JWB is the most destructive disease of Chinese jujube; it kills at least three to five percent of all the jujube trees each year in many orchards, mostly due to the lack of effective control methods (M. Liu, Zhao, & Zhou, 2010).

Fig. 10.1 (**A**, **B**) Cultivated plantation and fruiting conditions of Chinese jujube in Xinjiang, China. General view of the orchard (**A**) and close-up on fruits (**B**)

Phytoplasma are wall-less bacteria that live and reproduce in the phloem of their plant hosts and are transmitted by phloem feeders, mostly leafhoppers (Perilla-Henao & Casteel, 2016). Phytoplasma belong to the class Mollicutes and have been associated with several hundreds of diseases worldwide and are a severe threat to agriculture in many countries (Bertaccini, Duduk, Paltrinieri, & Contaldo, 2014). The inability to grow these bacteria in cell-free medium limits our understanding on their biology. Phytoplasma taxonomy is based mainly on restriction fragment length polymorphism (RFLP) analysis of their 16S rDNA and phytoplasma were classified into 19 groups (Bertaccini & Duduk, 2009). JWB has been associated with '*Ca.* P. ziziphi' (Jung et al., 2003). The 16S rDNA sequences of 15 JWB isolates from different regions in China were all clustered into the 16Sr V group and showed over 99.8% homology, indicating that JWB phytoplasmas are highly conserved (Bu, Peng, Liu, & Zhao, 2016).

JWB phytoplasmas are transmitted by phloem-feeding insects, such as *Hishimonus sellatus* Uhler (Hemiptera: Cicadellidae), *Hishimonoides aurifacialis* Kuoh (Hemiptera: Cicadellidae) and *Typhlocyba sp.* Germar (Hemiptera: Cicadellidae), or by grafting, root tillering and parasitic plants like *Cuscuta chinensis* Lam. (Convolvulaceae) (Dai, Viraktamath, & Zhang, 2010; Kusunoki et al., 2002; Weintraub & Beanland, 2006; Zhu, Hadidi, Gundersen, Lee, & Zhang, 1997).

JWB disease in jujube trees exhibit typical phytoplasma disease symptoms (Fig. 10.2). After infection with JWB phytoplasma, the vegetative growth of the

Fig. 10.2 (**A–D**) Symptoms of JWB in different organs of JWB-infected Chinese jujube tree. (**A**) Fascicular offsprings from roots; (**B**) Phyllody in flowers (close-up in the left top corner of the picture is the normal flower); (**C**) Condensed small yellowish leaves on branches; (**D**) Dried diseased leaves on infected trees during winter and leafless healthy trees

infected trees is stimulated, producing abundant and dense shoots and branches (witches' broom) and small leaves (Fig. 10.2B, C). Flowers shows phyllody symptoms and dense root suckers are generated on the ground (Fig. 10.2A, B). Fruits are deformed and production losses in quantity and quality can range from 30 to 80%. Sustained infection for 3 to 5 years can result in the death of the tree. Diseased trees can be asymptomatic at the initial stages of the infection. The concentration of phytoplasma varied with the growing seasons, and with the types of tissues as well as within the tissues. In lightly infected trees, phytoplasma are usually found in the trunk and roots located at the same side and/or in the prolongation of the diseased branches. The uneven distribution of phytoplasma is more often observed in lightly infected trees as compared to highly infected trees (Zhao, Liu, Zhou, & Dai, 2006). Some natural factors, orchard management and intercropping species are also closely related to the occurrence of JWB. For example, JWB occurrence is usually more serious in orchards that are poorly managed and/or grown in acid soil and sunny slope, and near pine, cypress and paulownia trees. In winter, there is still a large amount of JWB phytoplasma in the diseased branches (Zhao et al., 2006). Jujube trees are deciduous fruit trees that defoliate during winter, but diseased jujube trees do not defoliate in winter, leaving the dense diseased branches and leaves to dry and die on the trees (Fig. 10.2D). Because of the perceived incurability of the disease, JWB is often called 'jujube cancer'.

JWB phytoplasmas, whose sizes are approximately 200–500 nm, were observed solely in sieve tubes of diseased phloem and not in any other organs of the diseased tree (Fig. 10.3A, B).

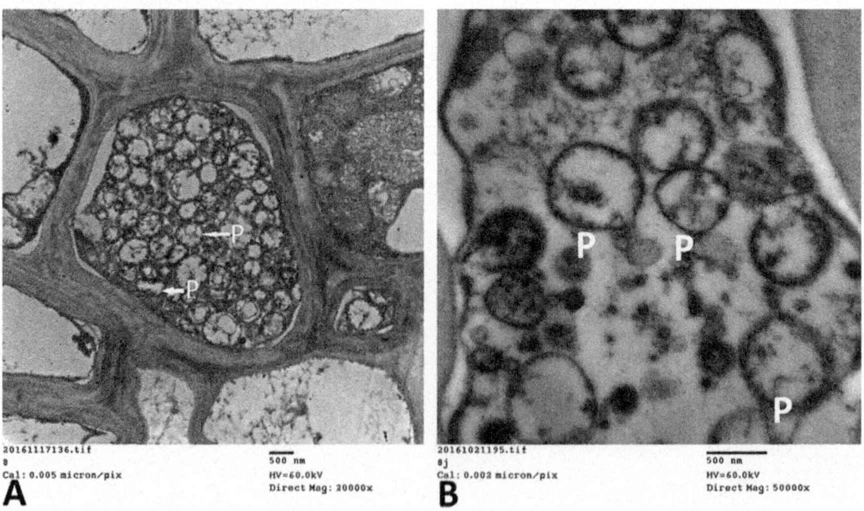

Fig. 10.3 (**A**–**B**) JWB phytoplasmas in sieve tubes of JWB-infected jujube trees, observed by Transmission Electron Microscopy (TEM). (**A**) Dense amounts of phytoplasmas observed in diseased sieve tubes. (**B**) Observation of 200–500 nm JWB phytoplasmas in diseased sieve tubes. *P* phytoplasmas

Phytoplasmas are phloem-limited and very difficult to control directly (Weintraub, Wilson, & Jones, 2010). In general, phytoplasma disease management relies on prophylaxis (use of clean plant materials, use of techniques to cure infected materials and/or destruction/roguing of diseased plants) and vector control using insecticides. However, except for the use of cryopreservation, prophylaxis methods used to eradicate JWB phytoplasma in diseased jujube trees have yet to be developed (R. Wang et al., 2015). Also, prophylaxis techniques can be expensive and time consuming. Moreover, insecticides are not environment-friendly, might not be registered for the infected crops and can lack efficiency in preventing the insect vectors to transmit the disease (Weintraub, 2007).

10.2 JWB Disease Management

10.2.1 Control with Antibiotics

Trunk injection is the main method used to cure fruiting trees from JWB disease. In diseased branches, the concentration of phytoplasma increased gradually with the rising of the temperature after bud sprouting in the months of April and May, then increased sharply to reach a peak in July and August, and then decreased before and during winter (Jin Zhao et al., 2006). Therefore, the optimal injection period to ensure treatment efficacy, is after sprouting and before blooming (May, Fig. 10.4). We developed two effective drug combinations (3 g/L Oxytetracycline hydrochloride +1% $MgSO_4$ + 2% Citric acid; 3 g/L Tetracycline hydrochloride +1% $MgSO_4$) through both indoor and field experiments. For both antibiotic combinations, the curing rate (no detection of phytoplasma and no reappearance of the disease for the next 2–3 years) and effective rate (alleviating the symptoms significantly) of one trunk injection before blooming were 85% and 100%, respectively. In addition, removing all the diseased branches can not only reduce the amount of pathogen in the tree but can also cure the diseased trees when the trees are newly infected and phytoplasmas still concentrated in the symptomatic branches. Moreover, it is recommended to remove all the symptomatic trees producing no economical yield (very young and very old trees as well as severely symptomatic trees) in order to remove pathogen inoculum from the orchard.

10.2.2 The Screening of Cultivars with High Resistance to JWB

In China, there are more than 900 jujube cultivars (Li, Guo, Fu, Liu, & Zhao, 2018; M. Liu & Zhao, 2008), most of them sensitive to JWB (M. Liu & Zhao, 2008). The large number of jujube germplasms in China provided a great opportunity to screen

Fig. 10.4 Trunk injection of antibiotics occurring during the month of May

for resistance to JWB. Based on literature search, survey information obtained from growers and various institutes as well as field observations, 29 cultivars demonstrated strong potential for JWB resistance and were collected from all over China for further screening (J. Zhao, Liu, Zhou, & Zhou, 2006).

Screening can be achieved by grafting diseased bark on healthy cultivar candidates or grafting healthy scions from cultivar candidates on diseased trees, and observing subsequent symptoms (M. Liu, Zhou, & Zhao, 2004). The 29 cultivar candidates were screened using the healthy scion grafting on diseased trees, as this method was found to be more effective for selecting highly-resistant germplasm as compared to the traditional method of grafting diseased bark onto healthy germplasm (J Zhao et al., 2006). Briefly, scions were cut from healthy jujube tree in spring and grafted on diseased stock in the field. There were at least 9 replications

Fig. 10.5 (**A–B**) (**A**) Resistant (winter defoliation) and susceptible (no winter defoliation) cultivars grafted onto severe diseased jujube trees. (**B**) Crown reconstruction of a diseased tree using resistant cultivar scions

for each cultivar. Symptoms were then observed at 40, 60, 80 and 110 days after grafting and scored based on a branch grading system. Phytoplasma presence was tested with fluorescence microscopy and PCR tests (M. Liu et al., 2004).

Interestingly, scions collected from resistant cultivars and grafted on severely diseased trees, sprouted, grew normally and became fully defoliated in winter. Scions collected from susceptible cultivars and grafted on severely diseased trees showed typical JWB-associated symptoms and did not defoliate in winter (Figs. 10.5A and 10.6A–C). After eight years of screening, four lines out of the 29 different cultivars were selected as being highly resistant to JWB. Moreover, the screening tests and field observations demonstrated that different lines of the same cultivar showed differential level of resistance to JWB (Qiao, Zhao, Wu, Yang, & Liu, 2008), which might be caused by genetic variation. Those strains are being used in breeding programs to produce JWB-resistant cultivars to reconstruct the crowns of diseased jujube trees (Fig. 10.5B).

The strain 'Xingguang' was identified among the 23 lines of the JWB-resistant cultivar 'Junzao'. 'Xingguang' developed mild symptoms shortly after the grafting and recovered within 40–80 days (Fig. 10.6D–F). Phytoplasmas were present in the resistant scions only for the short period of 40–80 days after grafting, indicating that the resistance of 'Xingguang' could be induced and stimulated by phytoplasma present in the diseased trees (Fig. 10.6). Moreover, 'Xingguang' was found to have fruits with excellent economic qualities. Indeed, 'Xingguang' produced fruits with a weight averaging 22.9 g (M. Liu et al., 2006). Moreover, the 'Xingguang' fruits are highly suitable for desiccation, with drying rate of 56.4% (M. Liu et al., 2006).

Fig. 10.6 (**A–F**) Growth condition of susceptible cultivar (**A, B, C**) and 'Xingguang' (**D, E, F**) grafted on diseased trees, 50 days after grafting (**A, D**),100 days after grafting (**B, E**) and 200 days after grafting (**C, F**)

The strain 'Xingguang' is currently being used successfully to cure diseased trees by crown reconstruction with top-grafting of 'Xingguang' scions, that induce a total recovery of the diseased trees with normal fruit production and no disease relapse (Z. Liu, Wang, Xiao, Zhao, & Liu, 2014). 'Xingguang' is heavily included in the jujube breeding program and being crossed with other susceptible cultivars.

10.2.3 The Jujube Genes Involved in Responses to Phytoplasma Infection

A suppression subtractive hybridization library (SSH) of resistant cultivar 'Xingguang' was constructed under phytoplasma disease stress to identify genes related to JWB resistance (Z. Liu et al., 2014). Seventy-seven of 200 unique expressed sequence tags had significant sequence homologies to known proteins and were classified into 10 functional groups. The most abundant group was disease/defense proteins (20.8%), which was consistent with the biotic stress associated with phytoplasma infection. These differentially expressed genes (such as *GST*, *TLP*, *PR10*, *HSP70*, *ERF* and *kinase-related protein*) provide the groundwork for addressing the plant-phytoplasma interaction (Peng, Zhao, Zhao, Zhao, & Liu, 2014; Xie, Yuan, Zhang, Zhao, & Liu, 2016).

Mitogen-activated protein kinases (MAPK) cascades and WRKYs are highly universal signal transduction modules and play crucial roles in regulating biotic responses in plants (Beckers et al., 2009; Pitzschke, Schikora, & Hirt, 2009). Transcriptome data and RT-qPCR demonstrated that a subset of *ZjMAPKs* and *ZjWRKYs* was activated in response to phytoplasma infection. Most *ZjMPK* and *ZjMKK* genes exhibited significant up-regulation or down-regulation upon phytoplasma infection. Compared with the control, *ZjMPK2*, *ZjMKK2* and *ZjMKK4* under phytoplasma stress exhibited significant differences in expression(Z. Liu et al., 2017), indicating that these genes might play important roles in jujube-phytoplasma interactions. Moreover, our recent studies have shown that the expression of more than 20 *ZjWRKYs* changed following phytoplasma infection, suggesting that they are also involved in signalling pathways that respond to phytoplasma stress.

Based on these results, we propose a model for the phytoplasma disease resistance of cultivar. Firstly, phytoplasma infection strongly induces the expression of a series of signal transduction genes. Then, protein kinase interacts with transcription factors and up-regulates the expression of relevant genes. Finally, transcription factors interact with the promoters of pathogenesis-related (PR) genes to promote the high expression of PR genes and synthesize PR proteins. Then, synthetized PR proteins interact with phytoplasma leading to a containment and eventual elimination of phytoplasma in the tissue.

10.2.4 The Novel Photosynthetic Response Pattern of the Resistant Cultivar to Phytoplasma Infection

Yellowing is a typical symptom of JWB-infected trees and diseased jujube trees in the field die 3–4 years after the beginnings of symptom expression. However, the diseased jujube seedlings can be subcultured and can survive for many years under a continuous nutrient supply in the lab. The long-term culture of diseased jujube seedlings can be performed *in vitro* (Fig. 10.7A, B), indicating that providing adequate nutrients is crucial to their growth. We found that phytoplasma infection inhibited photosynthesis, stimulated amino acids and energy metabolisms, and led to abnormal carbohydrate accumulation in the jujube diseased leaves (Xue et al., 2018). Simultaneous inhibition of the photosynthetic process and stimulation of the carbohydrate and energy metabolisms lead to "exhausted" trees that die a few years post-infection. Therefore, the effect of phytoplasma infection on these metabolisms is critical.

To further elucidate the mechanism of plant-phytoplasma interactions, we compared the effects of phytoplasma infection on photosynthetic pigments and activities, between the resistant cultivar 'Xingguang' and a susceptible cultivar. Total chlorophyll and carotenoid levels were significantly decreased in the susceptible

Fig. 10.7 (**A, B**) JWB-diseased jujube seedlings cultivated in vitro (**A**), with close up of the seedlings (**B**)

cultivar at later stages of the infection, but were remarkably increased in the resistant cultivar at the initial stages (Z. Liu, Zhao, & Liu, 2016). Compared to uninfected plant, a significant decrease in the main photochemical parameters (F_v/F_m, Φ_{PSII} and qP) was recorded at the initial stages of the infection in the resistant cultivar but occurred at later stages in the susceptible cultivar (Table 10.1).

Meanwhile, the expression patterns of key photosynthesis-related genes (such as *ZjGluTR*, *ZjCBP*, *ZjRubisco* and *ZjRCA2*) were similar in uninfected resistant and susceptible cultivars, but up-regulated in resistant cultivar and down-regulated in susceptible cultivar, 12 weeks after grafting diseased scions (Z. Liu et al., 2016). Collectively, it indicated that, in response to phytoplasma infection, the resistant cultivar 'Xingguang' undergoes an early decrease of photosynthetic activity (inhibiting phytoplasma multiplication) followed by a rapid increase of photosynthetic activity, helping jujube recovery, while the susceptible cultivar displays a late irreversible decrease in photosynthetic activity. The novel photosynthetic response pattern of the resistant cultivar may contribute to its stronger immunity to phytoplasma infection, which provides new insights into jujube-phytoplasma interactions.

Table 10.1 Changes of Chl fluorescence parameters in resistant and susceptible cultivars under phytoplasma stress

Parameters	Treatment	6 WAG	9 WAG	12 WAG	15 WAG	18 WAG
F_v/F_m	R-H	0.781 ± 0.013	0.693 ± 0.012	0.771 ± 0.013	0.739 ± 0.018	0.651 ± 0.010
	R-D	0.664 ± 0.010 **	0.689 ± 0.015	0.769 ± 0.012	0.736 ± 0.010	0.652 ± 0.010
	S-H	0.747 ± 0.009	0.746 ± 0.025	0.764 ± 0.016	0.749 ± 0.004	0.701 ± 0.015
	S-D	0.739 ± 0.014	0.745 ± 0.011	0.762 ± 0.006	0.673 ± 0.012 **	0.599 ± 0.012 **
F_0	R-H	0.055 ± 0.003	0.135 ± 0.004	0.100 ± 0.004	0.073 ± 0.003	0.110 ± 0.007
	R-D	0.100 ± 0.005 **	0.107 ± 0.003 **	0.050 ± 0.002 **	0.128 ± 0.003 **	0.107 ± 0.007
	S-H	0.082 ± 0.005	0.056 ± 0.003	0.064 ± 0.003	0.076 ± 0.002	0.117 ± 0.004
	S-D	0.094 ± 0.003 *	0.085 ± 0.004 **	0.116 ± 0.005 **	0.166 ± 0.004 **	0.159 ± 0.006 **
F_m	R-H	0.251 ± 0.007	0.439 ± 0.009	0.438 ± 0.010	0.280 ± 0.010	0.316 ± 0.013
	R-D	0.297 ± 0.010 **	0.345 ± 0.008 **	0.219 ± 0.005 **	0.486 ± 0.013 **	0.308 ± 0.012
	S-H	0.325 ± 0.010	0.223 ± 0.011	0.273 ± 0.011	0.302 ± 0.013	0.391 ± 0.011
	S-D	0.361 ± 0.009 *	0.332 ± 0.013 **	0.488 ± 0.012 **	0.506 ± 0.012 **	0.396 ± 0.011
Φ_{PSII}	R-H	0.679 ± 0.016	0.612 ± 0.016	0.655 ± 0.020	0.649 ± 0.024	0.553 ± 0.013
	R-D	0.509 ± 0.009 **	0.615 ± 0.010	0.654 ± 0.009	0.656 ± 0.018	0.549 ± 0.007
	S-H	0.645 ± 0.013	0.636 ± 0.016	0.654 ± 0.024	0.642 ± 0.005	0.619 ± 0.019
	S-D	0.641 ± 0.010	0.631 ± 0.007	0.651 ± 0.016	0.529 ± 0.002 **	0.492 ± 0.011 **
Abs	R-H	0.865 ± 0.009	0.858 ± 0.012	0.821 ± 0.017	0.851 ± 0.016	0.842 ± 0.027
	R-D	0.864 ± 0.008	0.866 ± 0.013	0.820 ± 0.015	0.845 ± 0.020	0.835 ± 0.022
	S-H	0.809 ± 0.012	0.879 ± 0.017	0.824 ± 0.005	0.857 ± 0.019	0.852 ± 0.017
	S-D	0.804 ± 0.014	0.875 ± 0.011	0.823 ± 0.015	0.785 ± 0.021 *	0.804 ± 0.016 *
q_p	R-H	0.972 ± 0.010	0.966 ± 0.012	0.941 ± 0.019	0.948 ± 0.025	0.959 ± 0.019
	R-D	0.914 ± 0.005 **	0.961 ± 0.023	0.939 ± 0.013	0.952 ± 0.013	0.957 ± 0.018
	S-H	0.923 ± 0.009	0.958 ± 0.024	0.965 ± 0.012	0.957 ± 0.011	0.976 ± 0.006
	S-D	0.923 ± 0.011	0.952 ± 0.015	0.963 ± 0.019	0.911 ± 0.013 *	0.946 ± 0.010 *
NPQ	R-H	0.248 ± 0.006	0.135 ± 0.006	0.246 ± 0.010	0.144 ± 0.006	0.127 ± 0.005
	R-D	0.291 ± 0.007 **	0.121 ± 0.007	0.248 ± 0.007	0.139 ± 0.010	0.129 ± 0.008
	S-H	0.131 ± 0.003	0.200 ± 0.019	0.286 ± 0.008	0.233 ± 0.010	0.160 ± 0.008
	S-D	0.131 ± 0.005	0.196 ± 0.019	0.293 ± 0.010	0.232 ± 0.010	0.155 ± 0.010

Note: R-D: Resistant cultivar grafted on diseased rootstock; R-H: Resistant cultivar grafted on healthy rootstock. S-D: Susceptible cultivar grafted on diseased rootstock; S-H: Susceptible cultivar grafted on healthy rootstock; WAG: weeks after grafting; Abs, absorption coefficient; F0, dark adapted minimal fluorescence; Fm, dark adapted maximum fluorescence; Fv/Fm, maximum quantum efficiency of PSII; ΦPSII, quantum efficiency of PSII; qp, photochemical quenching coefficient; NPQ, non-photochemical quenching coefficient. Values shown are means of 3 replicates ± standard error. Asterisk within rows indicates significantly different between the R-D or S-D treatment and their control (*P < 0.05; **P < 0.01)

In the field, the accumulation of photosynthetic product in susceptible diseased trees contributed to their vigorous and excessive growth leading to the witches' broom during growing season. However, inhibition of the photosynthetic process occurring with increase of the carbohydrate and energy metabolism, "exhausted" the susceptible diseased tree, leading to dried-up diseased tissues left on the tree and causing the typical symptoms of lack of defoliation during winter.

10.3 Conclusion

JWB is a devastating phytoplasma disease, causing serious economic losses in jujube production in China. The development of two effective antibiotic combinations as trunk injection treatment and the selection of four resistant lines extensively used in the jujube breeding programs might contribute to a successful JWB disease management in the future. Molecular studies showed that defence-related genes, such as ZjWRKYs and ZjMAPKs, are involved in responses to phytoplasma infection, and that earlier photosynthetic response pattern of the resistant cultivar likely contributes to its stronger immunity to phytoplasma infection. More studies on the jujube-phytoplasma interaction would provide new insights into the resistant mechanism and beneficial clues for effective preventive measures.

References

Beckers, G. J., Jaskiewicz, M., Liu, Y., Underwood, W. R., He, S. Y., Zhang, S., … Conrath, U. (2009). Mitogen-activated protein kinases 3 and 6 are required for full priming of stress responses in *Arabidopsis thaliana*. *The Plant Cell, 21*(3), 944–953.

Bertaccini, A., & Duduk, B. (2009). Phytoplasma and phytoplasma diseases: A review of recent research. *Phytopathologia Mediterranea, 48*(3), 355–378.

Bertaccini, A., Duduk, B., Paltrinieri, S., & Contaldo, N. (2014). Phytoplasmas and phytoplasma diseases: A severe threat to agriculture. *American Journal of Plant Sciences, 5*(12), 1763.

Bu, J., Peng, L., Liu, M., & Zhao, J. (2016). 16S rDNA sequence analysis of witches' broom phytoplasma isolates from Chinese jujube in North China. *Australasian Plant Pathology, 45*(1), 119–122.

Dai, W., Viraktamath, C. A., & Zhang, Y. (2010). A review of the leafhopper genus *Hishimonoides Ishihara* (Hemiptera: Cicadellidae: Deltocephalinae). *Zoological Science, 27*(9), 771–781.

Gao, Q.-H., Wu, C.-S., & Wang, M. (2013). The jujube (*Ziziphus jujuba* Mill.) fruit: A review of current knowledge of fruit composition and health benefits. *Journal of Agricultural and Food Chemistry, 61*(14), 3351–3363.

Jung, H.-Y., Sawayanagi, T., Kakizawa, S., Nishigawa, H., Wei, W., Oshima, K., … Namba, S. (2003). 'Candidatus Phytoplasma ziziphi', a novel phytoplasma taxon associated with jujube witches'-broom disease. *International Journal of Systematic and Evolutionary Microbiology, 53*(4), 1037–1041.

Kusunoki, M., Shiomi, T., Kobayashi, M., Okudaira, T., Ohashi, A., & Nohira, T. (2002). A leafhopper (*Hishimonus sellatus*) transmits phylogenetically distant phytoplasmas: Rhus yellows and Hovenia witches' broom phytoplasma. *Journal of General Plant Pathology, 68*(2), 147–154.

Li, S., Guo, M., Fu, P., Liu, H., & Zhao, X. (2018). Genetic diversity and population structure of Chinese jujube (*Ziziphus jujuba* Mill.) and sour jujube (*Ziziphus acidojujuba* Mill.) using inter-simple sequence repeat (ISSR) Markers. *PeerJ Preprints, 6*(e27088v1), 28.

Liu, M. (2006). Chinese jujube: Botany and horticulture. *Horticultural Reviews, 32*, 229–298.

Liu, M., Zhao, J., & Zhou, J. (2010). Jujube witches' broom disease. *China Agriculture Press, Beijing (China), 12*, 50–51.

Liu, M., & Zhao, Z. (2008). Germplasm resources and production of jujube in China. *Acta Horticulturae, 840*, 25–32.

Liu, M., Zhou, J., & Zhao, J. (2004). Screening of Chinese jujube germplasm with high resistance to witches' broom disease. *Acta Horticulturae, 663*, 575–580.

Liu, M., Zhou, J., Zhao, J., Wang, J., Liu, P., Dai, L., & Wang, Y. (2006). An excellent new cultivar of Chinese jujube with high resistance to jujube witches broom disease 'Xingguang'. *Acta Horticulturae Sinica, 33*(3), 687.

Liu, Z., Wang, Y., Xiao, J., Zhao, J., & Liu, M. (2014). Identification of genes associated with phytoplasma resistance through suppressive subtraction hybridization in Chinese jujube. *Physiological and Molecular Plant Pathology, 86*, 43–48.

Liu, Z., Zhang, L., Xue, C., Fang, H., Zhao, J., & Liu, M. (2017). Genome-wide identification and analysis of MAPK and MAPKK gene family in Chinese jujube (*Ziziphus jujuba* Mill.). *BMC Genomics, 18*(1), 855.

Liu, Z., Zhao, J., & Liu, M. (2016). Photosynthetic responses to phytoplasma infection in Chinese jujube. *Plant Physiology and Biochemistry, 105*, 12–20.

Peng, L., Zhao, Y., Zhao, Z., Zhao, J., & Liu, M. (2014). Cloning and expression of a tau class glutathione S-transferase (ZjGSTU1) from Chinese jujube in response to phytoplasma infection. *Acta Physiologiae Plantarum, 36*(11), 2905–2913.

Perilla-Henao, L. M., & Casteel, C. L. (2016). Vector-borne bacterial plant pathogens: Interactions with hemipteran insects and plants. *Frontiers in Plant Science, 7*, 1163.

Pitzschke, A., Schikora, A., & Hirt, H. (2009). MAPK cascade signalling networks in plant defence. *Current Opinion in Plant Biology, 12*(4), 421–426.

Qiao, Y., Zhao, J., Wu, T., Yang, Y., & Liu, M. (2008). Genetic diversity of twenty three strains of *Ziziphus jujuba* Mill. CV. Junzaov with different levels of resistance to jujube witches' broom disease. *Acta Horticulturae, 840*, 391–398.

Wang, L., & Hu, R. (2016). Technical and cost efficiency of jujube growers in Henan Province, China. *Custos e agronegocio, 12*(2), 279–297.

Wang, R., Mou, H., Gao, X., Chen, L., Li, M., & Wang, Q. (2015). Cryopreservation for eradication of jujube witches' broom phytoplasma from Chinese jujube (*Ziziphus jujuba*). *Annals of Applied Biology, 166*(2), 218–228.

Weintraub, P. G. (2007). Insect vectors of phytoplasmas and their control-an update. *Bulletin of Insectology, 60*(2), 169.

Weintraub, P. G., & Beanland, L. (2006). Insect vectors of phytoplasmas. *Annual Review of Entomology, 51*, 91–111.

Weintraub, P. G., Wilson, M. R., & Jones, P. (2010). Control of phytoplasma diseases and vectors. *Phytoplasmas: genomes, plant hosts and vectors* (pp. 233–249). Wallingford, UK: CABI.

Xie, N., Yuan, Z., Zhang, L., Zhao, J., & Liu, M. (2016). Molecular cloning and expression of a novel eukaryotes elongation factor1A gene (ZjeEF-1α) from Chinese jujube in response to phytoplasma infection. *Physiological and Molecular Plant Pathology, 96*, 101–108.

Xue, C., Liu, Z., Dai, L., Bu, J., Liu, M., Zhao, Z., … Zhao, J. (2018). Changing host photosynthetic, carbohydrate, and energy metabolisms play important roles in Phytoplasma infection. *Phytopathology, 108*(9), 1067–1077.

Yang, L., Ding, X., Liu, X., Li, P., & Eneji, A. E. (2016). Impacts of long-term jujube tree/winter wheat–summer maize intercropping on soil fertility and economic efficiency—A case study in the lower North China plain. *European Journal of Agronomy, 75*, 105–117.

Yi, C., & La, Y. (1973). Mycoplasma-like bodies found in the phloem elements of jujube trees infected with witches-broom disease. *Research Report Forest Research Institute, Korea, 20*, 111–114.

Zhang, W. (2015). *China rural statistical yearbook*. Retrieved January 18, 2019, from China Statistics Press http://cdi.cnki.net/Titles/SingleNJ?NJCode=N2016010146

Zhao, J., Liu, M., Zhou, J., & Dai, L. (2006). Distribution and year-round concentration variation of jujube Witches' Broom (JWB) phytoplasma in the plant of Chinese jujube. *Scientia Silvae Sinicae, 42*(8), 144.

Zhao, J., Liu, M., Zhou, J., & Zhou, H. (2006). Screening and application of Chinese jujube germplasm with high resistance to witches' broom disease. *Journal of Plant Genetic Resources, 7*(4), 398–403.

Zhu, S. F., Hadidi, A., Gundersen, D., Lee, I.-M., & Zhang, C. (1997). Characterization of the phytoplasmas. Associated with cherry lethal yellows and jujube witches'-broom diseases in China. *Acta Horticulturae, 472*, 701–714.

Chapter 11
Integrated Management of Coconut Lethal Yellowing Phytoplasma Disease in Mozambique: Current Challenges and Future Perspectives

J. Bila, A. Mondjana, B. Samils, L. Santos, and N. Hogberg

Abstract The coconut palm (*Cocos nucifera*), an importance cash and subsistence crop, is widely grown in the coastal regions of the tropics, and contributes to a large extend to the livelihood and food security of millions of rural inhabitants in Mozambique. However, coconut production has been challenged by the occurrence of an invasive and emerging pathogen, the causal agent of lethal yellowing disease (LYD). Mozambique was traditionally the top coconut producer in Africa and is now ranked fourth below Nigeria partially due to coconut LYD. The current LYD epidemic status in Mozambique expanded the typical plant-insect vector-phytoplasma triangle to a plant-insect vector-phytoplasma-beetle relationship. For LYD management, more effort must be directed towards preventive measures, since the infection is usually lethal. However, several measures aiming at maintaining the LYD pressure at low levels, can be combined. The integrated LYD strategies described in this chapter include the use of all available management strategies to maintain disease level under control, while minimising the use of chemicals. Currently, the top options for LYD management in Mozambique involve combination of plant resistance and cultural practices. The LYD field tolerant Mozambican tall (MZT) variety is widely used to replace the succumbed LYD-infected coconut palms. Black's approach for LYD management could be improved by including early field detection using LAMP technology to ensure that not only symptomatic palms, but also infected symptomless plants are felled and burnt. The integrated LYD management approach, targeting the LYD pathosystem and the coconut beetle *Oryctes monoceros* as the major coconut pest, is likely to successfully control the current epidemics in Mozambique.

J. Bila (✉) · A. Mondjana · L. Santos
Universidade Eduardo Mondlane, Maputo, Mozambique

B. Samils · N. Hogberg
Swedish University of Agricultural Sciences, Uppsala, Sweden
e-mail: Berit.Samils@slu.se; nils.hogberg@slu.se

C. Y. Olivier et al. (eds.), *Sustainable Management of Phytoplasma Diseases in Crops Grown in the Tropical Belt*, Sustainability in Plant and Crop Protection 12,
https://doi.org/10.1007/978-3-030-29650-6_11

Keywords Coconut · Coconut beetle · Lethal yellow disease · Plant resistance · Cultural practices · Phytoplasma disease management

11.1 Introduction

The coconut palm (*Cocos nucifera*), an import cash and subsistence crop, is widely grown in the coastal regions of the tropics, and contributes to a large extend to the livelihood and food security of millions of rural inhabitants in Mozambique (Bila, Högberg, Mondjana, & Samils, 2015; Bila et al., 2016). However, coconut production has been challenged by the occurrence of an invasive and emerging disease, known as lethal yellowing (LY). The disease, caused by phytoplasmas, has been threatening coconut production throughout the Americas and the Caribbean regions (Harrison & Oropeza, 2008; Tymon, Jones, & Harrison, 1998). Diseases with symptoms similar to those caused by LY occur in West and Eastern Africa and are collectively called lethal yellowing-like diseases (LYD) (Eden-Green, 1997; Mpunami, Tymon, Jones, & Dickinson, 1999; Tymon et al., 1998). LYD is also known as Coconut Lethal Yellowing or Lethal Yellowing Type Syndromes which reflect the typical symptoms caused by this complex of phytoplasma-associated coconut diseases (Gurr et al., 2016). In Tanzania, a similar disease is known as Lethal Disease (LD) (Mpunami et al., 1999).

Like most infectious diseases, no single symptoms would diagnose LYD accurately, but sequential development of symptoms need to be considered. Typical LYD symptoms progression start with fruit abortion, necrosis of inflorescences (even in unfolded inflorescences), followed by progressive yellow discoloration of the canopy (from the older to youngest leaves), rotting of the spear leaves, wilting and collapse of the palm crowns leading to palms resembling telephone poles (Fig. 11.1) (Dery & Philippe, 1997; Harrison & Elliot, 2005).

Eden-Green (1997) reported that records of a coconut disease with symptoms similar to those caused by LYD, were made in the mid-1950s in Mozambique, but the etiology was never established, and the disease did not reach epidemic status. Hence, it was only by the 1970s that the etiology of LYD was first established (Gurr et al., 2016). In 1992, a new coconut LYD outbreak, was detected in the Cabo Delgado province, in the northeast of Mozambique, which reached the epidemic level. Since then, a succession of outbreaks has been reported throughout the coastal areas of Northern and Central Mozambique. In 2010, a new outbreak was confirmed in the Inhambane Province in Southern Mozambique (Bila, 2016).

With advances in DNA-based technology, LYD phytoplasma taxonomy has been re-evaluated. Phylogenetic studies done by Tymon et al. (1998) showed that phytoplasmas associated with coconut belong to three different clades, namely the (1) Americas and Caribbean, (2) East and (3) West African types. Based on 16S rDNA Restriction Fragment Length Polymerase (RFLP) and sequence analyses, Mpunami

Fig. 11.1 Typical LYD symptoms. Fruit abortion (a), progressive yellow discoloration (from the older to youngest leaves of the canopy (b), wilting and collapse of the palm crown (c), rotting of the spear leaves (d), leading to telephone poles (e) (Photo courtesy: João Bila)

et al. (1999) showed that the LYD causal phytoplasma strains from Mozambique and Ghana were more similar to each other than to LD strains from East Africa. With progress in DNA-based identification and classification technologies, updated differentiation of phytoplasma groups and subgroups associated with LYD in Africa have been established. The Mozambican and Nigerian LYD phytoplasma group have been described as '*Candidatus* Phytoplasma palmicola', belonging to16S rRNA group XXII-A, while the closest group '*Ca.* P. palmicola'-related strains from Ghana and Côte d'lvoire belongs to 16S rRNA group XXII- B (Harrison et al., 2014). Tymon et al. (1998) have previously shown that the LYD in Ghana and Nigeria was either caused by the same or very closely related phytoplasmas. Apart from '*Ca.* P. palmicola', the major groups associated with LYD in Mozambique are the Tanzanian LD phytoplasma and a novel type closely related to '*Ca.* P. pini' (Bila, Högberg, et al., 2015). This suggests that at least three phytoplasma types are associated with coconut in Mozambique. So far, this is among the highest LYD phytoplasma diversity observed in a single country (Harrison et al., 2014; Mpunami et al., 1999; Nkansah-Poku, Philippe, Quaicoe, Dery, & Ransford, 2009). The presence of the Tanzanian LD phytoplasma group in Mozambique was further confirmed during the search for potential insect vectors of LYD phytoplasma in the country (unpublished data). The high LYD phytoplasma diversity observed in Mozambique may imply that different insect vectors and alternate hosts might be associated with their reservoir and dispersal, providing further challenges for the management of the disease. The nucleotide analysis of the Mozambican '*Ca.* P. palmicola' group, detected a genetic bottleneck followed by expansion and purifying selection, which was consistent with the observed LYD epidemic spread (Bila, Högberg, et al., 2015). Under high phytoplasma diversity, multiple infections can be observed (Arocha et al., 2016; Bila, Högberg, et al., 2015) which may affect the severity of the disease. For instance, co-infection of the coconut LYD phytoplasma with other phytoplasma groups have been reported in Mozambique (Bila et al., 2015) which may speed up the pathogen evolution since there may be opportunities for recombination among phytoplasma clades (Danet et al., 2011).

11.2 Socio-economic Implications of LYD in Mozambique

A significant part of Mozambicans depends on coconut for their livelihood. Apart from farmers' self-consumption, and endless uses of the different parts of the palm tree, the cultivation, marketing and processing of coconut are a major source of income, staple food, potable water and job opportunities. This is especially true for

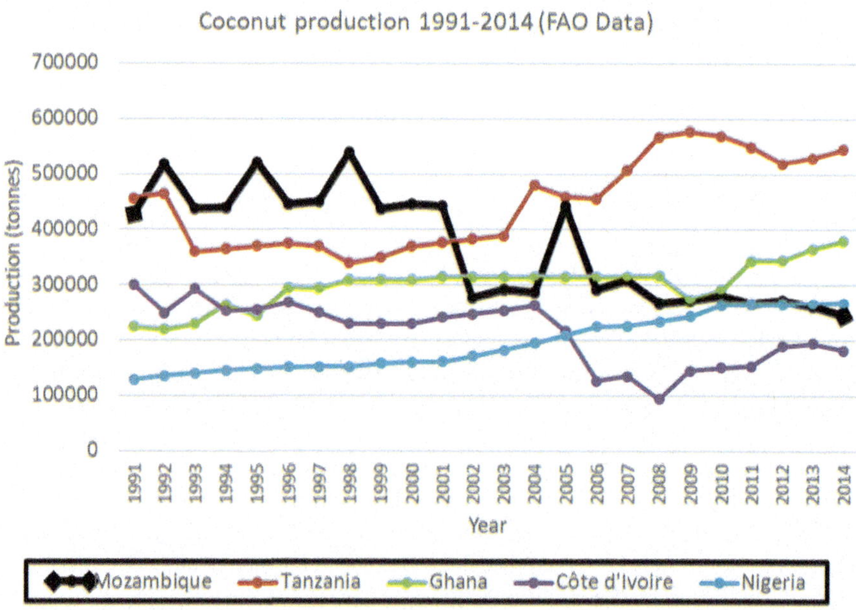

Fig. 11.2 Production trends (1991–2014) in the top five African coconut producer countries, also affected either with LYD (Mozambique, Ghana, Côte d'Ivoire, Nigeria) or Lethal disease (Tanzania) (Food and Agriculture Organisation of the United Nations, 2016)

resource poor families, commonly found in rural areas. Moreover, coconut timber and canopy are used in building houses, fences, furniture as well as fuel, and are a source of raw material for several value-added manufactured products.

Outbreaks of LYD have caused succession of epidemics and losses of millions of coconut palms, threatening the livelihood of a significant number of Mozambicans, mainly those living along the coastal belt. Mozambique was traditionally the top coconut producer in Africa, but is now ranked fourth below Nigeria (Fig. 11.2), due to coconut LYD (Food and Agriculture Organisation of the United Nations, 2016). The impact of LYD in Mozambique is not only felt by smallholder and subsistence coconut farmers but also by the commercial farmers and coconut industry investors and workers, the latter due to lack of raw material.

In the main coconut-growing areas of the Zambezia province where the disease is now endemic, 100% losses can be observed in several productions sites. After a disastrous epidemic in central Mozambique, which forced the biggest coconut business related companies to a closure and thousands of workers to be redundant, a new LYD outbreak is currently challenging Inhambane, in the southeast of Mozambique. This province used to be the second leading producer of coconut but is now number one and the current status of the LYD epidemic is scaring new investors in the coconut sector. Coconut losses due to LYD are still increasing in Mozambique, and a worst scenario is likely to occur in the near future. This prediction is supported by spatial and spatiotemporal pattern analysis of coconut LYD (Bonnot, De Franqueville, & Lourenço, 2010).

11.3 Lethal Yellowing Disease Cycle

Similarly to other phytoplasma disease, the LYD cycle (Fig. 11.3) resemble a plant-insect-phytoplasma matrimonial triangle relationship (Christensen, Axelsen, Nicolaisen, & Schulz, 2005; Garnier et al., 2001; Oshima et al., 2011). Knowledge of every aspect of the LYD cycle, such as host range, disease transmission and environmental conditions, is crucial for a successful disease management.

11.3.1 Host Range

Bila, Högberg, et al. (2015) reported several phytoplasma groups associated with LYD in Mozambique, including a novel strain closely related to '*Ca.* P. pini'. The discovery of the '*Ca.* P. pini'- related type in Mozambican coconut palms suggested that *Pinus* sp. could be a potential alternate host for LYD in Mozambique (Bila, Högberg, et al., 2015). Screening for alternative hosts provided *Pinus* sp. samples that were PCR positive for LYD, but sequencing failed to confirm their potential alternate host status (Bila, Mondjana, Samils, & Högberg, 2015).

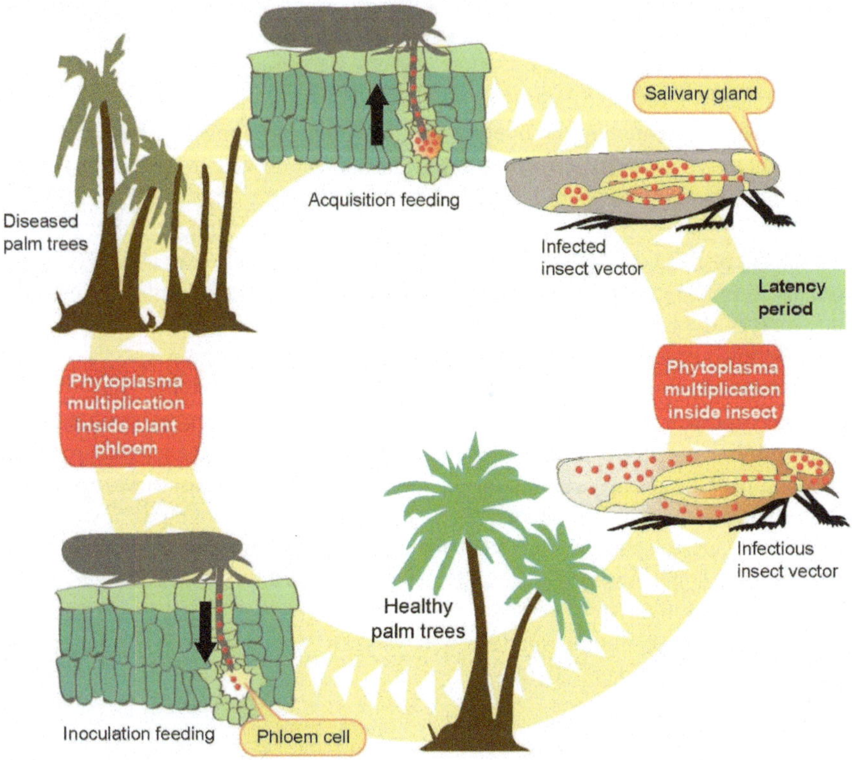

Fig. 11.3 Coconut lethal yellowing disease cycle. Phytoplasma within insect tissues or plant phloem are represented by red dots

African oil palm (*Elaeis guineensis* Jacq.) and fan palm (*Borassus aethiopum* Mart.) (Fig. 11.4) were found harbouring LYD phytoplasma in Mozambique, and assigned as alternate hosts for '*Ca.* P. palmicola' (Bila, Mondjana, et al., 2015). In Nigeria, where LYD is caused by the same phytoplasma strain as in Mozambique, Eziashi et al. (2013) found several weed species hosting potential LYD insect vectors. Interestingly, molecular screening demonstrated that some of the suspected weed species in Nigeria were reported as alternate hosts of '*Ca.* P. palmicola'-related strains belonging to the 16S rRNA group XXII-B, in Côte d'Ivoire (Arocha et al., 2016). However, more studies are needed to confirm that the suspected weed species are alternate hosts.

In West Africa, plant species from the families Poaceae (*Paspalum vaginatum* Sw., *Pennisetum pedicillatum* Trin.), Verbenaceae (*Stachytarpheta indica* (L.) Vahl), Plantaginaceae (*Scoparia dulcis* L.), Phyllanthaceae (*Phyllantus muelleria-nus* (Kuntze) Excell), Cyperacea (*Diplacrum capitatum* (Willd.) Boeckeler), Fabaceae (*Desmodium adscendus* (Sw.) DC) and Euphorbiaceae (*Manihot esculenta* Crantz) were also reported as alternate hosts for '*Ca.* P. palmicola'-related

Fig. 11.4 Alternate Hosts of coconut LYD phytoplasma in Mozambique. (**a**) African fan palm (*Borassus aethiopum*) showing progressive brown discoloration of the old leaves; (**b**) African oil palm (*Elaeis guineensis*) exhibits skirt shaped brown discoloration of the older leaves and (**c**) collapse of the necrotic crown. (Photo courtesy: João Bila)

strain belonging to the16Sr XXII-B, the other LYD causal agent in West Africa (Arocha et al., 2016; Danyo, 2011). This finding suggests that the host range of the LYD phytoplasma is expanding, which may accelerate the LYD phytoplasma evolution and consequently increasing the challenge of developing disease management strategies. Studies are also needed to determine if the reported LYD phytoplasma alternate hosts in Mozambique and West Africa are dead-end-hosts, or can act as reservoir for the pathogen which may result in new disease outbreaks.

11.3.2 Disease Transmission

LYD phytoplasmas are primarily spread through the movement of infected plants or insect vectors. Although LYD phytoplasmas have been detected in coconut embryos, seed transmission was not demonstrated (Cordova, Jones, Harrison, & Oropeza, 2003). However, recently, Oropeza et al. (2017) demonstrated for the first time, that the causal agent of LY (a 16SrIV group phytoplasma) could be transmitted from embryos to plantlets. So far, the known potential vectors associated with LY, LYD or LD coconut diseases are planthoppers (Derbidae, Cixiidae, Acrididade), leafhoppers (Cicadelidae) and Pentatomidae (Bila et al., 2017; Dollet, Macome, Vaz, & Fabre, 2011; Mpunami, Tymon, Jones, & Dickinson, 2000; Philippe et al., 2007). Based on molecular screening, different Hemiptera species were assigned as potential LYD insect vectors in different parts of the world (Bila et al., 2017; Brown, Been, & McLaughlin, 2006; Mpunami et al., 2000; Philippe et al., 2007) but their status as vectors has yet to be demonstrated through successful transmission trials. The planthopper *Diostrombus mkurungai* Wilson (Wilson, 1987) and the pentatomid bug *Platacantha lutea* were found harboring LYD phytoplasma DNA and labelled as potential LYD phytoplasma vector in Mozambique (Bila et al., 2017; Dollet et al., 2011). Based on both molecular screening and transmission trials, the planthopper *Haplaxius crudus* (formerly *Myndus crudus*) was confirmed as vector of LY in the Caribbean and Florida (Howard, Norris, & Thomas, 1983), while in Tanzania, LD phytoplasma might be transmitted by the planthoppers *D. mkurungai* and *Meenoplus sp.* (Mpunami et al., 2000).

11.3.3 The Impact of Oryctes monoceros *Beetle on LYD Epidemic in Mozambique*

The heavy epidemic status of LYD in Central and Northern Mozambique, is partially inflated by the high prevalence of the *Oryctes monoceros* (Oliver) beetle. The beetle attacks are weakening the palm tree defense response, providing conducive environment for other infections, such as LYD (Eden-Green, 2006). After LYD epidemics, the high numbers of fallen stems on the ground and still standing dead trees, increase the beetle population and threaten large scale replanting programs. Since 2007, Mozambique has been involved in large scale coconut replanting programs which ended up to be unsuccessful probably because of the beetle prevalence (Smith, 2015).

Decaying empty coconut stems left after crown collapse, are ideal sites for egg laying and larvae development of *O. monoceros* (Fig. 11.5a). The emerging adult beetles are more harmful to the young coconut palms, as they produce galleries in mature palms but also in the soft apical section of the seedlings and young palms (Fig. 11.5b) (Allou et al., 2012; Allou, Morin, Kouassi, N'klo, & Rochat, 2006; Bedford, 2014). Symptoms of typical beetle attack include truncated or broken

Fig. 11.5 The empty stems left over due to LYD attack provide favourable breeding sites for the beetle (**a**) and the emerging adult beetle feeds on spear leaves (**b**) which may result in truncated or broken fronds with wedge-shaped gaps or cuts (**c**). (Photo courtesy: João Bila (**a**) & FAEF-UEM (**b**, **c**))

fronds with wedge-shaped gaps or cuts (Fig. 11.5c), snapped-off spear leaves as well as holes in petioles and stalk (Bedford, 2014). Under the combined attack of the beetle and LYD phytoplasma, which is quite common in the LYD affected regions in central Mozambique, the real cause of the palm death is difficult to pinpoint. Overall, LYD is more likely to kill the mature palms while the coconut beetles usually kill the young palms. Apart from productivity reduction and palm's death, beetle damage may also be an entry point for other pathogens or saprophytes, hence further stressing the plant immunity. Eden-Green (2006) observed that significant numbers of the local Mozambique tall (MZT) coconut cultivar can survive continued LYD phytoplasma field exposure. However, survivors are usually weakened or killed by large population of the adult *O. monoceros*. In Mozambique, Smith (2015) observed that the current coconut LYD epidemic was sharing similar tendency with the coconut beetle damage.

Rather than the typical plant-insect-phytoplasma matrimonial triangle relationship, the current LYD epidemic status in Central and Northern Mozambique expanded the matrimonial triangle to a plant-insect vector-phytoplasma-beetle relationship. Any effective control strategy against the disease must therefore stop the disease cycle development, either by targeting the host, the insect vector or the conducive environment including the beetles.

11.4 Integrated Management of Coconut Lethal Yellowing Disease

Integrated disease management (IDM) is a disease control approach that endorses the combination of plant resistance as well as cultural, physical, biological, and chemical control strategies to keep disease pressures below an economic injury threshold (Agrios, 2005). The economic threshold level for LYD remains to be established.

The integrated LYD strategies described below include the use of all available management strategies to maintain disease level under control, while minimising the use of chemicals. For LYD management, more effort must be directed towards preventive measures, since the infection is usually lethal. However, several measures can be applied aiming at maintaining the LYD pressure at low levels. Among them, the Black's approach seems to be the more effective (Myrie et al., 2011). Black's approach named after Michael Black, is an integrated pest and LYD management method made of very extensive open tools box, including on-farm quarantine, weekly surveillance, felling and burning of palms with LYD symptoms, resistant variety and whole-farm weed control as well as a good fertilization regime (Myrie et al., 2011).

11.4.1 Plant Resistance

The use of resistant varieties has, so far, been the top option and most successful strategy against LYD all over the world (Eden-Green, 2006; Gurr et al., 2016; Myrie et al., 2011; Nipah, 2000; Quaicoe et al., 2009). Some coconut varieties are known to be less susceptible to LYD infection. For example, the Sri Lanka Green Dwarf and Vanatu tall varieties, are saving the coconut palm in Ghana (Quaicoe et al., 2009). In Mozambique, the Giant green variety, also known as Mozambican tall (MZT), is widely used to replace the succumbed LYD-infected coconut palms. The MZT variety is regarded as LYD tolerant, since it survives longer to LYD exposure and/or infection, as compared to the other varieties. In Mozambique, Bila et al. (2016) observed that dwarf coconut varieties were more vulnerable to LYD as compared to tall varieties. Similarly, the local MZT coconut variety has previously been observed to be "field resistant" for LYD phytoplasma infection in Mozambique (Eden-Green, 2006). Resistant/tolerant varieties identified through field observations are now being tested along with other cultivars using proper field assays. In 2007, two field trials, with four replicates each, were established in Zambezia Province, Central Mozambique, to screen coconut varieties for their resistance or tolerance against LYD in selected high LYD pressure sites. The trials consisted of 30 randomized pure and hybrid promising germplasms that were selected locally and abroad. The trial is still ongoing, and the evaluation results have yet not been published.

Early detection of infection with the use of specific primers and the loop mediated isothermal amplification (LAMP) technique, associated with the presence of resistant varieties is successfully saving the "tree of life" in Ghana against LYD (Quaicoe et al., 2009). After successful completion of the screening trials in the Zambezia Province, similar varietal trials should be conducted in other coconut growing sites, as variety resistance may be affected by several factors such as site specificity, environmental factors, phytoplasma strain and vector diversity as well as interactions between factors (Quaicoe et al., 2009).

Identification of the LYD resistant varieties, coupled with certification of the purity of the resistant lines could be an important approach to be considered within the IDM package against LYD. Specific markers for validation of the purity of breeding material resistant to Ghanaian LYD phytoplasma has been proposed (Swarbrick, Yankey, Nipah, Quaicoe, & Dickinson, 2013).

11.4.2 Cultural Practices

The prompt removal of infected trees, followed by replanting with a resistant or tolerant variety has been the primary approach after infection. Destruction and burning of the LYD felled palms is not a common practice in many affected areas, including Mozambique, since the felled and chopped palm would quickly loose turgidity and no longer be a preferred host for the putative insect vector (Nkansah-Poku et al., 2009). However, since the current LYD epidemic in Mozambique is partially accelerated by the plant-insect-phytoplasma-beetle matrimonial relationship, destruction, chopping and burning of the felled palms including palm logs is highly recommended. In Nigeria, Eziashi et al. (2013) reported that abundance of decaying palm trunks and logs was likely to predispose the coconuts to succession of infections and possible dissemination of LYD, which may be associated with build-up of the beetle population. Black's approach for LYD management could be improved by including early field detection using LAMP technology to ensure that symptomatic and infected symptomless plants are felled and burnt (Myrie et al., 2011).

In Mozambique, Bila et al. (2016) observed that the presence of other palm species in coconut farms increased the likelihood of higher LYD incidence. This finding is supported by the previous detection of LYD phytoplasma in African fan palm (*Borassus aethiopum*) and African oil palm (*Elaeis guineensis*) (Bila, Mondjana, et al., 2015). Apart from palm species, detection of LYD phytoplasma has also been reported in weed species and recently in cassava (Arocha et al., 2016; Danyo, 2011). This imply that monitoring and removal of infected plant material should not be limited to only coconut palms.

In general, plant nutrition plays a role in plant susceptibility or tolerance to a range of pathogens (Amtmann, Troufflard, & Armengaud, 2008; Ghosh, Bhattacharya, & Char, 2017). Hence, a vigorous plant with a well-balanced level of nutrients, may better tolerate virulent pathogen attacks and other abiotic stress conditions than weaker plants. Gurr et al. (2016) observed that landscape and climate factors are more likely to influence the population and distribution of the vectors and not directly the LYD incidence. Climate change, along with global warming and the El niño phenomenon, is likely to increase stress conditions in plants and therefore the severity and new outbreaks of LYD (Gurr et al., 2016). For example, in the Caribbean, a higher LYD disease incidence has been reported after hurricanes (Gurr et al., 2016). Similarly, resistant or tolerant coconut varieties have succumbed to LYD exposure under stressed conditions (Hunt, 2002). Also, Bila et al. (2016) observed that the practice of cutting the coconut roots, which subject the plants to stress, increased the risk of high LYD incidence. Rajagopal, Patil, and Sumathykuttyamma (1986) have found that palms with root wilt disease have compromised stomatal regulation activity which led to excessive water loss and leaflet looseness. In order to tolerate adverse conditions, it is very important to well maintain the coconut farms keeping the nutrients in the palms at optimal levels. Even though the ecology of the phytoplasma and putative insect vector of LYD in Mozambique is still unknown, it is also important to weed the farm routinely to prevent phytoplasma reservoir built-up and to avoid fertilizer loss to weeds (Bila et al., 2016).

Several studies have shown that young coconut plants are likely to be more vulnerable to LYD than mature plants, which may be related to the vector feeding and phytoplasma group-subgroup specificities (Bila et al., 2016; Harris & Maramorosch, 2013; Nipah, 2000). However, Howard (2001) reported that the insect vector of LY was commonly found in mature coconut palms and hardly ever in immature coconut plants. Gurr et al. (2016) suggested that plantations of varieties expressing a wide range of tolerance, should reduce the plant resistance break down, due to ecological adaptation of the pathogen or vector population. Intercropping different coconut growth stages, coupled with a range of tolerant or resistant varieties, would minimize the risk of widespread coconut death due to LYD (Bila et al., 2016; Harris & Maramorosch, 2013; Nipah, 2000). Such combinations would also have an effect on the plant-insect-phytoplasma-beetle relationship, since in Mozambique, younger coconut palms are preferred hosts for the adult *O. monoceros*.

11.4.3 Biological Control

Knowledge of the biology and ecology of the LYD phytoplasma and their potential insect vectors is still lacking, meaning that biological control against LYD has yet to be developed. However, some reasonable advances have been made on biopesticide application against the coconut beetle. Due to specific biology and ecology factors of both the host and the pest, chemical sprays do not control efficiently the beetle (Huger, 2005). Alternative methods for successful control of the *Oryctes* beetle should be biological and cultural control. Cultural practices would mostly aim at manipulating the environment of the breeding sites of the beetle, while the biological and mechanical control would mainly target the different developmental stages of the beetle. Several entomopathogens have been described as promising biological control agents against the *Oryctes* beetle, ranging from bacteria (*Bacillus thuringiensis* and *B. popilliae*), virus (Nudi virus or Baculo virus) and fungus (*Metarhizium anisopliae* and *Beauari abassiana*); the latter two (virus and fungus) already with successful widespread use (Bedford, 2014; Huger, 2005; Marschall & Ioane, 1982). *In vitro* trials conducted in Mozambique revealed that both *Metarhizium anisopliae* and *Beauaria bassiana* are promising biopesticides against the *Oryctes* beetle (Guilherme, 2013).

11.4.4 Chemical Control

Chemical spraying against the LYD putative insect vector, combined with felling of infected coconut palms, could have an impact on the disease spread by minimising both the source of inoculum and the vector of the pathogen. However, in Mozambique there is no record of chemical spray against LYD. A study conducted in Ghana (Nkansah-Poku, Dery, & Philippe, 2005) has shown that the coconut survival rate was not significantly higher when spraying was combined with felling infected palms, as compared to felling alone. The latter treatment was considered better due to its environmentally friendly aspect. Elucidating the biology of the putative insect vectors in Africa, might be crucial for the success of the insecticide spraying. In Florida, where the biology of the LY vector was known, chemical spraying significantly reduced the spread of the disease (Been, 1995).

Field trials conducted in Mozambique for evaluating the efficiency of mechanical, cultural and chemical measures against the coconut beetle (*O. monoceros*) revealed that chemical treatments were significantly more efficient than mechanical and cultural measures (Guilherme, 2013). Among the chemicals, Imidacloprid was better than all others (Guilherme, 2013). Yet, these experimental results require further confirmation, before being endorsed for widespread use. Where beetle populations are larger, the adults can also be collected through pheromone trapping (Bedford, 2014). Ethyl 4-methyloctanoate, an *Oryctes* female beetle attracting pheromone is suitable for different types of traps and is also available in slow-release dispenser (Bedford, 2014).

11.5 Future Perspectives

For more effective and efficient LYD management, additional biological information on either the plant-insect-phytoplasma triangle or the plant-insect-phytoplasma-beetle relationships needs to be generated. Advances in biotechnology, such as discovery of phytoplasma growth on axenic media (Contaldo, Bertaccini, Paltonieri, Windsor, & Windsor, 2012), the development of specific primer for LYD from less conservative regions, which combined with the use of LAMP, is allowing early-field detection of LYD phytoplasmas in palms, insects and other hosts, and phytoplasma group and subgroup classification, will provide the basic knowledge for multidisciplinary research towards a more robust and sustainable LYD integrated management package.

Advances in coconut breeding is urgent in order to develop resistant varieties which can suit the local conditions better than the present varieties. Fundamental knowledge needed to develop coconut varieties resistant to LYD is increasingly becoming available. However, breeding using conventional methods is a very time-consuming process which is not compatible with the desired response period. The breeding period could be reduced through molecular marker assisted breeding, which has been widely used in other crops. Successful breeding program can be followed by large scale replanting program in Zambezia and Nampula provinces. Care needs to be taken to intercrop the coconut with other suitable annual crop that can provide food and income while waiting for coconut production.

In Mozambique, LYD is associated with several phytoplasma species, which calls for more studies to analyse the co-infections using multiplex PCR and to understand the role of the high phytoplasma diversity in the current epidemic and future scenarios. The discovery of seed transmission of the 16S rRNA group IV phytoplasma (Oropeza et al., 2017), demonstrated the possibility that other LYD phytoplasma associated groups may also be seed borne. Seed transmission of LYD has tremendous impact, especially on the large-scale replanting program. This emphasizes the need for more effort and commitment to the currently ongoing field trials in the Zambezia Province and study the risk of seed transmission by '*Ca.* P. palmicola'.

The LYD vector status can only be conclusively stated after successful transmission trials. Research on biology and phytoplasma transmission for suspected species are urgently needed, since the understanding of the vector status is very important to develop reliable and sustainable LYD management. LYD transmission trials can be done through insect feeding assays followed by PCR testing for indirect phytoplasma detection on the feeding medium (Tanne et al., 2001). The number of identified plant host range for LYD is increasing, from palm to weed species and now to cassava (Arocha et al., 2016; Bila, Mondjana, et al., 2015). Since LYD alternate plant host species are also present in Mozambique, the search for LYD alternate hosts has been widened to other plant species and is now an ongoing research in Mozambique.

Black's approach for LYD management, targeting the LYD pathosystem and the coconut beetle *Oryctes monoceros* as a major coconut pest, is likely to successfully control the current epidemics in Mozambique. Research on future management strategies also needs to consider the interaction of LYD with the beetle, and to understand the biology of the beetle as well.

References

Agrios, G. (2005). *Plant pathology* (5th ed., p. 952). Amnsterdam, The Netherlands: Elsevier Academic Press Publication.

Allou, K., Issali, A. E., Lekadou, T., Konan, K., Zakra, N., Kouassi, K., … Saraka, Y. D. M. (2012). Comparative synergetic effect of coconut palm (*Cocos nucifera* L) slices and bunches residue of oil palm (*Elaeis guineensis* Jacq.) associated with two kinds of pheromone traps on *Oryctes monoceros* Olivier trapping in Côte d'Ivoire. *International Journal of Emerging Technology and Advanced Engineering, 2*(6), 1–6.

Allou, K., Morin, J.-P., Kouassi, P., N'klo, F. H., & Rochat, D. (2006). *Oryctes monoceros* trapping with synthetic pheromone and palm material in Ivory Coast. *Journal of Chemical Ecology, 32*(8), 1743–1754.

Amtmann, A., Troufflard, S., & Armengaud, P. (2008). The effect of potassium nutrition on pest and disease resistance in plants. *Physiologia Plantarum, 133*(4), 682–691.

Arocha, Y., Diallo, H. A., Konan Konan, J., Kouamé, A., Séka, K., Kra, K., … Beugré, N. (2016). Detection and identification of the coconut lethal yellowing phytoplasma in weeds growing in coconut farms in Côte d'Ivoire. *Canadian Journal of Plant Pathology, 38*(2), 164–173.

Bedford, G. O. (2014). Advances in the control of rhinoceros beetle, *Oryctes rhinoceros* in oil palm. *Journal of Oil Palm Research, 26*(13), 183–194.

Been, B. O. (1995). Integrated pest management for the control of lethal yellowing: Quarantine, cultural practices and optimal use of hybrids. In C. Oropeza, F. W. Howard, & G. Ashburner (Eds.), *Lethal yellowing: Research and practical aspects* (Vol. 5, pp. 101–109). Dordrecht, The Netherlands: Kluwer Academic Publishers/Springer.

Bila, J. (2016). *Coconut lethal yellowing phytoplasma disease in Mozambique. Diversity, host range, and the impact of farming practices on disease incidence.* PhD, Faculty of Forest Sciences, Uppsala, Sweden.

Bila, J., Högberg, N., Mondjana, A., & Samils, B. (2015). African fan palm (*Borassus aethiopum*) and oil palm (*Elaeis guineensis*) are alternate hosts of coconut lethal yellowing phytoplasma in Mozambique. *African Journal of Biotechnology, 14*(52).

Bila, J., Mondjana, A., Samils, B., & Högberg, N. (2015). High diversity, expanding populations and purifying selection in phytoplasmas causing coconut lethal yellowing in Mozambique. *Plant Pathology, 64*(3), 597–604.

Bila, J., Mondjana, A., Samils, B., Högberg, N., Wilson, M. R., & Santos, L. (2017). First report of '*Candidatus* Phytoplasma palmicola'detection in the planthopper *Diostrombus mkurangai* in Mozambique. *Bulletin of Insectology, 70*(1), 45–48.

Bila, J., Persson, I., Högberg, N., Mondjana, A., Manuel, L., Tostao, E., … Santos, L. (2016). Effect of farming practices and farm history on incidence of coconut lethal yellowing in Mozambique. *African Crop Science Journal, 24*(2), 167–178.

Bonnot, F., De Franqueville, H., & Lourenço, E. (2010). Spatial and spatiotemporal pattern analysis of coconut lethal yellowing in Mozambique. *Phytopathology, 100*(4), 300–312.

Brown, S., Been, B., & McLaughlin, W. (2006). Detection and variability of the lethal yellowing group (16Sr IV) phytoplasmas in the *Cedusa* sp.(Hemiptera: Auchenorrhyncha: Derbidae) in Jamaica. *Annals of Applied Biology, 149*(1), 53–62.

Christensen, N. M., Axelsen, K. B., Nicolaisen, M., & Schulz, A. (2005). Phytoplasmas and their interactions with hosts. *Trends in Plant Science, 10*(11), 526–535.

Contaldo, N., Bertaccini, A., Paltonieri, S., Windsor, H. M., & Windsor, G. D. (2012). Axenic culture of plant pathogenic phytoplasmas. *Phytopathologia Mediterranea, 51*(3), 607–617.

Cordova, I., Jones, P., Harrison, N., & Oropeza, C. (2003). *In situ* PCR detection of phytoplasma DNA in embryos from coconut palms with lethal yellowing disease. *Molecular Plant Pathology, 4*(2), 99–108.

Danet, J. L., Balakishiyeva, G., Cimerman, A., Sauvion, N., Marie-Jeanne, V., Labonne, G., … Škorić, D. (2011). Multilocus sequence analysis reveals the genetic diversity of European fruit tree phytoplasmas and supports the existence of inter-species recombination. *Microbiology, 157*(2), 438–450.

Danyo, G. (2011). Review of scientific research into the Cape Saint Paul wilt disease (CSPWD) of coconut in Ghana. *African Journal of Agricultural Research, 6*, 4567–4578.

Dery, S., & Philippe, R. (1997). Preliminary study on the epidemiology of Cape St Paul wilt disease of coconut in Ghana (November 1995). In S. Eden-Green & F. Ofori (Eds.), *Proceedings of an international workshop on lethal yellowing-like diseases of coconut* (pp. 255–260). Elmina, Ghana: NRI, Chatham, UK.

Dollet, M., Macome, F., Vaz, A., & Fabre, S. (2011). Phytoplasmas identical to coconut lethal yellowing phytoplasmas from Zambesia (Mozambique) found in a pentatomide bug in Cabo Delgado province. *Bulletin of insectology, 64*, S139–S140.

Eden-Green, S. (1997). History, distribution and research on coconut lethal yellowing-like diseases of palms. In S. Eden-Green & F. Ofori (Eds.), *Proceedings of the international workshop on lethal yellowing-like diseases of coconut (November 1995)* (Vol. 25, pp. 9–25). Elmina, Ghana: NRI, Chatham, UK.

Eden-Green, S. (2006). *An assessment of coconut Lethal Yellowing-type Disease (LYD) in Mozambique*. Consultant's report by EG Consulting, UK. Retrieved from http://www.clubofmozambique.com/solutions1/solutions/social_development/lethal_yellow_disease.pdf

Eziashi, E., Omamor, I., Aisueni, N., Aisagbonhi, C., Airede, C., Ikuenobe, C., … Osagie, I. (2013). Potential weed species as alternate hosts of insect vectors of the lethal yellowing disease (LYD) of coconut palms (*Cocos nucifera* L.) in Nigeria. *British Journal of Applied Science & Technology, 3*(1), 123–130.

Food and Agriculture Organisation of the United Nations. (2016). FAOSTAT. Crops. Retrieved August 25, 2018, from FAOSTAT Statistics Division. http://www.fao.org/faostat/en/#data/QC

Garnier, M., Foissac, X., Gaurivaud, P., Laigret, F., Renaudin, J., Saillard, C., & Bové, J. (2001). Mycoplasmas, plants, insect vectors: A matrimonial triangle. *Comptes Rendus de l'Académie des Sciences-Series III-Sciences de la Vie, 324*(10), 923–928.

Ghosh, P., Bhattacharya, A., & Char, B. (2017). Manipulating disease and pest resistance pathways in plants for enhanced crop improvement. *Bioscience Biotechnology Research Communications, 10*(4), 631–644.

Guilherme, P. (2013). *Avaliação de diferentes métodos de controlo do escaravelho rinoceronte Oryctes monoceros (coleoptera: scarabaeidae: dynastinae) do coqueiro.* (Mestrado em Protecção Vegetal), Universidade Eduardo Mondlane, Maputo,

Gurr, G. M., Johnson, A. C., Ash, G. J., Wilson, B. A., Ero, M. M., Pilotti, C. A., … You, M. S. (2016). Coconut lethal yellowing diseases: A phytoplasma threat to palms of global economic and social significance. *Frontiers in Plant Science, 7*, 1521.

Harris, K. F., & Maramorosch, K. (2013). *Vectors of plant pathogens*. Boston, MA: Academic.

Harrison, N., Davis, R. E., Oropeza, C., Helmick, E. E., Narvaez, M., Eden-Green, S., … Konan Konan, J. L. (2014). *Candidatus* phytoplasma palmicola': A novel taxon associated with a lethal yellowing-type disease (LYD) of coconut (*Cocos nucifera* L.) in Mozambique. *International Journal of Systematic and Evolutionary Microbiology, 64*(6), 1890–1899.

Harrison, N., & Elliot, M. (2005). *Lethal yellowing (LY) of palm*. Retrieved from Plant Pathology Department, University of Florida, USA. http://edis.ifas.ufl.edu

Harrison, N., & Oropeza, C. (2008). Coconut lethal yellowing. In N. Harrison, G. Rao, & C. Marcone (Eds.), *Characterization, diagnosis and management of phytoplasmas* (pp. 219–248). Houston, USA: Studium Press.

Howard, F. W. (2001). The animal class Insecta and the plant family Palmae. In F. W. Howard, D. Moore, R. Giblin-Davis, & R. Abad (Eds.), *Insects on palms* (pp. 1–32). New York, NY: Cabi Publishing.

Howard, F. W., Norris, R., & Thomas, D. (1983). Evidence of transmission of palm lethal yellowing agent by a planthopper, *Myndus crudus* (Homoptera, Cixiidae). *Tropical Agriculture, 60*(3), 168–171.

Huger, A. M. (2005). The Oryctes virus: Its detection, identification, and implementation in biological control of the coconut palm rhinoceros beetle, *Oryctes rhinoceros* (Coleoptera: Scarabaeidae). *Journal of Invertebrate Pathology, 89*(1), 78–84.

Hunt, P. (2002). *Abiotic factors affecting coconuts in relation to lethal yellowing*. Proceedings of the expert consultation on sustainable coconut production through control of lethal yellowing disease, CFC technical paper no. 18. Kingston: Common Fund for Commodities. P 77-79.

Marschall, K. J., & Ioane, I. (1982). The effect of re-release of *Oryctes rhinoceros* baculovirus in the biological control of rhinoceros beetles in Western Samoa. *Journal of Invertebrate Pathology, 39*(3), 267–276.

Mpunami, A., Tymon, A., Jones, P., & Dickinson, M. (1999). Genetic diversity in the coconut lethal yellowing disease phytoplasmas of East Africa. *Plant Pathology, 48*(1), 109–114.

Mpunami, A., Tymon, A., Jones, P., & Dickinson, M. (2000). Identification of potential vectors of the coconut lethal disease phytoplasma. *Plant Pathology, 49*(3), 355–361.

Myrie, W., Oropeza, C., Saenz, L., Harrison, N., Roca, M. M., Córdova, I., … Douglas, L. (2011). Reliable improved molecular detection of coconut lethal yellowing phytoplasma and reduction of associated disease through field management strategies. *Bull. Insectol, 64*, S203–S204.

Nipah, J. O. (2000). Some characteristics of the Sri Lanka Green dwarf (SGD): A variety resistant to the Cape Saint Paul wilt disease of coconut. *Journal of the Ghana Science Association, 2*(2), 88–93.

Nkansah-Poku, J., Dery, S. K., & Philippe, R. (2005). Reduction of spread of cape St Paul wilt disease (CSPWD) of coconut by insecticidal hot-fogging and removal of diseased palms. *Ghana journal of agricultural science*(1), 193-198.

Nkansah-Poku, J., Philippe, R., Quaicoe, R. N., Dery, S., & Ransford, A. (2009). Cape Saint Paul wilt disease of coconut in Ghana: Surveillance and management of disease spread. *Oléagineux Corps gras Lipides, 16*(2), 111–115.

Oropeza, C., Cordova, I., Puch-Hau, C., Castillo, R., Chan, J., & Sáenz, L. (2017). Detection of lethal yellowing phytoplasma in coconut plantlets obtained through *in vitro* germination of zygotic embryos from the seeds of infected palms. *Annals of Applied Biology, 171*(1), 28–36.

Oshima, K., Ishii, Y., Kakizawa, S., Sugawara, K., Neriya, Y., Himeno, M., … Yamaji, Y. (2011). Dramatic transcriptional changes in an intracellular parasite enable host switching between plant and insect. *PLoS One, 6*(8), e23242.

Philippe, R., Nkansah, J. P., Fabre, S., Quaicoe, R., Pilet, F., & Dollet, M. (2007). Search for the vector of Cape Saint Paul wilt (coconut lethal yellowing) in Ghana. *Bulletin of Insectology, 60*(2), 179.

Quaicoe, R. N., Dery, S. K., Philippe, R., Baudouin, L., Nipah, O., Nkansah Poku, J., … Pilet, F. (2009). Resistance screening trials on coconut varieties to Cape Saint Paul wilt disease in Ghana. *OCL. Oléagineux Corps gras Lipides, 16*(2), 132–136.

Rajagopal, V., Patil, K., & Sumathykuttyamma, B. (1986). Abnormal stomatal opening in coconut palms affected with root (wilt) disease. *Journal of Experimental Botany, 37*(9), 1398–1405.

Smith, J. (2015). *Fera report to ABT associate Inc. on impact evaluation of farmer income support project.* Food and Environment Research Institute. Sand Hutton, Fera.

Swarbrick, P., Yankey, E., Nipah, J., Quaicoe, R., & Dickinson, M. (2013). Identification of receptor like kinase genes in coconut and development of a marker for validation of breeding materials resistant to a phytoplasma disease in Ghana. *African Journal of Biotechnology, 12*(45), 6347–6357.

Tanne, E., Boudon-Padieu, E., Clair, D., Davidovich, M., Melamed, S., & Klein, M. (2001). Detection of phytoplasma by polymerase chain reaction of insect feeding medium and its use in determining vectoring ability. *Phytopathology, 91*(8), 741–746.

Tymon, A., Jones, P., & Harrison, N. (1998). Phylogenetic relationships of coconut phytoplasmas and the development of specific oligonucleotide PCR primers. *Annals of Applied Biology, 132*(3), 437–452.

Wilson, M. R. (1987). African Derbidae (Homoptera, Fulgoroidea): Taxonomic notes with descriptions of new species collected mainly from coconut. *Journal of Natural History, 21*(3), 567–595.

Chapter 12
Impact and Management of Major Phytoplasma Diseases in Brazil

Ivan Paulo Bedendo and João Roberto Spotti Lopes

Abstract In Brazil, phytoplasmas belonging to seven distinct groups and occurring in over 50 hosts belonging to commercial crops, weeds and wild plant species. The present chapter focuses on three economically relevant diseases occurring in three crops extensively grown in Brazilian fields: corn, brassicas and cassava. Brazil is one of the largest world producers of corn. The first epidemics of corn stunt diseases occurred in the beginning of 1990s and since this decade, the diseases has been reported as a major limiting factor of corn production, due its high level of incidence. General recommendations for control include the use of genetically resistant varieties and appropriate planting periods, eradication of volunteer plants and insecticides treatments. Brassicas, mainly cabbage, cauliflower and broccoli, are extensively cultivated and consumed by Brazilian population. Diseases associated with phytoplasmas are responsible for significant damage that reduce yield and quality of production and cause serious financial losses for growers. Prevention and control measures include the use of heathy seedlings, elimination of weeds from adjacent areas of crops, chemical applications in the first stages of crop development and maintenance of fields free of weeds after harvest. Cassava is an important source of carbohydrate for inhabitants of tropical areas, including Brazil. A disease known as witches'-broom is occurring with high incidence in the main producer regions, causing underdevelopment of roots and, consequently, low starch production per area. The most usual and efficient control is the use of healthy propagative materials for planting or resistant varieties.

Keywords Corn stunt · Cassava witches'-broom · Brassica stunt · Phytoplasma · Brazil

I. P. Bedendo (✉) · J. R. S. Lopes
University of São Paulo – Agricultural College, Piracicaba, SP, Brazil
e-mail: ibedendo@usp.br; jrslopes@usp.br

C. Y. Olivier et al. (eds.), *Sustainable Management of Phytoplasma Diseases in Crops Grown in the Tropical Belt*, Sustainability in Plant and Crop Protection 12,
https://doi.org/10.1007/978-3-030-29650-6_12

12.1 Introduction

The first studies on phytoplasma-associated diseases in Brazil were done in the 1970s, with the detection of these pathogens in tissues of diseased plants by transmission electron microscopy (Costa, Kitajima, & Arruda, 1971; Kitajima & Costa, 1970a, 1970b, 1979). A second wave of studies was initiated in the beginning of 2000s using molecular techniques, which allowed not only the detection, but also the identification and characterization of the associated phytoplasmas in the current system of groups and subgroups (Amaral Mello, Bedendo, & Camargo, 2006; Bedendo, Davis, & Dally, 2000; Junqueira, Bedendo, & Pascholati, 2004; Montano et al., 2001). Since then, several studies on phytoplasmas have been conducted in various pathosystems in Brazil, widening the knowledge about their host plants, genetic diversity and physiological aspects of plant-pathogen interactions (e.g. biochemical alterations in infected host plants), and allowing the description of members of new subgroups. Other studies have focused on identification of insects that are vectors or potential vectors of phytoplasmas, and on other aspects related to ecology and epidemiology.

In Brazil, phytoplasmas belonging to groups (and subgroups) 16SrI (B), 16SrII (O), 16 SrIII (B, J, Z), 16SrVII (B, D), 16SrIX (A, H), 16SrXIII (E, H), and 16SrXV (A) have been reported in association with several diseases (Montano, Cunha, & Pimentel, 2011; Pérez-López, Luna-Rodríguez, Olivier, & Dumonceaux, 2016). In addition, other phytoplasmas representing groups-subgroups 16SrIII-A (De Souza, Da Silva, Bedendo, & Carvalho, 2014), 16SrIII-U (Amaral Mello, Eckstein, Flores, Kreyci, & Bedendo, 2011), 16SrVII-C (Flôres, Amaral Mello, Massola Junior, & Bedendo, 2013), 16SrVII-F (da Silva Fugita et al., 2017) and 16SrXIII-J (de Aquino Melo et al., 2018) were identified. Over 50 botanical species found in crops, weeds and wild plant species were identified as hosts of phytoplasmas. Some are of particular economic importance, such as fruit trees (plum, citrus, papaya and passion-fruit), vegetables (eggplant, broccoli, chayote, cauliflower, strawberry, cabbage and tomatoes), ornamental plants (begonias, poinsettia, celosia, hibiscus, bougainvillea and periwinkle), major crops (sugarcane, cassava, corn and soybeans) (Montano et al., 2011; Pérez-López et al., 2016) and forage grasses (da Silva Fugita et al., 2017; Haas, 2005). Despite the large diversity of phytoplasmas and host plants, most diseases associated with phytoplasmas in Brazil are of minor economic importance and not well studied, hence lacking scientific information about yield losses, epidemiology, vectors, inoculum sources and management.

Therefore, this review is focused on three major diseases associated with phytoplasmas in economically-important and widely grown crop plants in Brazil: corn stunt, brassica stunt and cassava witches'-broom. The research progress obtained in these three pathosystems is summarized, providing a solid basis for development of efficient management strategies of the diseases, aiming at reducing losses and increasing profitability in the affected crops.

12.2 Corn Stunt

Corn (*Zea mays* L., Poaceae) is one of the most important cereals cultivated and consumed by man with Brazil having a prominent position as one of the largest world producers. The crop is widely distributed in the country, but the main producing areas are located in the southern, southeastern and center-west regions, corresponding to 77% of the corn cultivated area and 78% of the total volume of the Brazilian production (Companhia Nacional De Abastecimento – CONAB, 2015).

In the last three decades, areas traditionally cultivated with corn have been used for the cultivation of soybean due to its higher profitability (Coelho, Landau, & Sabato, 2017). Therefore, in various areas, the corn crops began to be planted late (mid to late summer), after harvesting soybean. This production system, known as "second crop season" or "safrinha", became economically relevant and routinely practiced in wide areas, representing currently, 60% of the Brazilian corn production (Coelho et al., 2017). Later, the practice of "second crop season" was also adopted in many areas in which corn was planted at the normal period (spring), thus widening the time period of this culture in the field. Interestingly, in the last years, the increase of irrigation has resulted in the production of 2–3 crop seasons per year, further extending the permanence of this crop in the field.

The introduction of the "second crop season" system, although profitable for corn growers, has caused important changes in the behavior of corn diseases. The incidence of viral and phytoplasmal diseases has increased significantly due to the increase of the vector populations, represented by aphids and leafhoppers. These populations are favored by climatic factors, especially by high temperatures, which occur during the late period (mid-late summer) of corn planting, and by the availability of hosts in the field during almost the entire year (Oliveira et al., 1998). Epidemiologically, successive crop seasons considerably favor the maintenance of pathogen inoculum and vector populations, contributing to high levels of disease incidence.

In the 1950s, two types of stunting diseases were recognized in corn and denominated "maize bushy stunt" and "corn stunt". Both diseases were supposedly caused by virus and differentiated based on symptomatology (Maramorosch, 1955). Later, it was demonstrated that a phytoplasma, named maize bushy stunt phytoplasma (MBSP), was associated with maize bushy stunt, while a spiroplasma was the causal agent of corn stunt (Nault, 1980). With the advance of molecular techniques, the phytoplasma associated with maize bushy stunt was identified as a representative of the 16SrI-B subgroup (Lee, Gundersen-Rindal, Davis, & Bartoszyk, 1998), while the spiroplasma associated with corn stunt was classified as *Spiroplasma kunkelii* (Granados, Maramorosch, & Shikata, 1968). Both diseases have caused epidemics in many countries, including the United States (Bradfute, Tsai, & Gordon, 1981), Argentina (Lenardon et al., 1993; Pecci et al., 2002), Nicaragua (Hruska, Gladstone, & Obando, 1996) and Brazil (Oliveira et al., 1998). In Brazil, both stunting forms were known since the 1970s and were considered of secondary importance (Costa et al., 1971). However, at the time, there were already warnings that these diseases could cause serious damage if corn was planted late (Costa et al., 1971). The molecular identification revealed that the maize bushy stunt phytoplasma is affiliated with subgroup 16SrI-B (Bedendo et al., 2000). Devastating epidemics of corn stunt diseases in Brazilian fields emerged in the 1990s, coinciding with the gradual increase of late plantings of the second crop season. Later, the diseases expanded to crops grown in the traditional planting period (spring), with reported average incidences ranging from 65.3% to 100% and resulting in low to zero yield production (Oliveira et al., 1998). Studies revealed that the severity of the symptoms and damage levels were directly proportional to the number of infected leafhoppers present in the crops (Toffanelli & Bedendo, 2002). Positive correlations between damage level and stunting incidence revealed that a reduction of 0.8% in grain production occurred for each 1% increase in disease incidence. Moreover, yield losses are also directly correlated with the infection timing, i.e., the sooner the infection, the higher the losses with reduction in grain production reaching values of up to 50–60% (Massola Júnior, Bedendo, Amorim, & Lopes, 1999).

12.2.1 Symptomatology

The symptoms of stunting exhibited by diseased plants (Fig. 12.1) vary according to many factors, such as the hybrid or variety of corn, the period of sowing, the climatic conditions and the vector population size (Sabato, 2017). In MBSP-infected plants, initial symptoms are chlorosis occurring on the edges of the leaves, followed by reddening of the affected areas (Nault, 1980). The reddening, which occurs in the form of stripes between the leaf veins, can expand and occupy the whole leaf blade. The occurrence of thin stems with short internodes and reduced floral tassels are common. Various corn cobs may be produced by the infected plant; however, these cobs are small, malformed, and have very few kernels. Despite being uncommon, tillers can emerge in the axillary buds of the leaves and at the base of the plants. For the corn stunt spiroplasma (*S. kunkelii*), the first symptoms are expressed by marginal chlorosis on the leaves and small chlorotic spots diffused at the base of the foliar limb (De Leon, 1974; Maramorosch, 1955). Later, stripes and chlorotic areas begin to take large portions of the foliar blade, contrasting with the normal green color. In addition to these typical symptoms, the plant can present internode shortening, thin stems, reduced growth and, in some cases, tillers at the base of the stem and axillary buds (Sabato, 2017). Due to the infection, malformed cobs with a few underdeveloped kernels loosely fixed to the cob, can proliferate (Sabato, 2017). The interaction between phytoplasma and corn induces a series of physiological changes to the host, such as increase in the levels of reducing sugars, proteins and phenolic compounds, as well as a significant reduction in the content of chlorophyll, which probably interferes with the photosynthetic capacity of the plant (Junqueira et al., 2004).

Fig. 12.1 Typical symptom of leaf reddening induced by phytoplasma in maize and malformed small cobs with very few kernels. (Photo courtesy of Dr. Elizabeth de Oliveira – EMBRAPA/CNPMS – Brazil)

12.2.2 Leafhopper Transmission

Some Deltocephalinae leafhoppers (Hemiptera: Cicadellidae) such as *Dalbulus maidis* (DeLong & Wolcott), *D. elimatus* (Ball) and *Graminela nigrifrons* (Forbes) can transmit both the phytoplasma (MBSP) and *S. kunkelii*, whereas *Exitianus exitiosus* (Uhler), *Exitianus obscurinervis* (Stål) and *Stirellus bicolor* (van Duzee) transmit only *S. kunkelii* (Carloni et al., 2011; Nault, 1980). The corn leafhopper, *D. maidis,* is the main vector species of both pathogens under natural conditions.

The corn leafhopper is the only *Dalbulus* species that occurs in Brazil (de Oliveira, Lopes, Dias, & Nault, 2004), where it is commonly found in the corn plant whorl, frequently in large numbers (Waquil, 1997). This leafhopper species is highly efficient in transmitting phytoplasma and spiroplasma associated with corn stunting, and is highly specific to the plant host. Large populations of *D. maidis* are found in the major corn growing regions of Brazil, where it is the only known vector for both MBSP and *S. kunkelii* (Waquil, 1997). The vector transmits these stunting agents in a persistent and propagative manner, with acquisition, latency and inoculation periods that vary from 4 to 7 days, 13 to 24 days and 2 to 7 days, respectively (Nault, 1980). Once acquired, both MBSP and *S. kunkelii* are transmitted by the leafhopper throughout its life (Alivizatos & Markham, 1986). The vector is thought to migrate long distances efficiently disseminating these pathogens from old crops to new crops, and thus promoting new disease cycles (de Oliveira, Molina, Albres, & Lopes, 2002).

12.2.3 Host Plants

Many weeds and wild species of plants may serve as hosts of insect vectors and act as intermediate hosts of plant pathogens transmitted by vectors. Regarding the corn stunt pathogens, the available studies indicate a very narrow host range, with large specificity of both host and vector (Sabato, 2017). In Brazil, experimental evidence demonstrates that a few species of grasses (Poaceae) such as *Brachiaria decumbens* Stapf., *Brachiaria Plantagenet* (Link) Hitchc. and *Panicum maximum* Jacq., are hosts of MBSP (Haas, 2005). However, the epidemiological role of these grasses as inoculum sources for the survival and dissemination of the phytoplasma is still unknown. The vector *D. maidis*, is host specific, with immatures developing only on maize (*Z. mays*) and close relatives (Nault, 1980). Therefore, maize appears to be the major (if not the only) source of inoculum for both primary and secondary spread of these stunting pathogens.

12.2.4 Control

Plant diseases result from the interaction between susceptible host plants, pathogenic agents and occurrence of favorable environmental factors. In the case of corn stunting diseases, one more element, the vector, must be considered. The reduction of losses caused by these diseases requires interference with the multiple components that generate favorable conditions to the establishment and development of the disease. Therefore, based on the information available in this chapter, management of corn stunt must consider a series of combined measures that involve resistance of corn varieties, reduction of pathogen inoculum and manipulation of factors that contribute to the increase and dispersal of the leafhopper vector population (Sabato, 2017). Although the environmental conditions that occur during the "second crop season" are more favorable to the diseases, it is not possible to abandon this planting system, due to its high production size and profitability. Thus, alternatives must be proposed to minimize the damage caused by stunting diseases.

The use of genetically resistant varieties is one of the first requisites. The Brazilian market provides genetic material with different levels of resistance, recommending alternance in varieties to avoid resistance overcome. The elimination of old plantings and volunteer plants that remain in the field after harvest is an important measure to avoid the survival of the pathogen and vector in the corn during off-season and to reduce the sources of pathogen inoculum. For the same reason, successive corn plantings (2–3 corn crop per season) in the same area should be avoided. Choosing the planting period most commonly adopted in the region is another appropriate measure, since insect populations can migrate from older crops to new crops, introducing and disseminating the pathogens associated with the diseases. These cultural practices are particularly important to reduce the rate of primary spread from external sources of inoculum.

Seed treatment with insecticides has been widely used in Brazil to control corn pests in the early stages of the crop, but its efficacy to prevent inoculation of MBSP and *S. kunkelii* by infected *D. maidis* adults is limited to a few days after crop germination (de Oliveira et al., 2002). This information suggests that seed treatments are ineffective to control primary spread, particularly under moderate to high populations of immigrating infected adults. However, insecticide treatments (seed or foliar applications) may be effective to control secondary spread within the crop, as long feeding periods (several hours to a few days) are required for the vector to acquire MBSP or *S. kunkelii*. In addition, after acquisition by healthy leafhoppers, these pathogens show a latent period of 2–3 weeks (Nault, 1980), during which the insecticides will likely kill the vector and prevent the pathogen transmission. Considering that yield losses caused by corn stunt are substantially reduced when infection occurs 7 weeks post-germination (plants with 5 leaves or older) (Massola Júnior et al., 1999), strict vector control during the first 4–5 weeks of the crop may play an important role in reducing damage by secondary spread.

12.3 Brassica Stunt

In Brazil, brassicas such as cabbage (*Brassica oleracea* L. var. *capitate*), cauliflower (*B. oleracea* var. *botrytis*) and broccoli (*B. oleracea* var. *italica*) are vegetables consumed on a large scale (IEA, 2015). The consumer market has room for expansion due to the nutritional properties of these vegetables and the development of healthy eating habits. The cultivation of these species also contributes to economic and social stability in the producing regions, given that the crops are demanding in labor and present high profitability (IEA – Instituto de Economia Agrícola, 2015). In Brazil, these species are cultivated in areas with mild climates throughout the year, due to the development of varieties well adapted to the distinct seasons of the year. The green belt region in the city of São Paulo is the main production center and is constituted by small and large farms (IEA – Instituto de Economia Agrícola, 2015).

One of the main issues for these crops is the damage and significant losses, caused by fungal, bacterial and viral diseases as well as nematodes (Maringoni & Silva Júnior, 2016). Among the diseases, Brassica stunting caused by phytoplasma is prominent. The disease has been observed by Brazilian producers since the beginning of the 1990s and has become more frequent since then, causing decreases in crop yield and commercial value. High incidence levels have been reported, from 30 to 70% for cabbage (Amaral Mello, 2007), up to 90% for cauliflower (Rappussi et al., 2012) and from 3 to 10% for broccoli (Eckstein et al., 2013). The disease is a limiting factor in production, forcing farmers to eliminate the crop, substituting the Brassica species for other species, or even abandon the cultivation areas.

The symptoms displayed by infected cabbage plants includes stunting, dwarfism, leaves with reddened edges, the presence of a dark ring when the branch is cut transversally (vessel necrosis), and in some cases, sprouting at the base of the plant. In addition to the above mentioned symptoms, infected cauliflower and broccoli plants also expressed malformation of inflorescence, and head asymmetry, triggering a reduction in their commercial values (Amaral Mello et al., 2006) (Fig. 12.2). Initially, fungi and bacteria were the suspected possible disease agents. However, this was disregarded after exhaustive but unsuccessful fungal and bacterial isolation and inoculation procedures (Canale, Amaral Mello, Martins, & Bedendo, 2012). Attempts to find virus in the diseased plant tissues also generated negative results. Research eventually associated phytoplasmas with affected cabbage plants (Amaral Mello, 2007). Similarly, investigations conducted with cauliflower and broccoli revealed that phytoplasmas were associated with symptomatic plants observed in commercial fields (Eckstein et al., 2013; Rappussi et al., 2012).

A similar situation occurred in the USA where cabbage plants expressing similar symptoms than those observed in Brazilian crops were reported for the first time, at the end of the 1930s, in the state of California (USA) (Severin & Frazier, 1945). Initially, the disease was attributed to a virus based on symptom characteristics and the transmission of the pathogen by an insect vector (Freitag, 1967; Westdal & Richardson, 1969). In the beginning of the 1980s, phytoplasmas were detected by transmission electron microscopy in broccoli and cauliflower plants that showed foliar yellowing and inflorescence malformation (Bertaccini, Pisi, & Marani, 1983).

Fig. 12.2 Symptoms of phytoplasmas in brassicas characterized by leaves with reddened edges and malformation and head or inflorescence

In Brazil, molecular methods have resulted in the identification of phytoplasmas in the three Brassica species, with predominance of phytoplasmas belonging to the group 16SrIII. In cabbage, phytoplasmas affiliated with subgroups 16SrI-B and 16SrIII-J were found (Amaral Mello, 2007). In cauliflower, phytoplasmas belonging to subgroups 16SrIII-J (Rappussi et al., 2012), 16SrVII-B (Pereira et al., 2016) and 16SrXV-A (Canale & Bedendo, 2013) were identified, whereas in symptomatic broccoli plants, the presence of groups 16SrI, 16SrIII and 16SrXIII phytoplasmas was revealed (Eckstein et al., 2013). Phytoplasma belonging to group 16SrXIII detected in broccoli was later identified as belonging to subgroup 16SrXIII-H (Pérez-López et al., 2016). The occurrence of phytoplasma from different groups and subgroups demonstrates the genetic diversity of these prokaryotes in association with Brassicas stunting. In other countries, a higher frequency of occurrence of phytoplasma belonging to group 16SrI was observed in Brassica species. Phytoplasma affiliated to subgroup 16SrI-B was identified in cabbage plants cultivated in Italy (Bertaccini et al., 1998), Hungary (Fodor, Viczian, Mergenthaler, & Sule, 1999) and the United States (Lee et al., 2003) and expressing symptoms similar to those observed in Brazil. Phytoplasma belonging to 16SrI-B subgroup was also identified found in symptomatic broccoli plants grown in commercial crops in Italy (Marcone & Ragozzino, 1995) and Serbia (Duduk et al., 2007).

12.3.1 Vectors

It is known that the most important species of phytoplasma insect vectors are leafhoppers (Hemiptera: Cicadellidae) and planthoppers (Hemiptera: Fulgoroidea, e.g. Delphacidae and Cixiidae) (Weintraub & Beanland, 2006). The constant association between insect and phytoplasma indicates the potential capacity of the insect in transmitting the pathogen (Wolf, 2004). During investigations aiming at identifying potential phytoplasma vectors in Brazilian brassicas fields, leafhoppers and planthoppers were collected in cabbage, cauliflower and broccoli crops as well as in the field surroundings. The highest levels of incidence of diseased plants were observed at the edges of the cultivated areas, suggesting that potential vectors could be introducing phytoplasma into the crops from external sources of inoculum. Phytoplasmas belonging to groups 16SrI and 16SrIII were identified in specimens belonging to the Cicadellidae and Delphacidae families and collected in cabbage production fields (Amaral Mello, 2007). In broccoli crops, phytoplasmas were detected in broccoli plants and in specimens of the leafhoppers species *Atanus nitidus* Linnavuori (Hemiptera: Cicadellidae), *Balcluta hebe* Kirkaldy (Hemiptera: Cicadellidae), *Planicephalus flavicosta* Stål (Hemiptera: Cicadellidae) and *Scaphytopius fulginosus* Osborn (Hemiptera: Cicadellidae) (Eckstein et al., 2014). The higher frequency of phytoplasma belonging to groups 16SrI and 16SrIII in Brassica crops in Brazil can be related to the size of the populations of potential vectors, the degree of specificity between vector and phytoplasma, and the transmission efficiency of the vector.

12.3.2 Alternative Hosts

Phytoplasma from distinct groups can induce similar symptoms in different botanical species and/or distinct symptoms in same botanical species. In addition, some phytoplasma-infected plants can be symptomless hosts (R. E. Davis, 1995). Numerous reports have revealed that weed or wild species that grow spontaneously within the crop and in surrounding areas can host phytoplasma related to the diseases present in the crop of interest. In Brazil, many asymptomatic and symptomatic weed species were sampled in Brassica crops in order to investigate their role as possible alternative hosts of phytoplasmas. Representatives from group 16SrIII, which were identified with higher frequency in cabbages, cauliflower and broccoli, were also found in many weed species, such as *Erigeron bonariensis* L. (Asteraceae), *Emilia sonchifolia* (L.) DC. ex Wight (Asteraceae), *Sonchus oleraceae* L. (Asteraceae), *Bidens Pilosa* L. (Asteraceae), *Agerantum conyzoides* L. (Asteraceae), *Lepidium virginicum* L. (Brasscaceae), *Ricinus communis* L. (Euphorbiaceae), *Leonurus sibiricus* L. (Lamiaceae), *Crotalaria lanceolate* E.Mey (Fabaceae), *Sida rhombifolia* L. (Malvaceae), *Paulicourea marcgravii* Hill. (Rubiaceae) and *Nicandra physalodes* (L.) Gaertn. (Solanaceae) (Kreyci, Eckstein, Lopes, Ferreira, & Bedendo, 2018). Among the weed species harboring phytoplasmas, *Brassica rapa* L. was prominent, occurring at a high frequency and occupying extensive cultivated areas when the main crop was removed from the field. These results indicate that various weed species can host phytoplasma and probably act as primary sources of inoculum for the cultivated Brassicas. The occurrence of vectors and alternative sources of phytoplasma are important factors that can contribute to the introduction and dissemination of the pathogen in the crop, hence increasing the intensity of the disease in commercial fields. Results show strong evidence that weed plants can have an important role in the epidemiology of brassica stunting.

12.3.3 Epidemiological Analysis

The epidemiological aspects related to the temporal and spatial distribution of phytoplasma diseases are important to increase the efficacy of control measures. In Brazil, analyses of spatial distribution of the disease for cabbage, cauliflower and broccoli fields revealed that symptomatic plants are distributed in an aggregated manner in the field (Amaral Mello, 2007; Eckstein, 2010; Rappussi et al., 2012). The aggregated pattern indicates the participation of vectors in the pathosystem. Furthermore, the aggregated pattern suggests that the role of the vector is stronger than other factors characteristics of the pathogen or host (Bergamin Filho, Hau, Amorim, & Jesus Junior, 2004). Confirming field observations, the analysis of isopath areas for phytoplasma and brassica pathosystem revealed that the outbreaks of the disease appear at the edges of the cultivated areas. This indicates that the pathogen is introduced in the crop from external areas, with subsequent spread to the center of the cultivated area (Amaral Mello, 2007; Eckstein, 2010; Rappussi et al., 2012).

12.3.4 Control

This review presents information on stunting diseases that cause relevant damage to vegetables of great social and economic importance. The discovery of phytoplasma in brassicas has contributed to an increase in the knowledge on the role of this group of pathogens in Brazilian agroecosystems. The detection of the same group of phytoplasma in cauliflower, cabbage and broccoli, in weed species that spontaneously grow within or surrounding the crops and in some species of leafhoppers collected in the area, as well as the presence of an aggregated model of disease distribution in the field, strongly indicates the participation of insect vectors in this pathosystem.

In general, field observations by growers agree with the information obtained from researchers. Although not experimentally proven, it has been observed that the employment of certain cultural practices and insecticide applications are usually effective in reducing the damage caused by stunting in brassicas. Efficient measures adopted by growers include planting of seedlings produced in protected nurseries, which hinder the entry of possible vectors, chemical treatment aiming at controlling the vectors in the first stages of the crop development, eradication of diseased plants from the crop, hindering their possible role as source of inoculum for secondary spread, maintaining crop fields and surrounding areas free of weed species, hence avoiding possible sources of inoculum for primary spread and eliminating cultivated plants and infesting species after harvest, avoiding their role as sources of inoculum for the next crop.

12.4 Cassava Witches'-Broom

Cassava is one of the most important sources of carbohydrates for many countries located in tropical climate zones. Cassava crop is grown between the parallels 30° North latitude and 30° South latitude, but most cultivated areas are located between 20° North and 20° South latitudes, in altitude inferior to 2.300 m (Cock, 1982). The crop is extensively distributed over the regions of Brazil and is exploited mostly for the production of flour and starch (Conceição, 1987). In addition to its use as food for humans and animals, cassava has an important socio-economic role, generating thousands of direct jobs in the field and industry (Conceição, 1987).

Cassava is a plant originated from the central region of Brazil, with a large number of species and varieties (Nassar, 2000). The species *Manihot glaziovii* Müll. Arg. (Euphorbiaceae), used for the extraction of latex, and *Manihot esculenta* Crantz, used for the production of starch or *in natura* consumption, the latest being of high economic importance (Hershey, 1992). The largest cultivated areas in Brazil are located in the northern, northeastern and center-west regions (Hershey & Amaya, 1980). The plant is rustic and demands little cultural management, being tolerant to draught and low soil fertility (Conceição, 1987). Due to these characteristics, this crop is highly cultivated by subsistence farmers, in poor soils that are unsuitable for other crops.

The occurrence of many diseases negatively affects crop yield, causing damage to the plants and generating losses to the growers. Among the diseases that occur in this crop, witches'-broom, associated with phytoplasma, is prominent. The disease has been reported in Latin America (Brazil, Mexico, Venezuela, Peru, Cuba) and, more recently, southeastern Asia, in countries such as Vietnam, Thailand, Cambodia, Laos, Indonesia and Philippines (Massola Júnior, Bedendo, & Oliveira, 2016). Diseased plants show excessive sprouting of branches, stunting, shortened internodes, reduced and malformed leaves, and numerous underdeveloped roots (Flôres, Haas, Canale, & Bedendo, 2013).

12.4.1 Brief History

In Brazil, the first report of cassava witches'-broom occurred in 1944 (Silberschmidt & Campos, 1944). The diseased plants presented an excessive number of sprouts on the branches, reduced size, decrease in the number of leaves, and lightening of the natural green color (Fig. 12.3). At the beginning of the 1970s, the disease was transmitted from diseased plants to healthy plants through grafting. Transmission electron microscope examinations of plant tissue revealed the presence of pleomorphic corpuscles in the phloem vessels, demonstrating the association between phytoplasma and symptomatic plants (Kitajima & Costa, 1970b). At the end of the 1990s, the

Fig. 12.3 Plants of cassava exhibiting shoot proliferation associated with phytoplasma, typical symptom of the disease known by cassava witches'-broom. (Photos by M. Dita and S. Oliveira)

phytoplasma associated with the disease was molecularly characterized and classified as belonging to group 16SrIII (Barros, Kitajima, & Resende, 1998). Later, further analyses identified the witches'-broom agent as a representative of subgroup 16SrIII-B, which was transmitted under controlled conditions, by the parasitic plant *Cuscuta subinclusa* Durand & Hilg. (Convolvulaceae) to healthy plants, confirming its pathogenicity (Flôres, Haas, et al., 2013). Distinct phytoplasmas are associated with the disease, such as members of 16SrI in islands in the Pacific and Cuba (Arocha, Piñol, et al., 2009; R. Davis, Arocha, Jones, & Malay, 2005), and a representative of 16SrII in Uganda (Arocha, Echodu, et al., 2009).

12.4.2 Importance and Control

In Brazil, high incidences of witches´-broom disease have been observed in commercial cassava plantations, causing significant losses for root production. For example, a survey conducted in a state from the northeastern region revealed incidence of 85%, with yield losses of 95% (Fukuda et al., 1990; Lozano, 1992). High incidence levels of cassava witches'-broom are attributed to the vegetative propagation of the crop. The plants are multiplied by growing branches removed from adult plants. As phytoplasma can be systematically distributed throughout the plant, branches harvested for the future crop, can be infected by the pathogen. The disease easily can be controlled by using healthy propagative materials for planting (Takatsu & Fukuda, 1990), which must be removed from healthy or asymptomatic plants of 12–14 months of age (Lozano, 1992). A previous study showed that the simple selection of planting materials was efficient for reducing the infection by over 98% (Lozano, Cavalcante, Lima, & Fukuda, 1990). Another recommended measure is the eradication of affected plants from the crop. The presence of insect vectors capable of promoting secondary spread of the pathogen in the crop remains to be demonstrated. Harvested plants remains must be eliminated before the plantation of new crop (Lozano, 1992). Another control method is the use of resistant or tolerant material, when available in the region. In Brazil, due to the economic importance of the crop and the interest in reducing the damage caused by the disease, a few varieties with good resistance characteristics were developed.

Aspects related to the survival and dissemination of the phytoplasma in cultivated fields are important parameters for better understanding the epidemiology of the disease. The occurrence of insects involved in the transmission of the pathogen, as well as the role of weeds and native plants that can act as reservoir of phytoplasmas and serve as source of inoculum to the commercial crops must be investigated. Such information is necessary to develop effective and sustainable strategies for controlling cassava witches'broom.

12.5 Final Remarks

Phytoplasmas of groups 16SrI, 16SrIII, 16SrVII and 16SrXIII are among the most frequently identified phytoplasma in Brazilian agricultural crops, with predominance of the two first groups, both associated with the three pathosystems addressed in this review and considered to be the most economically important. Although several botanical species were described as alternative hosts for both groups of phytoplasmas, the epidemiological role of these plants is largely unknown. In addition, there is a lack of information about vectors for most diseases associated with phytoplasmas in Brazil, except corn stunt, for which the corn leafhopper is a well-studied vector.

Future studies should investigate aspects related to the cycle of host-pathogen relationships, especially, those related to the survival and spread of the phytoplasmas in each pathosystem. Knowledge on alternative host plants and insect vectors represent major gaps that should be addressed by research, because of its importance for establishing effective and sustainable management strategies. Regarding control methods, more studies are needed to obtain resistant varieties and to implement cultural practices promoting eradication of inoculum sources and reduction of vector populations in the crop fields and surrounding areas. The efficacy of insecticides and other vector control methods to prevent primary and/or secondary spread of phytoplasmas has not yet been assessed and should be properly investigated before their recommendation for disease management.

References

Alivizatos, A., & Markham, P. (1986). Acquisition and transmission of corn stunt spiroplasma by its leafhopper vector *Dalbulus maidis*. *Annals of Applied Biology, 108*(3), 535–544.

Amaral Mello, A. (2007). *Identificação molecular de fitoplasmas associados ao enfezamento do repolho e análise epidemiológica da doença*. PhD, Universidade de São Paulo, Sao Paulo, Brazil.

Amaral Mello, A., Bedendo, I. P., & Camargo, L. E. A. (2006). Sequence heterogeneity in the 16S rDNA of tomato big bud phytoplasma belonging to group 16SrIII. *Journal of Phytopathology, 154*(4), 245–249.

Amaral Mello, A., Eckstein, B., Flores, D., Kreyci, P. F., & Bedendo, I. P. (2011). Identification by computer-simulated RFLP of phytoplasmas associated with eggplant giant calyx representative of two subgroups, a lineage of 16SrIII-J and the new subgroup 16SrIII-U. *International Journal of Systematic and Evolutionary Microbiology, 61*(6), 1454–1461.

Arocha, Y., Echodu, R., Talengera, D., Muhangi, J., Rockefeller, E., Asher, O., … Tripathi, J. (2009). Occurrence of '*Candidatus* Phytoplasma aurantifolia'(16SrII group) in cassava and four other species in Uganda. *Plant Pathology, 58*(2), 390–390.

Arocha, Y., Piñol, B., Almeida, R., Acosta, K. I., Quiñones, M., Zayas, T., … Lucas, J. (2009). First report of phytoplasmas affecting organoponic crops in central and eastern Cuba. *Plant Pathology, 58*(4), 793–793.

Barros, T., Kitajima, E., & Resende, R. (1998). Diversidade de isolados brasileiros de fitoplasmas através da análise do 16S rDNA. *Fitopatologia Brasileira, 23*(4), 459–465.

Bedendo, I. P., Davis, R. E., & Dally, E. L. (2000). Detection and identification of the maize bushy stunt phytoplasma in corn plants in Brazil using PCR and RFLP. *International Journal of Pest Management, 46*(1), 73–76.

Bergamin Filho, A., Hau, B., Amorim, L., & Jesus Junior, W. (2004). Análise espacial de epidemias. In F. Vale, J. W. Jesus, & L. Zambolim (Eds.), *Epidemiologia aplicada ao manejo de doenças de plantas* (pp. 195–240). Minas Gerais, Brazil: Belo Horizonte: Editora Perfil.

Bertaccini, A., Pisi, A., & Marani, F. (1983). Virescenza e fillodia del cavolfiore e del broccolo. *Informatore Fitopatologico, 33*, 57–60.

Bertaccini, A., Vorácková, Z., Vibio, M., Fránová, J., Navrátil, M., Špak, J., & Nebesárová, J. (1998). Comparison of phytoplasmas infecting winter oilseed rape in the Czech Republic with Italian Brassica phytoplasmas and their relationship to the aster yellows group. *Plant Pathology, 47*(3), 317–324.

Bradfute, O., Tsai, J., & Gordon, D. (1981). Corn stunt spiroplasma and viruses associated with a maize disease epidemic in Southern Florida. *Plant Disease, 65*(10), 837–841.

Canale, M., Amaral Mello, A., Martins, T., & Bedendo, I. P. (2012). Patossistema fitoplasma-brássicas: identificação do agente causal em repolho, couve flor e brócolis, potenciais vetores e epidemiologia. *Revisão Anual de Patologia de Plantas, 20*, 151–176.

Canale, M., & Bedendo, I. P. (2013). 'Candidatus Phytoplasma brasiliense'(16SrXV-A subgroup) associated with cauliflower displaying stunt symptoms in Brazil. *Plant Disease, 97*(3), 419–419.

Carloni, E., Virla, E., Paradell, S., Carpane, P., Nome, C., Laguna, I., … Giménez Pecci, M. (2011). *Exitianus obscurinervis* (Hemiptera: Cicadellidae), a new experimental vector of *Spiroplasma kunkelii*. *Journal of Economic Entomology, 104*(6), 1793–1799.

Cock, J. H. (1982). Aspectos fisiológicos del crecimiento y desarrollo de la planta de yuca. In C. Dominguez (Ed.), *Yuca: investigación, producción y utilización.* (pp. 51–74): Cali, Colombia: Centro Internacional de Agricultura Tropical.

Coelho, A., Landau, E., & Sabato, E. (2017). Time of sowing corn and incidence of diseases caused by mollicutes. In M. Oliveira & E. Sabato (Eds.), *Diseases in maize: Insect vectors, mollicutes and viruses* (pp. 195–209). Brasilia, Brazil: Empresa Brasileira de Pesquisa Agropecuária – EMBRAPA.

Companhia Nacional De Abastecimento – CONAB. (2015). Acompanhamento da safra brasileira de Grãos 2014/2015. In (pp. 83–97). Monitoramento Agrícola Cultivos de verão, 2ª safra e de inverno – Safra 2014/15: Brasília: Conab, 2015.

Conceição, A. J. (1987). *A mandioca* (382p). São Paulo, Brazil: Nobel.

Costa, A., Kitajima, E., & Arruda, S. (1971). Moléstias de vírus e de micoplasma do milho em São Paulo. *Revista da Sociedade Brasileira de Fitopatologia, 4*(4), 39–41.

da Silva Fugita, J. M., Pereira, T. B. C., Banzato, T. C., Kitajima, E. W., da Souto, E. R., & Bedendo, I. P. (2017). Molecular characterization of a phytoplasma affiliated with the 16SrVII group representative of the novel 16SrVII-F subgroup. *International Journal of Systematic and Evolutionary Microbiology, 67*(8), 3122–3126.

Davis, R., Arocha, Y., Jones, P., & Malay, A. (2005). First report of the association of phytoplasmas with plant diseases in the territory of Wallis and Futuna. *Australasian Plant Pathology, 34*(3), 417–418.

Davis, R. E. (1995). Fitoplasmas: fitopatógenos procarióticos sem parede celular, habitantes de floema e transmitidos por artrópodes. *Revisão Anual de Patologia de Plantas, 3*(1), 1–27.

de Aquino Melo, L., Ventura, J. A., Costa, H., Kitajima, E. W., Ferreira, J., & Bedendo, I. P. (2018). Delineation of a novel subgroup 16SrXIII-J phytoplasma, a '*Candidatus* Phytoplasma hispanicum'-related strain, based on computer-simulated RFLP and phylogenetic analysis. *International Journal of Systematic and Evolutionary Microbiology, 68*(3), 962–966.

De Leon, C. (1974). *Maize diseases: A guide for field identification. Information bulletin 11. In.* Mexico: International Maize and Wheat Improvement Center – CIMMYT.

de Oliveira, C., Lopes, J. R., Dias, C. T. D. S., & Nault, L. R. (2004). Influence of latitude and elevation on polymorphism among populations of the corn leafhopper, *Dalbulus maidis*

(DeLong and Wolcott)(Hemiptera: Cicadellidae), in Brazil. *Environmental Entomology, 33*(5), 1192–1199.

de Oliveira, C., Molina, R. M., Albres, R. S., & Lopes, J. R. (2002). Disseminação de molicutes do milho a longas distâncias por *Dalbulus maidis* (Hemiptera: Cicadellidae). *Fitopatologia Brasileira, 27*, 91–95.

De Souza, A. N., Da Silva, F. N., Bedendo, I. P., & Carvalho, C. M. (2014). A phytoplasma belonging to a 16SrIII-A subgroup and dsRNA virus associated with cassava frogskin disease in Brazil. *Plant Disease, 98*(6), 771–779.

Duduk, B., Bulajic, A., Duduk, N., Calari, A., Paltrinieri, S., Krstic, B., & Bertaccini, A. (2007). Identification of phytoplasmas belonging to aster yellows ribosomal group in vegetables in Serbia. *Bulletin of Insectology, 60*(2), 341–342.

Eckstein, B. (2010). *Enfezamento do brócolis: identificação molecular de fitoplasmas, potenciais insetos vetores e hospedeiros alternativos, e análise epidemiológica da doença*. Piracicaba, Brazil: Universidade de São Paulo.

Eckstein, B., Barbosa, J., Kreyci, P., Zanol, K., Coelho, L., Gonçalves, A., … Bedendo, I. (2014). Identification of potential leafhoppers vectors of phytoplasmas (16SrIII group) associated with broccoli stunt disease in Brazil. *Australasian Plant Pathology, 43*(4), 459–463.

Eckstein, B., Barbosa, J. C., Kreyci, P. F., Canale, M., Brunelli, K. R., & Bedendo, I. P. (2013). Broccoli stunt, a new disease in broccoli plants associated with three distinct phytoplasma groups in Brazil. *Journal of Phytopathology, 161*(6), 442–444.

Flôres, D., Amaral Mello, A., Massola Junior, N., & Bedendo, I. P. (2013). First report of a group 16SrVII-C phytoplasma associated with shoot proliferation of sunn hemp (*Crotalaria juncea*) in Brazil. *Plant Disease, 97*(12), 1652–1652.

Flôres, D., Haas, I., Canale, M., & Bedendo, I. P. (2013). Molecular identification of a 16SrIII-B phytoplasma associated with cassava witches' broom disease. *European Journal of Plant Pathology, 137*(2), 237–242.

Fodor, M., Viczian, O., Mergenthaler, E., & Sule, S. (1999). Cabbage infected with phytoplasma from the aster yellows group in Hungary. *Acta phytopathologica et entomologica hungarica, 34*(1/2), 1–6.

Freitag, J. (1967). Interactions between strain of aster yellows vírus in the six-spotted leafhopper *Macrosteles fascifrons*. *Phytopathology, 57*, 1016–1024.

Fukuda, C., Filho, J., Sa, M., Carneiro, J., Paiva, A., Pereira, C., … Lima, H. (1990). Levantamento da ocorrência do superbrotamento da mandioca na microrregião de Ibiapaba, CE. In *Paper presented at the VI Congresso Brasileiro De Mandioca (July 16–21, 1990)*. Londrina: Brazil.

Granados, R. R., Maramorosch, K., & Shikata, E. (1968). Mycoplasma: Suspected etiologic agent of corn stunt. *Proceedings of the National Academy of Sciences of the United States of America, 60*(3), 841.

Haas, I. (2005). *Gramíneas forrageiras como potenciais hospedeiros alternativos para o fitoplasma do enfezamento vermelho do milho*. Doctoral dissertation, Universidade de São Paulo, Piracicaba, Brazil.

Hershey, C. (1992). *Manihot esculenta diversity*. Paper presented at the first international network for cassava genetic resources (18–23 August, 1992) Cali, Colombia.

Hershey, C., & Amaya, A. (1980). Germoplasma de yuca: Evolución, distribución y colección. In C. Dominguez (Ed.), *Manual de produccion de yuca* (pp. 51–72). Cali, Columbia: CIAT.

Hruska, A. J., Gladstone, S. M., & Obando, R. (1996). Epidemic roller coaster: Maize stunt disease in Nicaragua. *American Entomologist, 42*(4), 248–252.

IEA – Instituto de Economia Agrícola. (2015). Estatística da Produção Paulista. Retrieved 08 June 2017, from http://www.iea.sp.gov.br

Junqueira, A., Bedendo, I., & Pascholati, S. (2004). Biochemical changes in corn plants infected by the maize bushy stunt phytoplasma. *Physiological and Molecular Plant Pathology, 65*(4), 181–185.

Kitajima, E., & Costa, A. (1970a). Corpúsculos do tipo micoplasmas associados a diversas moléstias de plantas do grupo amarelo no estado de São Paulo. *Ciência e Cultura, 23*, 285–291.

Kitajima, E., & Costa, A. (1970b). Micoplasma: possível agente etiológico de certas moléstias de plantas. *Ciência e Cultura, 22*(4), 351–363.

Kitajima, E., & Costa, A. (1979). Microorganismos do tipo micoplasma associados a moléstias do tipo amarelo em algumas plantas cultivadas e ornamentais, no estado de São Paulo e no Distrito Federal. *Fitopatologia Brasileira, 4*, 317–327.

Kreyci, P. F., Eckstein, B., Lopes, J. R. S., Ferreira, J., & Bedendo, I. P. (2018). Transmission of '*Candidatus* Phytoplasma pruni'-related strain associated with broccoli stunt by four species of leafhoppers. *Journal of Phytopathology, 166*(7–8), 502–505.

Lee, I.-M., Gundersen-Rindal, D. E., Davis, R. E., & Bartoszyk, I. M. (1998). Revised classification scheme of phytoplasmas based on RFLP analyses of 16S rRNA and ribosomal protein gene sequences. *International Journal of Systematic and Evolutionary Microbiology, 48*(4), 1153–1169.

Lee, I.-M., Martini, M., Bottner, K., Dane, R., Black, M., & Troxclair, N. (2003). Ecological implications from a molecular analysis of phytoplasmas involved in an aster yellows epidemic in various crops in Texas. *Phytopathology, 93*(11), 1368–1377.

Lenardon, S. L., Laguna, I. G., Gordon, D., Truol, G., Gomez, G., & Bratfute, O. (1993). Identification of corn stunt spiroplasma in maize from Argentina. *Plant Disease, 77*(1), 100.

Lozano, J. (1992). Overview of integrated control of cassava diseases. *Fitopatologia Brasileira, 17*, 18–22.

Lozano, J., Cavalcante, M. L. S., Lima, H. A., & Fukuda, C. (1990). *Efeito da seleçao de material de plantio para o controle de micoplasma da mandioca na microrregiao de Ibiapaba, CE [resumen]*. Paper presented at the VI Congresso Brasileiro de Mandioca (July 16–21, 1990), Londrina, Brazil.

Maramorosch, K. (1955). The occurrence of two distinct types of corn stunt in Mexico. *Plant Disease Reporter, 39*(12), 896–898.

Marcone, C., & Ragozzino, A. (1995). Detection of phytoplasmas in *Brassica* spp. in southern Italy and their characterization by RFLP analysis. *Zeitschrift für Pflanzenkrankheiten und Pflanzenschutz/Journal of Plant Diseases and Protection, 102*(5), 449–460.

Maringoni, A., & Silva Júnior, T. (2016). Doenças das brássicas. In L. Amorim, J. Rezende, F. Bergamin, & L. Camargo (Eds.), *Manual de Fitopatologia: Doenças das Plantas Cultivadas* (Vol. 2, 5th ed., pp. 165–173). Ouro Fino, Brazil: Editora Agronômica Ceres.

Massola Júnior, N., Bedendo, I. P., Amorim, L., & Lopes, J. R. (1999). Quantificação de danos causados pelo enfezamento vermelho e enfezamento pálido do milho em condições de campo. *Fitopatologia Brasileira, 24*(2), 136–142.

Massola Júnior, N., Bedendo, I. P., & Oliveira, S. (2016). Doenças da Mandioca. In L. Amorim, J. Rezende, A. Bergamin Filho, & L. Camargo (Eds.), *Manual de Fitopatologia: Doenças das Plantas Cultivadas* (5th ed., pp. 515–522). Ouro Fino, Brazil: Editora Agronômica Ceres.

Montano, H. G., Cunha Jr, J., & Pimentel, J. P. (2011). Phytoplasmas in Brazil: An update. *Bulletin of Insectology, 64*(Suppl), S251–S252.

Montano, H. G., Davis, R. E., Dally, E. L., Hogenhout, S., Pimentel, J. P., & Brioso, P. (2001). 'Candidatus Phytoplasma brasiliense', a new phytoplasma taxon associated with hibiscus witches' broom disease. *International Journal of Systematic and Evolutionary Microbiology, 51*(3), 1109–1118.

Nassar, N. (2000). Wild cassava, Manihot spp.: Biology and potentialities for genetic improvement. *Genetics and Molecular Biology, 23*(1), 201–212.

Nault, L. R. (1980). Maize bushy stunt and corn stunt: A comparison of disease symptoms, pathogen host ranges, and vectors. *Phytopathology, 70*(7), 659–662.

Oliveira, E., Waquil, J., Fernandes, F., Paiva, E., Resende, R., & Kitajima, E. (1998). Enfezamento pálido e enfezamento vermelho na cultura do milho no Brasil Central. *Fitopatologia Brasileira, 23*, 45–47.

Pecci, M. G., Oliveira, E., Resende, R., Laguna, I., Conci, L., Avila, A., … Nome, C. (2002). Ocorrencia de Doenças causadas por molicutes e por vírus em milho nas províncias de Tucumán e de Córdoba na Argentina. *Fitopatologia Brasileira, 27*, 4.

Pereira, T., Dally, E., Davis, R., Banzato, T., Galvão, S., & Bedendo, I. (2016). Cauliflower is a new host of a subgroup 16SrVII-B phytoplasma associated with stunting disease in Brazil. *Plant Disease, 100*(5), 1007–1007.

Pérez-López, E., Luna-Rodríguez, M., Olivier, C. Y., & Dumonceaux, T. J. (2016). The underestimated diversity of phytoplasmas in Latin America. *International Journal of Systematic and Evolutionary Microbiology, 66*(1), 492–513.

Rappussi, M., Eckstein, B., Flôres, D., Haas, I., Amorim, L., & Bedendo, I. (2012). Cauliflower stunt associated with a phytoplasma of subgroup 16SrIII-J and the spatial pattern of disease. *European Journal of Plant Pathology, 133*(4), 829–840.

Sabato, E. (2017). Corn stunting diseases. In M. Oliveira & E. Sabato (Eds.), *Diseases in maize: Insect vectors, Mollicutes and viruses* (pp. 11–23). Brasilia, Brazil: Empresa Brasileira de Pesquisa Agropecuária – EMBRAPA.

Severin, H., & Frazier, N. (1945). California aster yellows on vegetable and seed crops. *Hilgardia, 16*, 573–596.

Silberschmidt, K., & Campos, A. (1944). Estudos relativos à doença superbrotamento ou envassouramento da mandioca. *Arquivos do Instituto Biológico, 15*, 1–26.

Takatsu, A., & Fukuda, S. (1990). *Current status of cassava diseases in Brazil.* Paper presented at the Integrated pest management for tropical root and tuber crops: proceedings of the workshop on the global status of and prospects for integrated pest management of root and tuber crops in the tropics (25–30 October 1987), Ibadan, Nigeria.

Toffanelli, C., & Bedendo, I. P. (2002). Efeito da população infetiva de *Dalbulus maidis* na produção de grãos e no desenvolvimento de sintomas do enfezamento vermelho do milho. *Fitopatologia Brasileira, 27*(1), 82–86.

Waquil, J. M. (1997). Amostragem e abundância de cigarrinhas e danos de *Dalbulus maidis* (DeLong & Wolcott)(Homoptera: Cicadellidae) em plântulas de milho. *Anais da Sociedade Entomológica do Brasil, 26*, 27–33.

Weintraub, P. G., & Beanland, L. (2006). Insect vectors of phytoplasmas. *Annual Review of Entomology, 51*, 91–111.

Westdal, P., & Richardson, H. (1969). The susceptibility of cereals and wild oats to an isolate of the aster yellows pathogen. *Canadian Journal of Botany, 47*, 755–760.

Wolf, T. (2004). Ecology of potential vectors of North American Grapevine Yellows. *American Journal of Enology and Viticulture, 55*(4), 431A.

The manufacturer's authorised representative in the EU is Springer Nature Customer Service Centre GmbH, Europaplatz 3, 69115 Heidelberg, Germany. If you have any concerns regarding our products, please contact ProductSafety@springernature.com

Printed and bound by CPI Group (UK) Ltd, Croydon, CR0 4YY

28/07/2026

02179230-0002